Lecture Notes in Computer Science 12150

More information about this series at http://www.springer.com/series/7407

Fabio Gadducci · Timo Kehrer (Eds.)

Graph Transformation

13th International Conference, ICGT 2020
Held as Part of STAF 2020
Bergen, Norway, June 25–26, 2020
Proceedings

 Springer

Editors
Fabio Gadducci 🆔
Università di Pisa
Pisa, Italy

Timo Kehrer 🆔
Humboldt-Universität zu Berlin
Berlin, Germany

ISSN 0302-9743   ISSN 1611-3349 (electronic)
Lecture Notes in Computer Science
ISBN 978-3-030-51371-9   ISBN 978-3-030-51372-6 (eBook)
https://doi.org/10.1007/978-3-030-51372-6

LNCS Sublibrary: SL1 – Theoretical Computer Science and General Issues

This Springer imprint is published by the registered company Springer Nature Switzerland AG
The registered company address is: Gewerbestrasse 11, 6330 Cham, Switzerland

Preface

This volume contains the proceedings of ICGT 2020, the 13th International Conference on Graph Transformation held during June 25–26, 2020. Due to the pandemic situation leading to COVID-19 countermeasures and travel restrictions, the conference was held online. ICGT 2020 was affiliated with STAF (Software Technologies: Applications and Foundations), a federation of leading conferences on software technologies, and it took place under the auspices of the European Association of Theoretical Computer Science (EATCS), the European Association of Software Science and Technology (EASST), and the IFIP Working Group 1.3 on Foundations of Systems Specification.

The ICGT series aims at fostering exchange and collaboration of researchers from different backgrounds working with graphs and graph transformation, either by contributing to their theoretical foundations or by highlighting their relevance in different application domains. Indeed, the use of graphs and graph-like structures as a formalism for specification and modeling is widespread in all areas of computer science as well as in many fields of computational research and engineering. Relevant examples include software architectures, pointer structures, state space graphs, control/data flow graphs, UML and other domain-specific models, network layouts, topologies of cyber-physical environments, and molecular structures. Often, these graphs undergo dynamic change, ranging from reconfiguration and evolution to various kinds of behavior, all of which may be captured by rule-based graph manipulation. Thus, graphs and graph transformation form a fundamental universal modeling paradigm that serves as a means for formal reasoning and analysis, ranging from the verification of certain properties of interest to the discovery of new computational insights.

ICGT 2020 continued the series of conferences previously held in Barcelona (Spain) in 2002, Rome (Italy) in 2004, Natal (Brazil) in 2006, Leicester (UK) in 2008, Enschede (The Netherlands) in 2010, Bremen (Germany) in 2012, York (UK) in 2014, L'Aquila (Italy) in 2015, Vienna (Austria) in 2016, Marburg (Germany) in 2017, Toulouse (France) in 2018, and Eindhoven (The Netherlands) in 2019, following a series of six International Workshops on Graph Grammars and Their Application to Computer Science from 1978 to 1998 in Europe and in the USA.

This year, the conference solicited research papers describing new unpublished contributions in the theory and applications of graph transformation as well as tool presentation papers that demonstrate new features and functionalities of graph-based tools. The Program Committee selected 22 out of 40 submissions for inclusion in the conference's program. Out of these 22 papers, 2 of them have been accepted as new ideas papers that were presented at the conference, without appearing in the proceedings. All submissions went through a thorough peer-review process and were discussed online. There was no preset number of papers to accept, and each of them has been evaluated and assessed based on its own strengths and weaknesses. The topics of the accepted papers range over a wide spectrum, from theoretical approaches to graph transformation to the application of graph transformation in specific domains.

The papers presented new results on the DPO/SPO dichotomy and their rule application conditions, as well as introducing novel rewriting formalisms and establishing tighter connections with bigraphical reaction systems; furthermore, model checking issues were explored and the use of graph transformation advances in contemporary application domains such as life sciences. In addition to the submitted papers and tool presentations, the conference program included an invited talk, given by Bob Coecke (University of Oxford, UK), on the use of graphical structures for the implementation of natural language on quantum hardware.

We would like to thank all who contributed to the success of ICGT 2020, the invited speaker Bob Coecke, the authors of the submitted papers, as well as the members of the Program Committee and all the reviewers for their valuable contributions to the selection process. We are grateful to Reiko Heckel, the chair of the Steering Committee of ICGT, for his fruitful suggestions; and to Adrian Rutle, the STAF 2020 general chair, for the close collaboration during the dynamic pandemic situation in early 2020.

May 2020

Fabio Gadducci
Timo Kehrer

Organization

Steering Committee

Paolo Bottoni	Sapienza University of Rome, Italy
Andrea Corradini	University of Pisa, Italy
Gregor Engels	University of Paderborn, Germany
Holger Giese	Hasso Plattner Institute, University of Potsdam, Germany
Reiko Heckel (Chair)	University of Leicester, UK
Dirk Janssens	University of Antwerp, Belgium
Barbara König	University of Duisburg-Essen, Germany
Hans-Jörg Kreowski	University of Bremen, Germany
Leen Lambers	Hasso Plattner Institute, University of Potsdam, Germany
Ugo Montanari	University of Pisa, Italy
Mohamed Mosbah	University of Bordeaux, France
Manfred Nagl	RWTH Aachen, Germany
Fernando Orejas	Technical University of Catalonia, Spain
Francesco Parisi-Presicce	Sapienza University of Rome, Italy
John Pfaltz	University of Virginia, USA
Detlef Plump	University of York, UK
Arend Rensink	University of Twente, The Netherlands
Leila Ribeiro	Federal University of Rio Grande do Sul, Brazil
Grzegorz Rozenberg	University of Leiden, The Netherlands
Andy Schürr	Technical University of Darmstadt, Germany
Gabriele Taentzer	Philipps University of Marburg, Germany
Jens Weber	University of Victoria, Canada
Bernhard Westfechtel	University of Bayreuth, Germany

Program Committee

Paolo Baldan	University of Padua, Italy
Gábor Bergmann	Budapest University of Technology and Economics, Hungary
Paolo Bottoni	Sapienza University of Rome, Italy
Andrea Corradini	University of Pisa, Italy
Juan De Lara	Autonomous University of Madrid, Spain
Juergen Dingel	Queen's University, Canada
Maribel Fernandez	King's College London, UK
Holger Giese	Hasso Plattner Institute, University of Potsdam, Germany
Reiko Heckel	University of Leicester, UK

Thomas Hildebrandt	University of Copenhagen, Denmark
Wolfram Kahl	McMaster University, Canada
Jean Krivine	CNRS, France
Barbara König	University of Duisburg-Essen, Germany
Leen Lambers	Hasso Plattner Institute, University of Potsdam, Germany
Yngve Lamo	Western Norway University of Applied Sciences, Norway
Detlef Plump	University of York, UK
Arend Rensink	University of Twente, The Netherlands
Leila Ribeiro	Federal University of Rio Grande do Sul, Brazil
Andy Schürr	Technical University of Darmstadt, Germany
Pawel Sobocinski	Tallinn University of Technology, Estonia
Gabriele Taentzer	Philipps University of Marburg, Germany
Matthias Tichy	University of Ulm, Germany
Uwe Wolter	University of Bergen, Norway
Steffen Zschaler	King's College London, UK

Additional Reviewers

Barkowsky, Matthias
Campbell, Graham
Conte, Alessio
Courtehoute, Brian
Echahed, Rachid
Ehmes, Sebastian
Farkas, Rebeka
Goetz, Stefan
Grochau Azzi, Guilherme
Gönczy, László
Kosiol, Jens
Kulcsár, Géza

Kögel, Stefan
Küpper, Sebastian
Luthmann, Lars
Maximova, Maria
Orejas, Fernando
Sakizloglou, Lucas
Stegmaier, Michael
Stoltenow, Lara
Treinen, Ralf
Viennot, Laurent
Witte, Thomas

Quantum Natural Language Processing (on Actual Quantum Hardware): (Abstract of Invited Paper)

Bob Coecke, Giovanni de Felice, Konstantinos Meichanetzidis, and Alexis Toumi

Cambridge Quantum Computing Ltd., Oxford University, Department of Computer Science

This work involves three different graphical structures:

- Those of categorical quantum mechanics (CQM) [1, 6, 9].
- hose of distributional compositional language meaning (DisCoCat) [11].
- The ZX-calculus, in particular applied to quantum circuits [7–9].

By Quantum Natural Language Processing (QNLP) we mean the canonical implementation of natural language on quantum hardware, where by canonical we mean that compositional language structure, including grammar, matches the manner in which quantum systems compose.

The DisCoCat model for natural language enables such a canonical embedding. One instance of this is the perfect match of grammatical structure in terms of pregroups [17] and the compositional quantum structure of bipartite entanglement [1], in that both form compact closed categories (a.k.a. string diagrams). In fact, DisCoCat was directly inspired by teleportation-alike behaviours that exploit this entanglement [5].

Besides vector spaces and inner-products, which are commonplace in modern Natural Language Processing (NLP), DisCoCat also employs several other quantum-theoretic features, such as projector spectra for representing meanings of adjectives, verbs and relative pronouns [10, 14, 15, 19], density matrices for representing linguistic ambiguity and lexical entailment [2, 18], and entanglement for representing correlated concepts [4], all of which 'exist' on quantum hardware. Therefore DisCoCat-QNLP deserves to be referred to as 'quantum-native'.

The first proposal to implement QNLP was put forward in [21]. A first major upshot of quantum implementation of DisCoCat is an exponential reduction of space resources as compared to implementations on classical hardware. Other initially mentioned upshots include the nativeness of density matrices, and the availability of quantum algorithms that provide an algorithmic quantum advantage for typical NLP tasks such as classification.

However, a first shortcoming in that proposal was the reliance on quantum RAM [13], which does not yet, and may never do. Also, one needs to provide hardware-dependent conversion of DisCoCat diagrams into e.g. quantum circuits. These shortcoming are addressed in:

- Quantum natural language processing on near-term quantum computers. arXiv: 2005.04147.

Recently, we performed QNLP in the form of question-answering on IBM quantum hardware. In fact, this was the first time any form of NLP had been done on quantum hardware. The main two resources for our implementation are:

- A medium blog describing the implementation that we did: https://medium.com/cambridge-quantum-computing/quantum-natural-language-processing-748d6f27b31d

- A github repository containing the implementations: https://github.com/oxford-quantum-group/discopy/blob/ab2b356bd3cad1dfb55ca6606d6c4b4181fe590c/notebooks/qnlp-experiment.ipynb

This work makes use of tool box components presented in:

- DisCoPy: Monoidal Categories in Python. arXiv:2005.02975.

One key change as compared to [21] is the use of variational quantum circuits [3] instead of qRAM. Here is such a parametrised quantum circuit that we used:

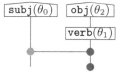

where the values of $\theta_0, \theta_1, \theta_2$ are learned using a small corpus. For reasons of simplicity, we used the verb structure that was studied in [16]. The task, rather than classification, is question answering [12], and we used a 1-dimensional sentence space. We made use of ZX-calculus [7] for easy translation between between DisCoCat diagrams and quantum circuits, and used CQC's t|ket> compiler and optimisation [20], which also relies on ZX-calculus.

References

1. Abramsky, S., Coecke, B.: A categorical semantics of quantum protocols. In: Proceedings of the 19th Annual IEEE Symposium on Logic in Computer Science (LICS), pp. 415–425 (2004). arXiv:quant-ph/0402130
2. Bankova, D., Coecke, B., Lewis, M., Marsden, D.: Graded entailment for compositional distributional semantics. J. Lang. Model. **6**(2), 225–260 (2019). arXiv:1601.04908.
3. Benedetti, M., Lloyd, E., Sack, S.: Parameterized quantum circuits as machine learning models. arXiv preprint arXiv:1906.07682 (2019)
4. Bolt, J., Coecke, B., Genovese, F., Lewis, M., Marsden, D., Piedeleu, R.: Interacting conceptual spaces i: grammatical composition of concepts. In: Kaipainen, M., Zenker, F., Hautamäki, A., Gärdenfors, P. (eds.) Conceptual Spaces: Elaborations and Applications. Synthese Library (Studies in Epistemology, Logic, Methodology, and Philosophy of Science), vol 405. Springer, Cham (2019). https://doi.org/10.1007/978-3-030-12800-5_9

5. Clark, S., Coecke, B., Grefenstette, E., Pulman, S., Sadrzadeh, M.: A quantum teleportation inspired algorithm produces sentence meaning from word meaning and grammatical structure. Malays. J. Math. Sci. **8**, 15–25 (2014). arXiv:1305.0556.
6. Coecke, B.: Kindergarten quantum mechanics. In: Khrennikov, A. (ed.) Quantum Theory: Reconsiderations of the Foundations III, pp. 81–98. AIP Press (2005). arXiv:quant-ph/0510032.
7. Coecke, B., Duncan, R.: Interacting quantum observables. In: Aceto, L., Damgård, I., Goldberg, L.A., Halldórsson, M.M., Ingólfsdóttir, A., Walukiewicz, I. (eds.) ICALP 2008. LNCS, vol. 5126. Springer, Berlin, Heidelberg (2008). https://doi.org/10.1007/978-3-540-70583-3_25
8. Coecke, B., Duncan, R.: Interacting quantum observables: categorical algebra and diagrammatics. New J. Phys. **13**, 043016 (2011). arXiv:quant-ph/09064725
9. Coecke, B., Kissinger, A.: Picturing Quantum Processes. A First Course in Quantum Theory and Diagrammatic Reasoning. Cambridge University Press (2017)
10. Coecke, B., Lewis, M., Marsden, D.: Internal wiring of cartesian verbs and prepositions. In: Lewis, M., Coecke, B., Hedges, J., Kartsaklis, D., Marsden, D. (eds.) Proceedings of the 2018 Workshop on Compositional Approaches in Physics, NLP, and Social Sciences. Electronic Proceedings in Theoretical Computer Science, vol. 283, pp. 75–88 (2018)
11. Coecke, B., Sadrzadeh, M., Clark, S.: Mathematical foundations for a compositional distributional model of meaning. In: van Benthem, J., Moortgat, M., Buszkowski, W. (eds.) A Festschrift for Jim Lambek. Linguistic Analysis, vol. 36, pp. 345–384 (2010). arxiv:1003.4394
12. De Felice, G., Meichanetzidis, K., Toumi, A.: Functorial question answering. arXiv:1905.07408 (2019)
13. Giovannetti, V., Lloyd, S., Maccone, L.: Quantum random access memory. Phys. Rev. Lett. **100**(16), 160501 (2008)
14. Grefenstette, E., Sadrzadeh, M.: Experimental support for a categorical compositional distributional model of meaning. In The 2011 Conference on Empirical Methods on Natural Language Processing, pp. 1394–1404 (2011). arXiv:1106.4058
15. Kartsaklis, D., Sadrzadeh, M.: A study of entanglement in a categorical framework of natural language. In: Proceedings of the 11th Workshop on Quantum Physics and Logic (QPL), Kyoto Japan (2014)
16. Kartsaklis, D., Sadrzadeh, M., Pulman, S., Coecke, B.: Reasoning about meaning in natural language with compact closed categories and Frobenius algebras. In: Logic and Algebraic Structures in Quantum Computing and Information. Cambridge University Press (2015). arXiv:1401.5980
17. Lambek, J.: From word to sentence. Polimetrica, Milan (2008)
18. Piedeleu, R., Kartsaklis, D., Coecke, B., Sadrzadeh, M.: Open system categorical quantum semantics in natural language processing. In: CALCO 2015 (2015). arXiv:1502.00831
19. Sadrzadeh, M., Clark, S., Coecke, B.: The Frobenius anatomy of word meanings I: subject and object relative pronouns. J. Logic Comput. **23**, 1293–1317 (2013). arXiv:1404.5278
20. Sivarajah, S., Dilkes, S., Cowtan, A., Simmons, W., Edgington, A., Duncan, R.: t|ket>: A retargetable compiler for NISQ devices. arXiv preprint arXiv:2003.10611 (2020)
21. Zeng, W., Coecke, B.: Quantum algorithms for compositional natural language processing. In: Proceedings SLPCS 2016, EPTCS, vol. 221, pp. 67–75 (2016)

Contents

Tool Presentations

Theoretical Advances

Conditional Bigraphs

Blair Archibald$^{(\boxtimes)}$ ⓘ, Muffy Calder ⓘ, and Michele Sevegnani ⓘ

School of Computing Science, University of Glasgow, Glasgow, UK
{blair.archibald,muffy.calder,michele.sevegnani}@glasgow.ac.uk

Abstract. Bigraphs are a universal graph based model, designed for analysing reactive systems that include spatial and non-spatial (*e.g.* communication) relationships. Bigraphs evolve over time using a rewriting framework that finds instances of a (sub)-bigraph, and substitutes a new bigraph. In standard bigraphs, the applicability of a rewrite rule is determined completely by a local match and does not allow any non-local reasoning, *i.e.* contextual conditions. We introduce conditional bigraphs that add conditions to rules and show how these fit into the matching framework for standard bigraphs. An implementation is provided, along with a set of examples. Finally, we discuss the limits of application conditions within the existing matching framework and present ways to extend the range of conditions that may be expressed.

Keywords: Bigraphs · Bigraphical reactive systems · Application conditions · Conditional rewriting

1 Introduction

Bigraphs are a universal mathematical model, introduced by Milner [15], for representing spatial and non-spatial relationships of physical or virtual entities. They have been applied to a wide range of systems including: mixed-reality systems [4], networking [6], Iot [2], security of cyber-physical systems [1], and biology [13].

Bigraphical reactive systems (BRS) augment bigraphs with a rewriting theory that allows models to evolve over time. The rewrite theory consists of a set of reaction rules $L \longrightarrow R$ that finds an *occurrence* of L in a larger bigraph B and replaces it with R. This form of rewriting only allows local reasoning through *matching* a pattern bigraph L exactly, but does not allow non-local reasoning that takes into account the context of a rule. We introduce *conditional* rules that use *application conditions* to specify contextual requirements within the rewrite system. Such conditional rules have proved invaluable in graph transformation systems [11] (GTS), a closely related formalism to bigraphs. However, it is important to note that although BRS and GTS are based on graph structures, the formalisms require completely different semantics for conditional rewriting. For example, in GTS there is a single context for the rules, whereas BRS feature a distinct *context* and a *parameter*, and so application conditions can be specified over either.

ⓒ Springer Nature Switzerland AG 2020
F. Gadducci and T. Kehrer (Eds.): ICGT 2020, LNCS 12150, pp. 3–19, 2020.
https://doi.org/10.1007/978-3-030-51372-6_1

A common requirement for conditional rules in BRS is to avoid the duplication of links between entities. As an example, consider the `createLink` rule shown in Fig. 1a (full details of this notation is in Sect. 2). Bigraphs consist of entities, A, B and L, shown as shapes that are related either by a nesting relationship – L *inside* A – or via the green hyperlinks. Sites (grey rectangles) represent parts of the system that have been abstracted away, *i.e.* other bigraphs may appear inside. Without an application condition, this rule allows *any* number of L-L links to be created between an A and a B; this is because the sites may contain any number of other entities, including existing L entities. If we wish to restrict to *single* L-L links between A-B pairs, we have to employ some sort of *tagging* scheme [6], often coupled with *rule priorities* [3], that can determine when a link does or does not exist. In practice, this requires an extra entity (for tagging) and additional (four) reaction rules. This inflates the model with non-domain specific rules, generates additional control-only steps in the resulting transition system, and, more importantly, obfuscates the purpose of the rule, which is to create non-duplicate links.

With conditional rules, we can achieve this goal in a *single* rewrite step. In the example, we create a conditional rule though the addition of a negative application condition, which is shown in Fig. 1b. This states that within the *parameter* of the rule, *i.e.* in the *sites*, we must not find an existing L-L link. If such a link is found then the rule does not apply. Consider application of the rule to the example bigraph in Fig. 1c. For the linked A, the existing L entities appear *inside* the two sites of the parameter (of the left-hand side of the rule). As the parameter negative condition forbids such a shape to appear in the left-hand side, no new link can be created. On the other hand, for the unlinked A, no L-L link is present in the parameter and so the negative condition is not satisfied, and a new link can be created.

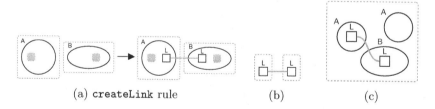

(a) `createLink` rule   (b)   (c)

Fig. 1. Negative application condition to avoid duplicate L-L links. (b) Negative application condition for the parameter. Given the bigraph in (c), the rule does not apply for the already linked A and B but does for the unlinked A and B. (Color figure online)

In Sects. 3–5 we show that conditional rules are possible within Milner's original BRS formalism, and we give our implementation in BigraphER [19]. In Sect. 6 we reflect on the fact that, due to how bigraph matching is defined, conditional rules are limited in the conditions that can be expressed. We discuss these limitations, with examples, and indicate possible extensions that include matching on names and patterns, spatial logics to encode spatial context, and matching with sorting schemes, site numbering, and nested application conditions.

We make the following contributions:

- We extend the original formalism of bigraphs to support non-local application conditions for bigraphical reactive systems.
- To show application conditions are both implementable and useful, we implement application conditions in BigraphER [19] and provide example models that highlight common uses of application conditions in practice.
- We show the limits of such an approach, based strictly on the existing matching/decomposition of bigraphs, and highlight areas for future exploration.

2   Bigraphs

A bigraph consists of two orthogonal structures: a *place graph* that describes the *nesting* of entities, *e.g.* a Phone inside a Room, and *link graph* that provides non-local hyperlinks between entities, *e.g.* allowing Phone entities to communicate regardless of location. In standard bigraphs place graphs are forests, however here we use bigraphs with sharing [18], that has place graphs as directed acyclic graphs, allowing entities to have multiple parents. Bigraphs feature an equivalent algebraic and an intuitive diagrammatic form, and we use this diagrammatic form where possible.

An example bigraph is in Fig. 1c. We draw entities as different (colored) shapes, often omitting the label when possible. Containment illustrates the spatial nesting relationship, *e.g.* L is contained by A, while green hyperedges represent non-spatial connections. Entities have a fixed *arity* (number of links), *e.g.* L has arity 1, but links may be disconnected/closed.

Each place graph has m *regions*, shown as the dashed rectangles, and n *sites*, shown as filled dashed rectangles. Regions represent parallel parts of the system, and sites represent abstraction, *i.e.* an unspecified bigraph (including the empty bigraph) exists there. Similarly, link graphs have a (finite) set of inner names, *e.g.* $\{z\}$ and outer names, *e.g.* $\{x, y\}$. For example, in Fig. 2b, C has an inner name x, d has outername x, and id_I has both an inner and outer name x (where both x's are distinct).

Bigraphs are compositional structures, that is, we can build larger bigraphs from smaller bigraphs. Composition of bigraphs consists of placing regions in sites, and connecting inner and outer-faces on like-names.

Algebraically we describe bigraphs using their *interfaces*, *e.g.* $B : \langle n, X \rangle \rightarrow \langle m, Y \rangle$, or more succinctly $B : I \rightarrow J$, where n is the number of sites, m number of regions, X a set of inner names, and Y a set of outer names. Composition of bigraphs is defined when the interfaces match, *i.e.* $B_1 \circ B_0$ is defined for $B_0 : I \rightarrow J$ and $B_1 : J \rightarrow K$. We use ϵ to refer to the empty interface $\langle 0, \emptyset \rangle$, and call bigraphs of the form $\epsilon \rightarrow I$ *ground*, *i.e.* bigraphs with no sites and no inner names. Figure 1a is non-ground as it contains two sites, while Fig. 1c is ground as it contains no sites or inner names.

We let $\mathrm{id}_I : \mathrm{id}_I : \langle m, X \rangle \to \langle m, X \rangle$ be the identity bigraph over an interface $I : \langle m, X \rangle$. id_I maps names in X to themselves and places m sites in m regions. This bigraph is particularly important for matching as it allows names and entities to move between the context and parameter of a match. With these definitions, bigraphs form a pre-category[1] with objects as interfaces and arrows as bigraphs.

While composition combines bigraphs *vertically*, we can also combine bigraphs *horizontally* through the tensor product $\otimes$. This tensor product extends both the sites/regions and name sets for the interfaces. For example, given $A : \langle 0, \{x\} \rangle \to \langle 1, \{y\} \rangle$ and $B : \langle 1, \emptyset \rangle \to \langle 2, \{z\} \rangle$, we can construct a new bigraph $A \otimes B : \langle 1, \{x\} \rangle \to \langle 3, \{y, z\} \rangle$ Note that $\otimes$ is only defined when the sets of interface names are disjoint.

Notation. When referring to a ground bigraph we use lower-case letters, while general bigraphs, that may or may not be ground, are denoted in upper-case. Where the identity of an interface is not required we use $\cdot$ as a placeholder for $I, J, \ldots$

2.1 Bigraphical Reactive Systems

Bigraphical reactive systems (BRS) equip bigraphs with a rewriting theory that allows models to evolve over time. Intuitively, applying a reaction rule $L \longrightarrow R$ to bigraph B finds an *occurrence* of L in B (if one exists) and replaces it with R to create B'. Most often we rewrite over *ground* bigraphs as these represent fully formed models, *e.g.* without holes/sites. Here we give the most general definitions possible, *i.e.* for arbitrary (including ground) bigraphs B, and specialise to ground bigraphs when necessary.

We work with a restricted version of reaction rules that are "well-behaved" where L is *solid*[2]. Solid bigraphs were introduced by Krivine *et al.* [13] to count unique occurrences for stochastic BRS.

Definition 1 (solid). *A bigraph is* solid *if:*

- *All roots contain at least one node, and all outer names are connected to at least one edge.*
- *No two sites or inner names are siblings*
- *No site has a root as a parent*
- *No outer name is linked to an inner name.*

Definition 2 (occurrence). *We say a bigraph P occurs in B, written $B \vDash P$, if there exists a decomposition $B = C \circ (P \otimes \mathrm{id}_I) \circ D$ for some context C and parameter D. That is, there is a match for P in B.*

Likewise, we say P does not occur in B, written $B \nvDash P$ if $\nexists C' \nexists D', B = C' \circ (P \otimes \mathrm{id}_I) \circ D'$. That is, there is no match for P in B.

---

[1] Bigraphs are not a full category as composition is not defined for non-disjoint *supports*. We do not discuss support here.

[2] The definition of solid for bigraphs with sharing differs slightly, see [17, Defn 3.6.1].

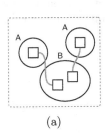

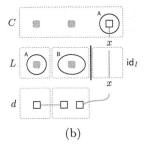

(a)   (b)

Fig. 2. Decomposition of ground bigraph $b = C \circ (L \otimes \mathsf{id}_I) \circ d$.

For a solid L, we gain the property that an occurrence $B \models L$ *uniquely identifies* a context C and parameter D.

We show graphically how occurrences are found in Fig. 2. Given the ground bigraph b shown in Fig. 2a, we show in Fig. 2b one (of the two possible) decompositions when matching against the rule given in Fig. 1a. The match L, by definition, has the same form as the left-hand-side of the reaction rule. The context C captures entities in the bigraph that do not lie within the match, while the parameter d, which must be ground as b is ground, provides the entities required to fill any sites/inner names in the match. To allow names (and entities in the case of sharing, *i.e.* for those sharing a parent in the match and context) to move between the parameter and the context, we allow an interface id_I next to the match. The use of id_I means the distinction between context and parameter is not always clear as both names and unmatched entities can move between the context and parameter as required, *e.g.* an entity from the context can appear in the parameter by extending id_I with an additional, trivial, region/site. Note that we only take L to be solid allowing these region/sites to be added as required.

Allowing entities to move between the context and parameter complicates the specification of application conditions that must determine if the condition is within the context of the parameter. We deal with this by forcing the parameter to be *minimal* such that it contains only entities that are within sites of the match, and all other entities move to the context. We note this is a choice and the theory also applies to systems that take the minimal context.

Definition 3 (reaction rule). *A reaction rule* R *is a pair of bigraphs,* R = (L, R), *defined over the same interface, and often written as* $L \longrightarrow R$, *with* L *solid. Applying a reaction rule consists of replacing an occurrence* $B \models L$, *in a given bigraph* B, *with* R.

Rewriting (over ground bigraphs) is shown graphically in the commuting diagram of Fig. 3. Given a ground bigraph a, we first find a decomposition $a = C \circ (L \otimes \mathsf{id}_I) \circ d$ and, should such a decomposition exist, rewrite it to obtain $a' = C \circ (R \otimes \mathsf{id}_I) \circ d$. Both the context C and parameter d are the same for both the left and right hand sides of the rewrite[3]. In Sect. 3 we show how checking application conditions corresponds to a further decomposition of C and d.

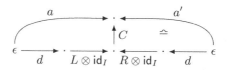

Fig. 3. Bigraph rewriting [15]. a rewrites to (a support equivalent) a' if there is a decomposition $a = C \circ (L \otimes \mathrm{id}_I) \circ d$ and rule $L \longrightarrow R$.

While the rules are specified for *abstract* bigraphs, *i.e.* they match any entity of the correct type, rewriting itself works on *concrete* bigraphs where entities have distinct identifiers. In general, we may rewrite into any *support equivalent* a', notated $\backsimeq$ where support equivalence allows renaming of entity identifiers and link-names while keeping the structure intact. For the rest of the paper we assume support equivalence without explicitly stating it.

Given a set of reaction rules we construct a BRS as follows.

Definition 4 (bigraphical reactive system (BRS)). *A bigraphical reactive system consists of a set of* ground *bigraphs $\mathcal{B}$ and set of reaction rules $\mathcal{R}$, defined over $\mathcal{B}$, of the form $L \longrightarrow R$. The reaction relation $\longrightarrow$ over ground bigraphs is the smallest such that $b \longrightarrow b'$ when $b = C \circ (L \otimes \mathrm{id}_I) \circ d$ and $b' = C \circ (R \otimes \mathrm{id}_I) \circ d$ for $b, b' \in \mathcal{B}$, reaction $(L \longrightarrow R) \in \mathcal{R}$, context C, and parameter d.*

That is, our system consists all possible (ground) bigraphs closed under $\longrightarrow$.

3   Application Conditions for Bigraphs

We show how application conditions for bigraphs are instances of the bigraph matching problem. We begin by defining application conditions, which can be viewed as stand-alone instances of the left-hand-side of a rule.

Application conditions include information such as if they are positive or negative and if they apply to the context or parameter of a match.

Definition 5 (application condition). *An application condition is a tuple $\langle t, P, l \rangle$ where $t \in \{+, -\}$ is the type of application condition, either positive or negative, P is a (non-ground, not necessarily solid) constraint bigraph, and $l \in \{\uparrow, \downarrow\}$ determines if the condition is over the context ($\uparrow$) or parameter ($\downarrow$).*

---

[3] Bigraphs allow the use of an instantiation map η [15, Defn 8.3] that specifies a mapping of sites in the left-hand side to those in the right-hand site. We do not consider instantiation maps here.

Checking the conditions is a matching problem and no *rewriting* is performed on the context/parameter, *i.e.* we do not reduce constraints. Finally, we define conditional reaction rules.

Definition 6 (conditional reaction rule). *A conditional reaction rule* R *consists of a reaction rule* R : L $\longrightarrow$ R, *and a set of application conditions* $\mathcal{A}$. *Unconditional rules are those where* $\mathcal{A} = \emptyset$.

A rule L $\longrightarrow$ R *applies to a if there is a decomposition* $a = C \circ (L \otimes \mathrm{id}_I) \circ d$, *and*

$$\forall \langle +, P, \uparrow \rangle \in \mathcal{A}, \ C \vDash P$$
$$\forall \langle -, P, \uparrow \rangle \in \mathcal{A}, \ C \nvDash P$$
$$\forall \langle +, P, \downarrow \rangle \in \mathcal{A}, \ d \vDash P$$
$$\forall \langle -, P, \downarrow \rangle \in \mathcal{A}, \ d \nvDash P$$

In a slight overload of notation, we use $C, d \vDash a$ *when an application condition* $a \in \mathcal{A}$ *is satisfied (positively or negatively) in context* C *and parameter* d.

We show graphically how application conditions of the form $\langle +, P, \uparrow \rangle$ and $\langle +, P', \downarrow \rangle$ are checked in Fig. 4. This diagram shows the left-hand side of Fig. 3, *i.e.* the decomposition of ground bigraph a, but with the context C and parameter d further decomposed. As C can be decomposed into three arrows and as P exists in the decomposition we know the application condition is met. The use of the additional parameter E allows for application conditions to have a different interface from the original match (shown here as an arbitrary $\cdot$). Without E all application conditions would be forced to have the same number of sites as regions of L, and matching outer/inner names, making it difficult to specify reusable conditions. Likewise we can check P' via the decomposition of d. For negative application conditions we instead show that no such decomposition exists for a given P, *i.e.* we cannot form the diagram in Fig. 4.

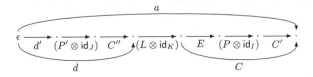

Fig. 4. Context and parameter decomposition over left-hand side of Fig. 3 to check positive application conditions

When multiple application conditions are specified, we check that matches exist (or do not exist) separately for each condition. That is, we take the conjunction of the conditions. Importantly, we do not force the decompositions to cover a unique set of entities, and allow the same entity to be matched for multiple application conditions, *i.e.* conditions can overlap. We discuss non-overlapping rules as an extension to this approach in Sect. 6.4.

Definition 7 (conditional bigraphical reactive system). *A conditional BRS consists of a set of* ground *bigraphs $\mathcal{B}$ and set of conditional reaction rules $\mathcal{R}_c$ of the form $(L \rightarrow R, \mathcal{A})$.*

*The reaction relation $\longrightarrow$ is the smallest such that $b \longrightarrow b'$ when $b = C \circ (L \otimes \mathrm{id}_I) \circ d$ and $b' = C \circ (R \otimes \mathrm{id}_I) \circ d$, for $b, b' \in \mathcal{B}$, reaction $(L \rightarrow R, \mathcal{A}) \in \mathcal{R}$, context C, and parameter d. **Additionally**, $\forall a \in \mathcal{A}, C, d \vDash a$.*

When $\mathcal{A} = \emptyset$, a conditional BRS is a standard BRS as in Definition 4.

4   Implementation

We have implemented application conditions in BigraphER [19] an open-source framework for bigraphs[4]. BigraphER supports bigraphs with sharing [18], including an efficient matching algorithm based on SAT that we use to decompose the bigraph to check application condition predicates.

We show the example rule for Fig. 1 in the BigraphER language in Listing 1.1. The rules are written as before, with the addition of an `if` clause that allows application conditions to be specified as arbitrary bigraphs. The structure of the `conds` production has the following BNF specification and appears as an optional statement of any reaction rule; the production ⟨bigraph_exp⟩ parses an arbitrary bigraph expression. As is common in programming languages, we use ! to represent negation, while `param` and `ctx` become reserved keywords that specify *where* we should search for the constraint specified by ⟨bigraph_exp⟩ – in the parameter (sites) or context respectively.

$$\begin{aligned}
\langle \text{place} \rangle &::= \texttt{param} \mid \texttt{ctx} \\
\langle \text{bang} \rangle &::= \texttt{!} \mid \epsilon \\
\langle \text{app\_cond} \rangle &::= \langle \text{bang} \rangle \, \langle \text{bigraph\_exp} \rangle \, \texttt{in} \, \langle \text{place} \rangle \\
\langle \text{app\_conds} \rangle &::= \langle \text{app\_cond} \rangle \mid \langle \text{app\_cond} \rangle , \langle \text{app\_conds} \rangle \\
\langle \text{conds} \rangle &::= \texttt{if} \, \langle \text{app\_conds} \rangle
\end{aligned}$$

5   Examples

We apply conditional rewriting to three typical examples: ensuring entities are unique, implementing a $\not\exists$ operator and performing counting in multi-sets, and replacing priorities/control with conditionals.

---

[4] Available, along with the example models of Sect. 5, at www.dcs.gla.ac.uk/~michele/bigrapher.html.

Listing 1.1: Specifying application conditions in BigraphER.

```
1   # Example from Fig. 1
2
3   # control <name> = <arity>
4   ctrl A = 0; # Circle
5   ctrl B = 0; # Ellipse
6   ctrl L = 1; # Square
7
8   react createLink =
9   A.id || B.id --> /x (A.(L{x} | id) || B.(L{x} | id))
10  if
11  !(/y (L{y} || L{y})) in param;
```

5.1   Uniqueness of Entities

Many applications have constraints on the uniqueness of particular entities. For example, in a networking application *e.g.* in [20], we want to disallow two devices having the same MAC address.

In bigraphs there is no general method to declare an entity as *unique*. However, with application conditions we can check, before an entity is created, that no identical entity exists in either the context *or* the parameter. Since there is no way to create a duplicate, if we use conditional reaction rules with appropriate conditions to generate a model, this will ensure entities are unique.

As an example, consider the rule `createUnique` shown in Fig. 5. This rule allows an entity Unique to be created in a given Place so long as no other entity Unique already exists in the model – either in the same place, *i.e.* the parameter, or anywhere in the context.

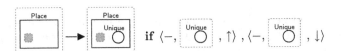

Fig. 5. `createUnique` – application conditions force uniqueness of entity Unique.

5.2   Non-existence and Counting in Multisets

Non-atomic entities in bigraphs can be considered multisets, *i.e.* they hold an arbitrary number of children including duplicates. An example bigraph used as a (multi-)set is in Fig. 6a.

Checking an entity is in the set is simple: we match on the entity of interest and use a site to allow other children (including the empty child) to be present, *i.e.* the rule Fig. 6b would apply to Fig. 6a. However, matching on the non-existence of a child is difficult. To ensure the entity of interest cannot not appear in a site, we must specify rules for every possible combination of other entities in the set, *e.g.* Fig. 6c. In this case the rule does not apply to Fig. 6a as the circle entity is present. It is often not practical to specify rules for all permutations of additional entities in the set as the number of rules increases factorially.

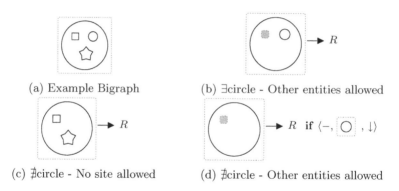

(a) Example Bigraph   (b) ∃circle - Other entities allowed

(c) ∄circle - No site allowed   (d) ∄circle - Other entities allowed

Fig. 6. Using negative application conditions for non-existence in multisets.

As shown in Fig. 6d, negative application conditions for the parameters allows sites to be used in non-existence checks by allowing anything in the parameter *except* the entity of interest – a concise and natural way of specifying the rule.

A similar issue of sites hiding too much information occurs when counting. For example, if we wish to match at most one T then without negative conditions we must enumerate all possible sets with a single T. With negative conditions, we can specify that a site exists but that the site does not contain more than one T.

5.3 Encoding Control Flow

Models often need to encode some control flow, for example, to implement turn-based control. Often this is achieved through the use of tagging, counters, and prioritised rules, e.g. [16], that determine when the algorithms should change state.

Application conditions can make it easier to encode elements of control without requiring counting, tagging, or priorities. Consider the system in Fig. 7 that uses turn-based control where entities, shown as circles, representing autonomous agents, cycle between a Move phase and a Act phase, with the current phase determined by a Controller. Each agent keeps track of its local state, *i.e.* what the last action it performed was.

The move and act rules (Figs. 7b and 7e) show example actions the agents can take. While the act shown only changes internal state (represented by the fill color) to say an action has been performed, in practice this would perform some meaningful step. Applications conditions on the rules ensure that agents only perform valid actions for the given controller state. It is possible to write a similar rule without application conditions by matching on Move in a separate region, however this does not allow the agents to be nested *under* a controller and obscures the meaning of the rule.

Rules switch$_1$ and switch$_2$ (Figs. 7c and 7d) toggle the controller state only once *all* agent has taken the appropriate action, *i.e.* it encodes fix-point behaviour. As the rules simply check for the non-existence of agents waiting to take an action, they work for *any* number of agents without needing to explicitly encode counting. Likewise, there is no need to introduce priorities to the rules as the conditions guard them from firing at the wrong time.

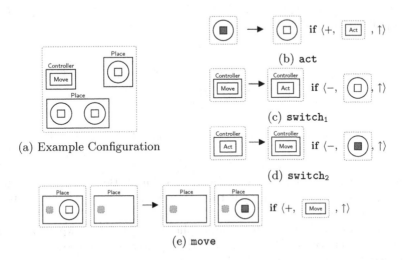

(a) Example Configuration

(b) act

(c) switch$_1$

(d) switch$_2$

(e) move

Fig. 7. Encoding turn-based control. (Color figure online)

6 Discussion and Limitations

The key advantage of the presented approach is that application conditions are defined solely in terms of the existing rewriting theory for bigraphs – allowing theory/tool reuse.

However, this approach does not capture all the application conditions we might wish to specify in practice. In this section we highlight the limitations of the current approach and discuss how, by moving away from standard bigraph theory, *i.e.* changing the semantics of matching, or utilising spatial logics, we can express a wider range of application conditions.

6.1 Matching on Names

Commonly, application conditions in GTS make use of graph edges to access the context for a *particular* node, *i.e.* the existence of a link identifies an entity of interest in the context. In bigraphs such an approach is not possible as link names *cannot* be used as identifiers.

For example, consider the reaction rule and application condition in Fig. 8. The intention of the rule is to state that an (empty) circle can be transformed into a red circle when it is connected to a square on the x link. However, this is not the interpretation because an open name in a reaction rule is not an identifier, but, much like sites, it signifies that there *may* be additional entities on the same link. As such, the x on the left-hand side and the x in the application condition are not considered to be the same link (*i.e.* the second x is not bound to the first x); we would get the same rewrite if we replaced, for example, the x in the application condition with y.

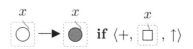

Fig. 8. Names are not identifiers in bigraphs. (Color figure online)

This issue has been observed elsewhere, for example Benford *et al.* [4] extend the matching semantics for bigraphs to "bigraph patterns" that allow matching on specific identifiers.

Utilising this type of matching for application conditions would allow the intended interpretation of Fig. 8, where x is scoped over both the left-hand side and the application condition. No implementation of this matching currently exists, and it remains unclear how this might affect other aspects of the bigraph theory.

6.2   Matches Can Be Too Large

Although we have shown application conditions as defined are useful for practical applications, care must be taken in their use. Currently an application condition allows its constraint to appear *anywhere* in the context/parameter. This is sometimes too strong. Consider our first example of avoiding duplicate links (Fig. 1). If the target bigraph contains an L-L link *within* A, as shown in Fig. 9, then the rule does not apply, even though there is no duplicate link *between* A and B.

Practically such cases are often not an issue as, for example, the `createLink` rule only ever creates links between an A and B – disallowing internal A–A links from ever being created. If internal links are needed, then a different entity type could be used to distinguish between external and internal links. Finally, a *sorting scheme* [15, Chapter 6] could be used to disallow invalid contexts/parameters from existing, however there is currently no automated tool support for checking a sorting scheme is satisfied.

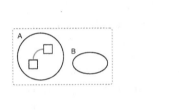
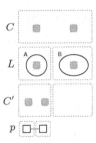

Fig. 9. Matches occur *anywhere* in the parameter, not only in specific regions. Decomposition shows further decomposition $d = C' \circ p$ for application condition constraint p (Fig. 1b) for rule `createLink`.

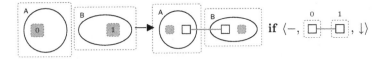

Fig. 10. Explicit parameter placement

6.3 Matching Specific Places

Another possible solution to the internal link issue above is to extend matching to allow application conditions to specify *where* in the context/parameter a particular entity should be found. For example, in Fig. 10, we add explicit indices to the sites allowing us to specify how the application condition should compose with the match – in this case, that the two L ends are in distinct sites. Matching routines that can check for the explicit placement of entities are not currently available, and as with name linking it remains unclear what affect this might have on the rest of the bigraph theory. As it would only be used for application condition matching it could potentially be defined as a special case matching routine. It is particularly unclear how to perform such matches if sharing is allowed *e.g.* as shares can merge regions in the parameter.

Explicit placement is also possible for conditions in the context, this time allowing us to specify that a particular region in the match occurs as a descendent of a site in the context. Figure 11 is an example of this where entity C must be a – not-necessarily direct – descendant of D in the condition.

$$\overset{0}{\underset{C}{\bigcirc}} \longrightarrow R \quad \text{if} \ \langle +, \ \overset{D}{\boxed{0}}, \ \uparrow \rangle$$

Fig. 11. Explicit context placement

6.4   Handling Overlaps

A conditional rule only applies if the conjunction of all application conditions is true, *e.g.* all matches are present. As each application condition is checked independently, the same entity can be used in multiple matches *i.e.* we allow overlaps between application conditions. For example, consider the rule in Fig. 12. The intended behaviour is to check there are *two* distinct square entities in the context – regardless of how many other entities *e.g.* the diamond, are in the nesting – however as we allow overlaps this matches even in the case a single square entity is in the context.

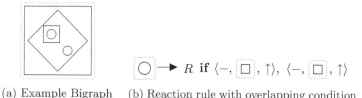

(a) Example Bigraph   (b) Reaction rule with overlapping condition

Fig. 12. Rule applies to both circle entities as overlaps are allowed

A solution for this in GTS is to use *nested conditions* [12] that allow further checks to be made on the context of the constraint *within* in the application condition.

We can define a form of nested application conditions for bigraphs by allowing further decomposition within the application condition match. That is, we first find a suitable context (parameter) and then decompose the context (parameter) into a new context/parameter and check nested conditions on these new context/parameter.

As it only requires additional decompositions, such an approach is possible within the existing bigraph matching framework. However, it requires matches to apply in a specific order *e.g.* closest match first, to ensure we know if the nested condition should be in the parameter or context. For example, in Fig. 12 if we first match on the outermost square then the next condition appears in the parameter. But, if we first match on the innermost square then the next condition is in the context.

6.5   Related Work

Sorting Schemes. It is possible to give a sorting scheme to bigraphs [15, Chapter 6] that determines when a bigraph is well formed, *e.g.* that a Room cannot be within a Person – much like a type graph for GTS. Sorting schemes compliment application conditions. The sorting scheme defines what *can* be in the context/parameter, while application conditions determine what *is* in the current context/parameter. While there is an existing theory for sorts, there is currently no tool support available.

Conditional Transformation Systems. Conditional rewriting is often found in rewriting logic [14] where rules take the form $r \rightarrow l$ if $u_1 \wedge u_2 \wedge \ldots$. Unlike application conditions, rewriting logic allows the terms in u_1 to be reduced over a set of equations.

Application conditions are also a common feature in GTS [10, Chapter 7] and are present in most graph rewriting frameworks. Intuitively, a GTS application condition is defined as the existence/non-existence of a match (graph morphism) between a non-local pattern and the current graph.

More general treatments of (nested) application conditions considers them at the level of adhesive categories [5,12]. Such categories are are closely linked with the DPO rewriting approach. The relationship between bigraphs and cospan categories/DPO rewriting has previously been explored [9], and such an input-linear variant of bigraphs[5], such as that of Sobociński [21], could fit such a framework.

Spatial Logics. Another approach to application conditions for bigraphs was explored by Tsigkanos *et al.* [22]. Here application conditions are specified using a spatial logic for closure spaces [7], of which graphs are an instance. The logic requires flattening the bigraph to a graph – losing the distinction between spatial and non-spatial links – but provides features such as matching links by name and specifying reachability constraints, *e.g.* a PC connects to a Printer through some path.

A more general spatial logic for bigraphs is BiLog [8], which could also implement application conditions while maintaining the orthogonality between space and linking. However, there is a lack of tool support and the decidability of the logic remains an open question.

Importantly, both logics require the user to specify constraints in a language separate to that of bigraphs, while our approach maintains the diagrammatic approach by having conditions as bigraphs.

7   Conclusion

Reactive modelling formalisms, such as bigraphical reactive systems, should make it as easy as possible to express how a system evolves over time. Whether or not a reaction rule is applicable often depends not only on a local match, but also on the surrounding context. Application conditions allow non-local reasoning to be added to reaction rules allowing the context to be interrogated to check the existence/non-existence of constraints.

We have extended the theory of bigraphical reactive systems with conditional reaction rules that allow application conditions to be specified. Unlike graph transformation systems that feature a single context, bigraphs have both a *context* (above the match) and a *parameter* (below the match). We show how these contexts can be further decomposed as additional instances of the bigraph matching problem, enabling the existing matching framework to be used to check

---

[5] Standard bigraphs are *output-linear*.

application conditions. To show this is useful in practice we implement conditional rules in BigraphER [19].

Unfortunately, such rules do not let us express all conditions of interest. For example we cannot track a name from the match into the context, or specify the exact location of entities *e.g.* do not apply a rule if entity A is a grandparent. Specifying these types of property require extensions to how bigraphs are matched, and potentially the use of spatial logics to provide exact specification of spatial constraints.

This paper paves the way for future work on application conditions for bigraphs, and, more generally, improvements to the matching algorithm that allow more expressive constraints to be described.

Acknowlegements. This work was supported EPSRC grant S4: Science of Sensor Systems Software (EP/N007565/1), and PETRAS SRF grant MAGIC (EP/S035362/1).

References

1. Alrimawi, F., Pasquale, L., Nuseibeh, B.: On the automated management of security incidents in smart spaces. IEEE Access **7**, 111513–111527 (2019)
2. Archibald, B., Shieh, M., Hu, Y., Sevegnani, M., Lin, Y.: BigraphTalk: verified design of IoT applications. IEEE Internet Things J. **7**(4), 2955–2967 (2020). https://doi.org/10.1109/JIOT.2020.2964026. ISSN 2372-2541
3. Baeten, J.C.M., Bergstra, J.A., Klop, J.W., Weijland, W.P.: Term-rewriting systems with rule priorities. Theor. Comput. Sci. **67**(2&3), 283–301 (1989)
4. Benford, S., Calder, M., Rodden, T., Sevegnani, M.: On lions, impala, and bigraphs: modelling interactions in physical/virtual spaces. ACM Trans. Comput. Hum. Interact. **23**(2), 1–56 (2016)
5. Bruggink, H.J.S., Cauderlier, R., Hülsbusch, M., König, B.: Conditional reactive systems. In: FSTTCS, pp. 191–203 (2011)
6. Calder, M., Koliousis, A., Sevegnani, M., Sventek, J.S.: Real-time verification of wireless home networks using bigraphs with sharing. Sci. Comput. Program. **80**, 288–310 (2014)
7. Ciancia, V., Latella, D., Loreti, M., Massink, M.: Specifying and verifying properties of space. In: Diaz, J., Lanese, I., Sangiorgi, D. (eds.) TCS 2014. LNCS, vol. 8705, pp. 222–235. Springer, Heidelberg (2014). https://doi.org/10.1007/978-3-662-44602-7_18
8. Conforti, G., Macedonio, D., Sassone, V.: Spatial logics for bigraphs. In: Caires, L., Italiano, G.F., Monteiro, L., Palamidessi, C., Yung, M. (eds.) ICALP 2005. LNCS, vol. 3580, pp. 766–778. Springer, Heidelberg (2005). https://doi.org/10.1007/11523468_62
9. Ehrig, H.: Bigraphs meet double pushouts. Bull. EATCS **78**, 72–85 (2002)
10. Ehrig, H., Ehrig, K., Prange, U., Taentzer, G.: Fundamentals of Algebraic Graph Transformation. Monographs in Theoretical Computer Science. An EATCS Series. Springer, Heidelberg (2006). https://doi.org/10.1007/3-540-31188-2
11. Habel, A., Heckel, R., Taentzer, G.: Graph grammars with negative application conditions. Fundam. Inform. **26**(3/4), 287–313 (1996)

12. Habel, A., Pennemann, K.-H.: Nested constraints and application conditions for high-level structures. In: Kreowski, H.-J., Montanari, U., Orejas, F., Rozenberg, G., Taentzer, G. (eds.) Formal Methods in Software and Systems Modeling, Essays Dedicated to Hartmut Ehrig on the Occasion of His 60th Birthday. LNCS, vol. 3393, pp. 293–308. Springer, Heidelberg (2005). https://doi.org/10.1007/978-3-540-31847-7_17

13. Krivine, J., Milner, R., Troina, A.: Stochastic bigraphs. Electron. Notes Theor. Comput. Sci. **218**, 73–96 (2008)

14. Meseguer, J.: Conditional rewriting logic as a unified model of concurrency. Theor. Comput. Sci. **96**(1), 73–155 (1992)

15. Milner, R.: The Space and Motion of Communicating Agents. Cambridge University Press, Cambridge (2009)

16. Muffy, C., Michele, S.: Modelling IEEE 802.11 CSMA/CA RTS/CTS with stochastic bigraphs with sharing. Form. Asp. Comput. **26**(3), 537–561 (2014). https://doi.org/10.1007/s00165-012-0270-3. ISSN 0934-5043

17. Sevegnani, M.: Bigraphs with sharing and applications in wireless networks. Ph.D. thesis, School of Computing Science, University of Glasgow (2012)

18. Sevegnani, M., Calder, M.: Bigraphs with sharing. Theor. Comput. Sci. **577**, 43–73 (2015)

19. Sevegnani, M., Calder, M.: BigraphER: rewriting and analysis engine for bigraphs. In: Chaudhuri, S., Farzan, A. (eds.) CAV 2016. LNCS, vol. 9780, pp. 494–501. Springer, Cham (2016). https://doi.org/10.1007/978-3-319-41540-6_27

20. Sevegnani, M., Kabác, M., Calder, M., McCann, J.A.: Modelling and verification of large-scale sensor network infrastructures. In: ICECCS, pp. 71–81 (2018)

21. Sobociński, P.: Deriving process congruences from reaction rules. Ph.D. thesis, Aarhus University (2004)

22. Tsigkanos, C., Kehrer, T., Ghezzi, C.: Modeling and verification of evolving cyber-physical spaces. In: ESEC/FSE, pp. 38–48 (2017)

Confluence up to Garbage

Graham Campbell[1]([⊠])[iD] and Detlef Plump[2][iD]

[1] School of Mathematics, Statistics and Physics, Newcastle University,
Newcastle upon Tyne, UK
`g.j.campbell2@newcastle.ac.uk`
[2] Department of Computer Science, University of York, York, UK
`detlef.plump@york.ac.uk`

Abstract. The transformation of graphs and graph-like structures is ubiquitous in computer science. When a system is described by graph-transformation rules, it is often desirable that the rules are both terminating and confluent so that rule applications in an arbitrary order produce unique resulting graphs. However, there are application scenarios where the rules are not globally confluent but confluent on a subclass of graphs that are of interest. In other words, non-resolvable conflicts can only occur on graphs that are considered as "garbage". In this paper, we introduce the notion of confluence up to garbage and generalise Plump's critical pair lemma for double-pushout graph transformation, providing a sufficient condition for confluence up to garbage by non-garbage critical pair analysis. We apply our results to language recognition by backtracking-free graph reduction, showing how to establish that a graph language can be decided by a system which is confluent up to garbage. We present two case studies with backtracking-free graph reduction systems which recognise a class of flow diagrams and a class of labelled series-parallel graphs, respectively. Both systems are non-confluent but confluent up to garbage.

Keywords: Graph transformation · Confluence · Graph languages · Decision procedures

1 Introduction

Rule-based graph transformation and graph grammars date back to the late 1960s. The best developed theoretical framework is the so-called double-pushout (DPO) approach to graph transformation [10,12]. When specifying systems in computer science by DPO graph transformation rules, it is often desirable that the rules are both terminating and confluent so that rule applications in an arbitrary order produce unique resulting graphs. However, there are application scenarios where the rules are not confluent but confluent on a subclass of graphs

G. Campbell—Supported by a Vacation Internship and a Doctoral Training Grant No. (2281162) from the Engineering and Physical Sciences Research Council (EPSRC) in the UK, while at University of York and Newcastle University, respectively.

F. Gadducci and T. Kehrer (Eds.): ICGT 2020, LNCS 12150, pp. 20–37, 2020.
https://doi.org/10.1007/978-3-030-51372-6_2

that are of interest. In other words, non-resolvable conflicts can only occur on graphs that are considered as "garbage".

In this paper, we introduce the notions of (local) confluence up to garbage and termination up to garbage in graph transformation. We generalise Plump's Critical Pair Lemma [26,28] and Newmann's Lemma [25] and thereby allow to check confluence up to garbage via non-garbage critical pair analysis. We apply our results to language recognition by backtracking-free graph reduction, showing how to establish that a graph language can be decided by a system which is confluent up to garbage. We present two case studies with backtracking-free graph reduction systems which recognise a class of flow diagrams and a class of labelled series-parallel graphs, respectively. Both systems are non-confluent but confluent up to garbage. Parts of this paper are based on Chapter 4 of an unpublished report [6], in turn developed from Campbell's BSc Thesis [5].

2 Preliminaries

We review some terminology for binary relations, the DPO approach to graph transformation, graph languages, and confluence checking.

2.1 Abstract Reduction Systems

An *abstract reduction system* (ARS) is a pair $(A, \to)$ where A is a *set* and $\to$ a *binary relation* on A. We say that:

1. y is a *successor* to x if $x \xrightarrow{+} y$, and a *direct successor* if $x \to y$;
2. x and y are *joinable* if there is a z such that $x \xrightarrow{*} z \xleftarrow{*} y$. We write $x \downarrow y$;
3. $\to$ is *confluent* if $y_1 \xleftarrow{*} x \xrightarrow{*} y_2$ implies $y_1 \downarrow y_2$;
4. $\to$ is *locally confluent* if $y_1 \leftarrow x \to y_2$ implies $y_1 \downarrow y_2$;
5. $\to$ is *terminating* if there is no infinite sequence $x_0 \to x_1 \to \dots$.

The principle of *Noetherian induction* is:

$$\frac{\forall x \in A, (\forall y \in A, x \xrightarrow{+} y \Rightarrow P(y)) \Rightarrow P(x)}{\forall x \in A, P(x)}$$

Theorem 1 (Noetherian Induction [1]). *Given an ARS $(A, \to)$, the principle of Noetherian induction holds if and only if $\to$ is terminating.*

Theorem 2 (Newman's Lemma [25]). *A terminating relation is confluent if and only if it is locally confluent.*

2.2 Labelled Graphs and Morphisms

We will be working with *directed labelled graphs* [15]. An *alphabet* is a pair $\Sigma = (\Sigma_V, \Sigma_E)$ of finite sets of node and edge labels from which a graph can be labelled. A *graph* (over Σ) is a tuple $G = (V, E, s, t, l, m)$ where V is a finite set of nodes, E is a finite set of edges, $s : E \to V$ is the source function, $t : E \to V$ is the target function, $l : V \to \Sigma_V$ is the node labelling function, and $m : E \to \Sigma_E$ is the edge labelling function. We may write the components as V_G, E_G, s_G, etc.

A *graph morphism* $g : G \to H$ is a pair $g = (g_V, g_E)$ of functions $g_V : V_G \to V_H$ and $g_E : E_G \to E_H$ such that $g_V \circ s_G = s_H \circ g_E$, $g_V \circ t_G = t_H \circ g_E$, $l_G = l_H \circ g_V$ and $m_G = m_H \circ g_E$. We say g is *injective* (*surjective, bijective*) if both functions g_V and g_E are. A graph H is a *subgraph* of G, denoted by $H \subseteq G$, if there exists an *inclusion morphism* $i : H \to G$ with $i(x) = x$ for all items x.

It is well known that graphs and morphisms over Σ form a category. Graph morphisms are bijective if and only if they are isomorphisms in the categorical sense. Given a graph G, we write $[G]$ for the isomorphism class of G and call $[G]$ an *abstract graph*. We denote by $\mathcal{G}(\Sigma)$ the set of all abstract graphs over Σ.

2.3   Double-Pushout Graph Transformation

A *rule* is a pair of inclusions $r = \langle L \leftarrow K \to R \rangle$, where L is the left-hand side (LHS), K the interface, and R the right-hand side (RHS). A *match* of r in a graph G is an injective morphism $L \to G$. An application of rule r to G with match $g : L \to G$ requires to construct two pushouts as in Fig. 1. We write $G \Rightarrow_{r,g} H$ for this application and call the diagram in Fig. 1 a *direct derivation*.

$$
\begin{array}{ccccc}
L & \longleftarrow & K & \longrightarrow & R \\
\downarrow{\scriptstyle g} & & \downarrow{\scriptstyle d} & & \downarrow{\scriptstyle h} \\
G & \longleftarrow & D & \longrightarrow & H
\end{array}
$$

Fig. 1. A direct derivation

Given r and the match $g : L \to G$, the direct derivation of Fig. 1 exists if and only if the *dangling condition* is satisfied: nodes in $g(L - K)$ must not be incident to edges in $G - g(L)$. In this case the graphs D and H are determined uniquely up to isomorphism [10]. We call the injective morphism h the *comatch* of the rule application.

Given a set of rules $\mathcal{R}$, we write $G \Rightarrow_{\mathcal{R}} H$ if H is obtained from G by applying any of the rules from $\mathcal{R}$. We write $G \Rightarrow_{\mathcal{R}}^{+} H$ if H is obtained from G by one or more rule applications, and $G \Rightarrow_{\mathcal{R}}^{*} H$ if $G \cong H$ or $G \Rightarrow_{\mathcal{R}}^{+} H$.

By pushout properties, the relation $\Rightarrow_{\mathcal{R}}$ can be lifted to abstract graphs. Hence we have an ARS $(\mathcal{G}(\Sigma), \Rightarrow_{\mathcal{R}})$. This view gives us the definition of (local) confluence and termination for graph transformation systems.

2.4   Graph Languages

A graph language is simply a set of graphs, in the same way that a string language is a set of strings. Just like we can define string languages using string grammars, we can define graph languages using graph grammars, where we rewrite some start graph using a set of graph transformation rules. Derived graphs are then defined to be in the language exactly when they are terminally labelled.

Given a *graph transformation system* $T = (\Sigma, \mathcal{R})$, a subalphabet of *non-terminals* $\mathcal{N}$, and a *start graph* S over Σ, then a *graph grammar* is a tuple $\mathcal{G} = (\Sigma, \mathcal{N}, \mathcal{R}, S)$. We say that a graph G is *terminally labelled* if $l(V) \cap \mathcal{N}_V = \emptyset$ and $m(E) \cap \mathcal{N}_E = \emptyset$. Thus, we can define the *graph language* generated by $\mathcal{G}$:

$$L(\mathcal{G}) = \{[G] \mid S \Rightarrow_{\mathcal{R}}^* G, G \text{ terminally labelled}\}.$$

Given $\mathcal{G} = (\Sigma, \mathcal{N}, \mathcal{R}, S)$, we have $G \Rightarrow_r H$ if and only if $H \Rightarrow_{r^{-1}} G$, for some $r \in \mathcal{R}$, by using the *comatch*. Moreover, $[G] \in L(\mathcal{G})$ if and only if $G \Rightarrow_{\mathcal{R}^{-1}}^* S$ and G is terminally labelled. So we have a non-deterministic membership checking.

2.5   Confluence Checking

In 1970, Knuth and Bendix showed that confluence checking of terminating term rewriting systems is decidable [18]. Moreover, it suffices to compute all *critical pairs* and check their joinability [1,17]. Unfortunately, for (terminating) graph transformation systems, confluence is not decidable in general, and joinability of critical pairs does not imply local confluence. In 1993, Plump showed that *strong joinability* of all critical pairs is sufficient but not necessary to show local confluence [26,28].

The derivations $H_1 \Leftarrow_{r_1,g_1} G \Rightarrow_{r_2,g_2} H_2$ are *parallelly independent* if $(g_1(L_1) \cap g_2(L_2)) \subseteq (g_1(K_1) \cap g_2(K_2))$. We say two *parallelly independent* derivations are a *critical pair* if additionally $G = g_1(L_1) \cup g_2(L_2)$, and if $r_1 = r_2$ then $g_1 \neq g_2$. Every graph transformation system has only finitely many critical pairs.

Let $G \Rightarrow H$ be a *direct derivation*. Then the *track morphism* is defined to be the partial morphism $tr_{G \Rightarrow H} = in' \circ in^{-1}$, where in and in' are the bottom left and right morphisms in Fig. 1, respectively. We define $tr_{G \Rightarrow^* H}$ inductively as the composition of track morphisms. The set of *persistent nodes* of a critical pair $\Phi : H_1 \Leftarrow G \Rightarrow H_2$ is $Persist_\Phi = \{v \in G_V \mid tr_{G \Rightarrow H_1}(\{v\}), tr_{G \Rightarrow H_2}(\{v\}) \neq \emptyset\}$. That is, those nodes that are not deleted by the application of either rule.

A critical pair $\Phi : H_1 \Leftarrow G \Rightarrow H_2$ is *strongly joinable* if it is *joinable* without deleting any of the persistent nodes, and the persistent nodes are identified when joining. That is, there exists a graph M and derivations $H_1 \Rightarrow_{\mathcal{R}}^* M \Leftarrow_{\mathcal{R}}^* H_2$ such that $\forall v \in Persist_\Phi, tr_{G \Rightarrow H_1 \Rightarrow^* M}(\{v\}) = tr_{G \Rightarrow H_2 \Rightarrow^* M}(\{v\}) \neq \emptyset$.

Theorem 3 (Critical Pair Lemma [26,28]). *A graph transformation system T is locally confluent if all its critical pairs are strongly joinable.*

The original proof of the Critical Pair Lemma needs the Commutativity, Clipping and Embedding Theorems, and some auxiliary definitions. We will need these intermediate results when we come to prove our generalised version.

Theorem 4 (Commutativity [11]). *If $H_1 \Leftarrow_{r_1,g_1} G \Rightarrow_{r_2,g_2} H_2$ are parallelly independent, then there is a graph G' and derivations $H_1 \Rightarrow_{r_2} G' \Leftarrow_{r_1} H_2$.*

Let the derivation $\Delta : G_0 \Rightarrow^* G_n$ be given by pushouts $(1), (1'), \ldots, (n), (n')$ and suppose there are pushouts $(\underline{1}), (\underline{1'}), \ldots, (\underline{n}), (\underline{n'})$ whose vertical morphisms are injective (Fig. 2). Then, the derivation $\Delta' : G_0' \Rightarrow^* G_n'$ consisting of the composed pushouts $(1 + \underline{1}), \ldots, (n' + \underline{n'})$ is an instance of Δ based on the morphism $G_0 \rightarrow G_0'$. Moreover, we define the subgraph UseΔ to be all items x such that there is some $i \geq 0$ with $G_0 \Rightarrow^* G_i(x) \in \text{Match}(G_i \Rightarrow G_{i+1})$ where $\text{Match}(G_i \Rightarrow G_{i+1})$ is the image of the associated rule's left hand side graph under the match $L \rightarrow G_i$.

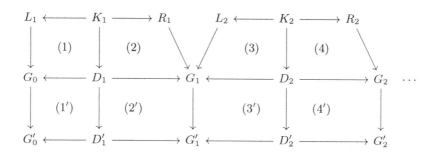

Fig. 2. Derivation instances

Theorem 5 (Clipping [27]). *Given a derivation $\Delta' : G' \Rightarrow^* H'$ and an injective morphism $h : G \rightarrow G'$ such that $\text{Use}\,\Delta' \subseteq h(G)$, there exists a derivation $\Delta : G \Rightarrow^* H$ such that Δ' is an instance of Δ based on h.*

Given a derivation $\Delta : G \Rightarrow^* H$ the subgraph of G, PersistΔ, consists of all items x such that $tr_{G \Rightarrow^* H}(x)$ is defined.

Theorem 6 (Embedding [27]). *Let $\Delta : G \Rightarrow^* H$ be a derivation, $h : G \rightarrow G'$ an injective graph morphism, B_Δ be the discrete subgraph of G consisting of all nodes x such that $h(x)$ is incident to an edge in $G' \setminus h(G)$. If $B_\Delta \subseteq \text{Persist}\Delta$, then there exists a derivation $\Delta' : G' \Rightarrow^* H'$ such that Δ' is an instance of Δ based on h. Moreover, there exists a pushout of $t : B_\Delta \rightarrow H$ along $h' : B_\Delta \rightarrow C_\Delta$ where $C_\Delta = (G' \setminus h(G)) \cup h(B_\Delta)$ and t is the restriction of $tr_{G \Rightarrow^* H}$ to B_Δ.*

3   Closedness and Confluence up to Garbage

In this section, we introduce (local) confluence and termination up to garbage, and closedness. We show that if we have closedness and termination up to garbage, then local confluence up to garbage implies confluence up to garbage: the Generalised Newmann's lemma. Moreover, we recap that closedness is undecidable in general, in the context of DPO graph transformation.

3.1   Closedness and Garbage

Definition 1. Let $T = (\Sigma, \mathcal{R})$ be a GT system, and $\mathcal{D} \subseteq \mathcal{G}(\Sigma)$ be a set of abstract graphs. Then, a graph G is called *garbage* if $[G] \notin \mathcal{D}$ and $\mathcal{D}$ is *closed* under T if for all G, H such that $G \Rightarrow_{\mathcal{R}} H$, if $[G] \in \mathcal{D}$ then $[H] \in \mathcal{D}$.

The idea is that a set of abstract graphs $\mathcal{D}$ represents the *good input*, and the *garbage* is the graphs that are not in this set. $\mathcal{D}$ need not be explicitly generated by a graph grammar. For example, it could be defined by some (monadic second-order [8]) logical formula.

Example 1. Consider the reduction rules in Fig. 3. The language of acyclic graphs is *closed* under the GT system $((\{\square\}, \{\square\}), \{r_1\})$, and the language of trees (forests) and its complement are both *closed* under $((\{\square\}, \{\square\}), \{r_2\})$.

Fig. 3. Example reduction rules

Definition 2 (Closedness Problem).

Input: A GT system $T = (\Sigma, \mathcal{R})$ and a graph grammar $\mathcal{G}$ over Σ.
Question: Is $\mathrm{L}(\mathcal{G})$ *closed* under T?

It turns out that closedness is undecidable in general, even if we restrict ourselves to recursive languages and terminating GT systems. In 1998, Fradet and Le Métayer showed the following result:

Theorem 7 (Undecidable Closedness [14]). *The closedness problem is undecidable in general, even for terminating GT systems T with only one rule, and $\mathcal{G}$ an edge replacement grammar.*

3.2   Confluence up to Garbage

We can now define (*local*) *confluence* and *termination* up to garbage, allowing us to say that, ignoring the garbage graphs, a system is (*locally*) *confluent*.

Definition 3. Let $T = (\Sigma, \mathcal{R})$, $\mathcal{D} \subseteq \mathcal{G}(\Sigma)$. Then:

1. if for all graphs G, H_1, H_2, such that $[G] \in \mathcal{D}$, $H_1 \Leftarrow_{\mathcal{R}} G \Rightarrow_{\mathcal{R}} H_2$ implies that H_1, H_2 are *joinable*, then T is *locally confluent* (up to garbage) on $\mathcal{D}$;
2. if for all graphs G, H_1, H_2, such that $[G] \in \mathcal{D}$, $H_1 \overset{*}{\Leftarrow}_{\mathcal{R}} G \overset{*}{\Rightarrow}_{\mathcal{R}} H_2$ implies that H_1, H_2 are *joinable*, then T is *confluent* (up to garbage) on $\mathcal{D}$;

3. if there is no infinite derivation sequence $G_0 \Rightarrow_{\mathcal{R}} G_1 \Rightarrow_{\mathcal{R}} G_2 \Rightarrow_{\mathcal{R}} \cdots$ such that $[G_0] \in \mathcal{D}$, then T is *terminating* (up to garbage) on $\mathcal{D}$.

The following is an immediate consequence of set inclusion:

Proposition 1. *Let* $T = (\Sigma, \mathcal{R})$, $\mathcal{D} \subseteq \mathcal{G}(\Sigma)$, $\mathcal{E} \subseteq \mathcal{D}$. *Then (local) confluence on* $\mathcal{D}$ *implies (local) confluence on* $\mathcal{E}$, *and similarly for termination.*

Example 2. Looking again at r_1 and r_2 from our first example, it is easy to see that r_1 is *terminating* and *confluent up to garbage* on the language of acyclic graphs, but is not *confluent* on all graphs. Similarly, r_2 is *terminating* and *confluent up to garbage* on the language of trees.

Example 3. Consider the rules in Fig. 4. Clearly they are terminating, since they are size reducing. Moreover, the language of all linked lists with edge labels a or b and its complement are closed under the rules. The rules are not locally confluent. To see this, consider the 3-cycle with edges labelled with a, a, b. It is possible for the cycle to be reduced to either the 2-cycle with edges a and a or the 2-cycle with edge a and b. Neither of these cycles can be reduced further, and so we have a counter example to confluence. These rules are locally confluent on linked lists. Moreover, an input graph G is a linked list if and only if it can be reduced using these rules to a length one linked list.

Fig. 4. List reduction rules

Theorem 8 (Generalised Newman's Lemma). *Let* $T = (\Sigma, \mathcal{R})$, $\mathcal{D} \subseteq \mathcal{G}(\Sigma)$. *If* T *is terminating on* $\mathcal{D}$ *and* $\mathcal{D}$ *is closed under* T, *then* T *is confluent on* $\mathcal{D}$ *if and only if it is locally confluent on* $\mathcal{D}$.

Proof. This can be seen by Noetherian Induction (Fig. 5), due to the fact that closedness ensures applicability of the induction hypothesis. □

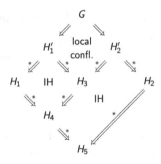

Fig. 5. Induction step diagram

4   Generalised Critical Pair Lemma

In this section, we generalise Plump's Critical Pair Lemma, providing a machine checkable sufficient condition for local confluence up to garbage. For this, we need to define a notion of subgraph closure and non-garbage critical pairs.

4.1   Subgraph Closure

In the proof of the traditional critical pair lemma for (hyper)graphs, the argument is that if a pair of derivations is not parallelly independent, then it must be the case that a critical pair can be embedded within it. In our new setting, the possible start graphs will be restricted, since some of the graphs will be *garbage*. We are only interested in those critical pairs with start graphs that can be embedded in non-garbage graphs. This is exactly the statement that the start graph of the critical pair is in the subgraph closure of the non-garbage graphs.

Definition 4. Let $\mathcal{D} \subseteq \mathcal{G}(\Sigma)$ be a set of abstract graphs. Then $\mathcal{D}$ is *subgraph closed* if for all graphs G, H, such that $H \subseteq G$, if $[G] \in \mathcal{D}$, then $[H] \in \mathcal{D}$. The *subgraph closure* of $\mathcal{D}$, denoted $\widehat{\mathcal{D}}$, is the smallest set containing $\mathcal{D}$ that is *subgraph closed*.

Proposition 2. *Given* $\mathcal{D} \subseteq \mathcal{G}(\Sigma)$, $\widehat{\mathcal{D}}$ *always exists, and is unique. Moreover,* $\mathcal{D} = \widehat{\mathcal{D}}$ *if and only if* $\mathcal{D}$ *is subgraph closed.*

Proof. The key observations are that the subgraph relation is transitive, and each graph has only finitely many subgraphs. Clearly, the smallest possible set containing $\mathcal{D}$ is just the union of all subgraphs of the elements of $\mathcal{D}$, up to isomorphism. This is the unique subgraph closure of $\mathcal{D}$.   □

$\widehat{\mathcal{D}}$ always exists, however it need not be decidable, even when $\mathcal{D}$ is! It is not obvious what conditions on $\mathcal{D}$ ensure that $\widehat{\mathcal{D}}$ is decidable. Interestingly, the classes of regular and context-free string languages are actually closed under substring closure [4].

Example 4. $\emptyset$, $\mathcal{G}(\Sigma)$, and the language of discrete graphs are subgraph closed.

Example 5. The subgraph closure of the language of trees is the language of forests. The subgraph closure of the language of connected graphs is the language of all graphs.

4.2   Non-garbage Critical Pairs

We now define non-garbage critical pairs, which allow us to ignore certain pairs, which if all are strongly joinable, will allow us to conclude local confluence up to garbage, even in the presence of (local) non-confluence on all graphs.

Definition 5. Let $T = (\Sigma, \mathcal{R})$, $\mathcal{D} \subseteq \mathcal{G}(\Sigma)$. A *critical pair* $H_1 \Leftarrow G \Rightarrow H_2$ is *non-garbage* if $[G] \in \widehat{\mathcal{D}}$.

Lemma 1. *Given a GT system $T = (\Sigma, \mathcal{R})$ and $\mathcal{D} \subseteq \mathcal{G}(\Sigma)$, then there are only finitely many non-garbage critical pairs up to isomorphism. Moreover, if $\widehat{\mathcal{D}}$ is decidable, then one can find them in finite time.*

Proof. There are only finitely many critical pairs for T, up to isomorphism, and there exists a terminating procedure for generating them. It then remains to filter out the garbage pairs, which can always be done if $\widehat{\mathcal{D}}$ is decidable.   □

Corollary 1. *Let $T = (\Sigma, \mathcal{R})$, $\mathcal{D} \subseteq \mathcal{G}(\Sigma)$ be such that T is terminating on $\mathcal{D}$ and $\widehat{\mathcal{D}}$ is decidable. Then, one can decide if all the non-garbage critical pairs are strongly joinable.*

Proof. By Lemma 1, we can generate all the pairs, but then since T is terminating on D, there are only finitely many successor graphs to be generated. We can then test each for strong joinability in finite time.   □

Theorem 9 (Generalised Critical Pair Lemma). *Let $T = (\Sigma, \mathcal{R})$, $\mathcal{D} \subseteq \mathcal{G}(\Sigma)$. If all its non-garbage critical pairs are strongly joinable, then T is locally confluent on $\mathcal{D}$.*

Proof. Our proof is a generalisation of Plump's original proof of the Critical Pair Lemma for (hyper)graphs (Theorem 3) [26,28]. We need to show that every pair of derivations $H_1 \Leftarrow_{r_1,g_1} G \Rightarrow_{r_2,g_2} H_2$ such that G is non-garbage can be joined. There are two cases to consider. Firstly, if the derivations are parallelly independent, then by Theorem 4, the result is immediate. Otherwise, we must consider the case that they are not parallelly independent.

By Theorem 5, we can factor out a pair $T_1 \Leftarrow S \Rightarrow T_2$. Since critical pairs are, by construction, the overlaps of rule left hand sides, it must be the case that this pair is actually a critical pair. Moreover, since $[G] \in \mathcal{D}$, then $[S] \in \widehat{\mathcal{D}}$ and so the critical pair must be non-garbage, and must be strongly joinable to U. We can now apply Theorem 6 to $T_1 \Rightarrow^* U$ and $T_2 \Rightarrow^* U$, separately, giving result graphs M_1 and M_2 (applicability of the theorem is a consequence of strong joinability). To see that M_1 and M_2 are isomorphic follows from elementary properties of pushouts along monomorphisms [28] (Fig. 6).   □

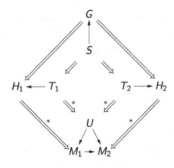

Fig. 6. Generalised critical pair lemma diagram

Corollary 2. *Let $T = (\Sigma, \mathcal{R})$, $\mathcal{D} \subseteq \mathcal{G}(\Sigma)$. If T is terminating on $\mathcal{D}$, $\mathcal{D}$ is closed under T, and all T's non-garbage critical pairs are strongly joinable, then T is confluent on $\mathcal{D}$.*

Proof. By the above theorem, T is *locally confluent up to garbage*, so by the Generalised Newman's Lemma (Theorem 8), T is *confluent up to garbage.*   □

Obviously, checking for local confluence up to garbage is undecidable in general, even when $\widehat{\mathcal{D}}$ is decidable and the system is terminating and closed. What is remarkable though, is that local confluence up to garbage is actually undecidable in general for a terminating non-length-increasing string rewriting systems and $\mathcal{D}$ a regular string language [7]!

4.3   Checking for Confluence up to Garbage

Given a GT system T and a language $\mathcal{D}$ (possibly specified by a grammar), the process is to:

1. Establish (by means of direct proof) that T is terminating on $\mathcal{D}$ and $\mathcal{D}$ is closed under T. If this is not true, one may want to restart with some language containing $\mathcal{D}$ to try to establish closedness.
2. Generate the finitely many non-garbage critical pairs of T.
3. Check if each generated pair is (strongly) joinable.

If all the pairs are strongly joinable, then we have confluence up to garbage due to Corollary 2. If all the pairs are joinable, but not all strongly, then we cannot draw any conclusions, but one may be able to construct a counter example to confluence by attaching context to nodes. Finally, if one of the pairs is not joinable at all, then we have a direct counter example to confluence, and we can conclude non-confluence up to garbage.

5   Language Recognition

In this section, we introduce a general notion of what it means to recognise a language, and what it means to be a confluent decider. We then demonstrate the applicability of our earlier results by showing that there are confluent deciders for Extended Flow Diagrams and Labelled Series-Parallel Graphs, even in the absence of confluence. We thus have algorithms, specified by reduction rules, that can check membership of these languages without needing to backtrack.

5.1 Confluent Recognition

One can think of graph transformation systems in terms of grammars that define languages. If they are terminating, then membership testing is decidable, but in general, non-deterministic in the sense that a deterministic algorithm must backtrack if it produces a normal form not equal to the start graph, to determine if another derivation sequence could have reached it. If the system is confluent too, then the algorithm becomes deterministic.

In general, the requirement of confluence is too strong, and one only requires confluence on the language we are recognising. Using the results from the last section, it is often possible to prove local confluence up to garbage using the Generalised Critical Pair Lemma, and then, in the presence of termination and closure, use the Generalised Newman's Lemma to show confluence up to garbage. Closedness and language recognition has actually been considered before by Bakewell, Plump, and Runciman, in the context of languages specified by reduction systems without non-terminals [3], but without the development of the theory we have provided.

Before continuing, we must provide a formal definition of what it means to recognise a language, and that grammars satisfy our definition by considering their rules in reverse, abstracting away from grammars, with a more general definition that accounts for the fact that reduction systems may need auxiliary symbols, not in the input, in the same way grammars can use non-terminals.

Definition 6 (Language Recognition). Let $T = (\Sigma, \mathcal{R})$ be a GT system, $\mathcal{I} \subseteq \Sigma$ an input alphabet, and $\mathcal{S}$ a finite set of graphs over Σ. Then we say that $(T, \mathcal{S})$ *recognises* a language $\mathcal{L}$ over $\mathcal{I}$ if for all graphs G over $\mathcal{I}$, $[G] \in \mathcal{L}$ if and only if $G \Rightarrow_{\mathcal{R}}^* S$ for some $S \in \mathcal{S}$.

Theorem 10 (Membership Checking). *Given a grammar* $\mathcal{G} = (\Sigma, \mathcal{N}, \mathcal{R}, S)$, $[G] \in \mathrm{L}(\mathcal{G})$ *if and only if* $G \Rightarrow_{\mathcal{R}^{-1}}^* S$ *and* G *is terminally labelled. That is,* $((\Sigma, \mathcal{R}^{-1}), \{S\})$ *recognises* $\mathrm{L}(\mathcal{G})$ *over* $\Sigma \setminus \mathcal{N}$.

Proof. The key is that rules and derivations are invertible, which means that if S can be derived from G using the reverse rules, then G can be derived from S using the original rules so is in the language. If S cannot be derived from G, then G cannot be in the language since that would imply there was a derivation sequence from S to G which we could invert to give a contradiction. □

We are now ready to define *confluent deciders*, and show that such systems can test for language membership without backtracking.

Definition 7 (Confluent Decider). Let $T = (\Sigma, \mathcal{R})$ be a GT system, $\mathcal{I} \subseteq \Sigma$ an input alphabet, and $\mathcal{S}$ a finite set of graphs over Σ. Then we say that $(T, \mathcal{S})$ is a *confluent decider* for a language $\mathcal{L}$ over $\mathcal{I}$ if $(T, \mathcal{S})$ recognises $\mathcal{L}$ over $\mathcal{I}$, T is terminating on $\mathcal{G}(\mathcal{I})$, and T is confluent on $\mathcal{L}$.

Theorem 11 (Confluent Decider Correctness). *Given a confluent decider $(T, \mathcal{S})$ for a language $\mathcal{L}$ over $\mathcal{I} \subseteq \Sigma$ and an input graph G over $\mathcal{I}$, the following algorithm is correct: Compute a normal form of G by deriving successor graphs using T as long as possible. If the result graph is isomorphic to S, the input graph is in the language. Otherwise, the graph is not in the language.*

Proof. Suppose G is not in $\mathcal{L}$. Then, since T is terminating on $\mathcal{G}(\mathcal{I})$ our algorithm must be able to find a normal form of G, say H, and because T recognises $\mathcal{L}$, it must be the case that H is not isomorphic to S, and so the algorithm correctly decides that G is not in $\mathcal{L}$.

Now, suppose that G is in $\mathcal{L}$. Then, because T is terminating, as before, we must be able to derive some normal form, H. But then, since T is both confluent on $\mathcal{L}$ and recognises $\mathcal{L}$, it must be the case that H is isomorphic to S, and so the algorithm correctly decides that G is in $\mathcal{L}$. $\qquad\qquad\Box$

What we really want is a version of Theorem 10 for instantiating confluent deciders. We really want is a Since both termination and confluence testing is undecidable in general, we cannot hope for an effective procedure, even for a terminating system, however the theory we introduced in the previous sections will help by automating local confluence checking. It just remains for us to choose a suitable set $\mathcal{D}$, and proceed in a similar way to as described in Subsect. 4.3. For the remainder of this section, we will look at two examples that demonstrate how we can use the Generalised Newman's Lemma and Generalised Critical Pair Lemma to show that we have a confluent decider for a language, given a grammar that generates the language.

5.2   Extended Flow Diagrams

In 1976, Farrow, Kennedy and Zucconi presented *semi-structured flow graphs*, defining a grammar with confluent reduction rules [13]. Plump has considered a restricted version of this language: *extended flow diagrams* (EFDs) [28]. The reduction rules for *extended flow diagrams* are a confluent decider for the EFDs, despite not being confluent.

Definition 8. The language of *extended flow diagrams* is generated by EFD $= (\Sigma, \mathcal{N}, \mathcal{R}, S)$ where $\Sigma_V = \{\bullet, \square, \Diamond\}$, $\Sigma_E = \{t, f, \square\}$, $\mathcal{N}_V = \mathcal{N}_E = \emptyset$, $\mathcal{R} = \{seq, while, ddec, dec1, dec2\}$, and $S = \bullet\!\!\rightarrow\!\!\square\!\!\rightarrow\!\!\bullet$ (Fig. 7).

In the next figure, the shorthand notation with the numbers under the nodes places such nodes in the interface graph of the rules. We assume that the interface graphs are discrete (have no edges).

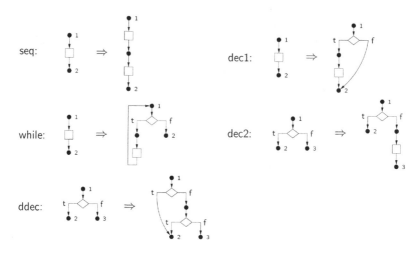

Fig. 7. EFD grammar rules

Lemma 2. *Every directed cycle in an EFD contains a t-labelled edge*

Proof. By induction. □

Theorem 12 (Confluent EFD Decider). *Let* $T = (\Sigma, \mathcal{R}^{-1})$. *Then* $(T, \{S\})$ *is a confluent decider for* L(EFD) *over* Σ.

Proof. By Theorem 10, T recognises L(EFD) over Σ, and one can see that it is terminating since each rule is size reducing.

We now proceed by performing critical pair analysis on T. There are ten critical pairs, all but one of which are strongly joinable apart from one (Fig. 8). Now observe that Lemma 2 tells us that EFDs cannot contain such cycles. With this knowledge, we define $\mathcal{D}$ to be all graphs such that directed cycles contain at least one t-labelled edge. Clearly, $\mathcal{D}$ is subgraph closed, and then by our Generalised Critical Pair Lemma (Theorem 9), we have that T is locally confluent on $\mathcal{D}$.

Next, it is easy to see that $\mathcal{D}$ is closed under T, so we can use Generalised Newman's Lemma (Theorem 8) to conclude confluence on $\mathcal{D}$ and thus, by Proposition 1, T is confluent on L(EFD).

Thus, T is a confluent decider for L(EFD) over Σ, as required. □

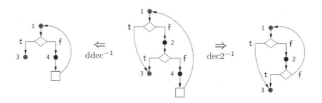

Fig. 8. Non-joinable EFD critical pair

5.3   Series-Parallel Graphs

Series-parallel graphs were introduced by Duffin [9] as a model of electrical networks. A more general version of the class was introduced by Lawler [23] and Monma and Sidney [24] as a model for scheduling problems.

Definition 9. *Series-parallel* graphs are inductively defined:

1. P is a series-parallel graph where s is the *source* and t the *sink*.
2. The class of series-parallel graphs is closed under *parallel composition* and *sequential composition*.

where $P = $ •→→• , parallel composition identifies the two sources and the two sinks, and sequential composition identifies the sink of one with the source of another.

Duffin showed that a graph is series-parallel if and only if it can be reduced to P by a sequence of series and parallel reductions. We can rephrase this in terms of a graph grammar.

Theorem 13 (SP Recognition [29]). *The class of series-parallel graphs is the language generated by grammar* $SP = ((\{\Box\}, \{\Box\}), (\emptyset, \emptyset), \{s, p\}, P)$ *(Fig. 9).*

Fig. 9. Series-parallel grammar rules

By traditional critical pair analysis, one can establish that the reversed rules are confluent, however, we run into a problem if we want to consider arbitrarily labelled graphs. Consider the case where the edge alphabet is of size 2, rather than size 1. The obvious modification to the rules is to use all combinations of labels in LHS graphs (Fig. 10), however Hristakiev and Plump [16] observed that when doing (the equivalent of) this in GP 2, we no longer have confluence. We exhibit a counter example to confluence in Fig. 11.

Fig. 10. Labelled series-parallel reduction rules

All is not lost, however, because we can use our new theory to show that, via non-garbage critical pair analysis, the new system is confluence up to garbage, and so we can show that we have a confluent decider. Moreover, our system for two edge labels can be easily generalised for any finite edge alphabet.

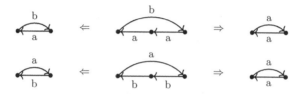

Fig. 11. Non-joinable labelled series-parallel pairs

Definition 10. The class of *labelled series-parallel graphs* (LSPs) is all series-parallel graphs, but with arbitrary edge labels chosen from $\Sigma_E = \{a, b\}$.

Theorem 14 (Confluent LSP Decider). *Let* $\Sigma = (\{\Box\}, \{a, b\})$, $T = (\Sigma, \{s_i, p_i \mid i \in I\})$ *(where* I *indexes label choice),* $P_a = \bullet \overset{a}{\to} \bullet$ *and* $P_b = \bullet \overset{b}{\to} \bullet$. *Then* $(T, \{P_a, P_b\})$ *is a confluent decider for the labelled series-parallel graphs over* Σ.

Proof. We denote by $\mathcal{L}$ the language of all labelled series-parallel graphs.

Our rules are structurally the same as the unlabelled rules, so because our LHS graphs are arbitrarily labelled, language recognition of $\mathcal{L}$ over Σ follows from Theorem 13. Termination follows from the fact that the combined metric of graph size plus number of b-labelled edges strictly decreases with each derivation.

We now proceed by performing critical pair analysis on T. We find that we have two non-isomorphic critical pairs that are not joinable (Fig. 11). These pairs have a cyclic start graph, but the series-parallel graphs are acyclic, so we can define $\mathcal{D}$ to be the language of acyclic graphs over Σ, thus classifying these two pairs as garbage. The remaining critical pairs are strongly joinable, so by Theorem 9, we have that T is locally confluent on $\mathcal{D}$.

We find that all the non-garbage critical pairs are strongly joinable. We have, up to isomorphism, two garbage critical pairs. These are not even joinable, which give us a counter example to (local) confluence, but since all our non-garbage pairs are strongly joinable, we can claim local confluence up to garbage.

Next, it is easy to see that $\mathcal{D}$ is closed under T, so we can use Theorem 8 to conclude confluence on $\mathcal{D}$ and thus, by Proposition 1, T is confluent on $\mathcal{L}$. $\Box$

6   Conclusion and Future Work

In this paper we have introduced (local) confluence and termination up to garbage for DPO graph transformation systems, and shown that Newmann's Lemma and Plump's Critical Pair Lemma can be generalised, providing us with machine checkable conditions for confluence up to garbage, using only critical pairs. Of course, confluence up to garbage of terminating graph transformation systems is undecidable in general, however, now we can detect more positive cases of confluence up to garbage using non-garbage critical pair analysis, where we previously would have been unable to draw a conclusion due to non-strong joinability of some critical pairs.

In particular, our results can be directly applied to recognition of languages, which we have demonstrated with extended flow diagrams and labelled series-parallel graphs. We have backtracking-free algorithms that apply reduction rules as long as possible, with correctness established via non-garbage critical pair analysis. We also anticipate there to be other applications, since there are many other reasons one would want to show confluence up to garbage, such as considering GT systems as computing functions where we restrict [15]. Indeed, one might only be interested in the non-garbage critical pairs themselves, and classification of conflicts [20, 22].

Confluence analysis of GT systems (and related systems) still remains a generally under-explored area. One obvious piece of future work is to investigate the connection to the work by Lambers, Ehrig and Orejas on *essential critical pairs* [21] and the continued work by others including Born and Taentzer [20]. It is also not obvious if there is a relation between confluence up to garbage and graphs satisfying negative constraints [19]. Moreover, developing a stronger version of the Generalised Critical Pair Lemma that allows for the detection of persistent nodes that need not be identified in the joined graph would allow conclusions of confluence up to garbage where it was previously not determined.

Future work also includes developing checkable sufficient conditions under which one can decide if a graph is in the subgraph closure of a language. Finally, applying our theory in a rooted context and to GP 2 is future work [2]. It is likely that the theory will be applicable there, since program preconditions correspond exactly to non-garbage input, and so it is only natural to be interested in confluence up to garbage, rather than confluence. We would also expect there to be analogues of our results for other kinds of rewriting systems such as string and term rewriting.

References

1. Baader, F., Nipkow, T.: Term Rewriting and All That. Cambridge University Press, Cambridge (1998)
2. Bak, C.: GP 2: efficient implementation of a graph programming language. Ph.D. thesis, Department of Computer Science, University of York, UK (2015). https://etheses.whiterose.ac.uk/12586/
3. Bakewell, A., Plump, D., Runciman, C.: Specifying pointer structures by graph reduction. In: Pfaltz, J.L., Nagl, M., Böhlen, B. (eds.) AGTIVE 2003. LNCS, vol. 3062, pp. 30–44. Springer, Heidelberg (2004). https://doi.org/10.1007/978-3-540-25959-6_3
4. Berstel, J.: Transductions and Context-Free Languages. Vieweg+Teubner, Stuttgart (1979). https://doi.org/10.1007/978-3-663-09367-1
5. Campbell, G.: Efficient graph rewriting. BSc thesis, Department of Computer Science, University of York, UK (2019). https://arxiv.org/abs/1906.05170
6. Campbell, G., Plump, D.: Efficient recognition of graph languages. Technical report, Department of Computer Science, University of York, UK (2019). https://arxiv.org/abs/1911.12884
7. Caron, A.-C.: Linear bounded automata and rewrite systems: influence of initial configurations on decision properties. In: Abramsky, S., Maibaum, T.S.E. (eds.)

CAAP 1991. LNCS, vol. 493, pp. 74–89. Springer, Heidelberg (1991). https://doi. org/10.1007/3-540-53982-4_5

8. Courcelle, B.: The monadic second-order logic of graphs: definable sets of finite graphs. In: van Leeuwen, J. (ed.) WG 1988. LNCS, vol. 344, pp. 30–53. Springer, Heidelberg (1989). https://doi.org/10.1007/3-540-50728-0_34

9. Duffin, R.J.: Topology of series-parallel networks. J. Math. Anal. Appl. **10**(2), 303–318 (1965). https://doi.org/10.1016/0022-247X(65)90125-3

10. Ehrig, H., Ehrig, K., Prange, U., Taentzer, G.: Fundamentals of Algebraic Graph Transformation. Monographs in Theoretical Computer Science. An EATCS Series. Springer, Heidelberg (2006). https://doi.org/10.1007/3-540-31188-2

11. Ehrig, H., Kreowski, H.-J.: Parallelism of manipulations in multidimensional information structures. In: Mazurkiewicz, A. (ed.) MFCS 1976. LNCS, vol. 45, pp. 284–293. Springer, Heidelberg (1976). https://doi.org/10.1007/3-540-07854-1_188

12. Ehrig, H., Pfender, M., Schneider, H.: Graph-grammars: an algebraic approach. In: Proceedings 14th Annual Symposium on Switching and Automata Theory, SWAT 1973, pp. 167–180. IEEE (1973). https://doi.org/10.1109/SWAT.1973.11

13. Farrow, R., Kennedy, K., Zucconi, L.: Graph grammars and global program data flow analysis. In: Proceedings 17th Annual Symposium on Foundations of Computer Science, SFCS 1976, pp. 42–56. IEEE (1976). https://doi.org/10.1109/SFCS. 1976.17

14. Fradet, P., Métayer, D.L.: Structured gamma. Sci. Comput. Program **31**(2–3), 263–289 (1998). https://doi.org/10.1016/S0167-6423(97)00023-3

15. Habel, A., Müller, J., Plump, D.: Double-pushout graph transformation revisited. Math. Struct. Comput. Sci. **11**(5), 637–688 (2001). https://doi.org/10.1017/S0960129501003425

16. Hristakiev, I., Plump, D.: Checking graph programs for confluence. In: Seidl, M., Zschaler, S. (eds.) STAF 2017. LNCS, vol. 10748, pp. 92–108. Springer, Cham (2018). https://doi.org/10.1007/978-3-319-74730-9_8

17. Huet, G.: Confluent reductions: abstract properties and applications to term rewriting systems. J. ACM **27**(4), 797–821 (1980). https://doi.org/10.1145/322217. 322230

18. Knuth, D., Bendix, P.: Simple word problems in universal algebras. In: Leech, J. (ed.) Computational Problems in Abstract Algebra, pp. 263–297. Pergamon Press, Oxford (1970). https://doi.org/10.1016/B978-0-08-012975-4.50028-X

19. Lambers, L.: Certifying rule-based models using graph transformation. Ph.D. thesis, Technical University of Berlin, Elektrotechnik und Informatik (2009). https://doi.org/10.14279/depositonce-2348

20. Lambers, L., Born, K., Orejas, F., Strüber, D., Taentzer, G.: Initial conflicts and dependencies: critical pairs revisited. In: Heckel, R., Taentzer, G. (eds.) Graph Transformation, Specifications, and Nets. LNCS, vol. 10800, pp. 105–123. Springer, Cham (2018). https://doi.org/10.1007/978-3-319-75396-6_6

21. Lambers, L., Ehrig, H., Orejas, F.: Efficient conflict detection in graph transformation systems by essential critical pairs. In: Proceedings Fifth International Workshop on Graph Transformation and Visual Modeling Techniques, GT-VMT 2006, Electronic Notes in Theoretical Computer Science, vol. 211, pp. 17–26. Elsevier (2008). https://doi.org/10.1016/j.entcs.2008.04.026

22. Lambers, L., Kosiol, J., Strüber, D., Taentzer, G.: Exploring conflict reasons for graph transformation systems. In: Guerra, E., Orejas, F. (eds.) ICGT 2019. LNCS, vol. 11629, pp. 75–92. Springer, Cham (2019). https://doi.org/10.1007/978-3-030-23611-3_5

23. Lawler, E.: Sequencing jobs to minimize total weighted completion time subject to precedence constraints. Ann. Discrete Math. **2**, 75–90 (1978). https://doi.org/10.1016/S0167-5060(08)70323-6
24. Monma, C., Sidney, J.: Sequencing with series-parallel precedence constraints. Math. Oper. Res. **4**(3), 215–224 (1979). https://doi.org/10.1287/moor.4.3.215
25. Newman, M.: On theories with a combinatorial definition of "equivalence". Ann. Math. **43**(2), 223–243 (1942). https://doi.org/10.2307/1968867
26. Plump, D.: Hypergraph rewriting: critical pairs and undecidability of confluence. In: Sleep, M.R., Plasmeijer, M.J., van Eekelen, M.C. (eds.) Term Graph Rewriting, pp. 201–213. Wiley, Chichester (1993)
27. Plump, D.: Computing by graph rewriting. Habilitation thesis, Universität Bremen, Fachbereich Mathematik und Informatik (1999)
28. Plump, D.: Confluence of graph transformation revisited. In: Middeldorp, A., van Oostrom, V., van Raamsdonk, F., de Vrijer, R. (eds.) Processes, Terms and Cycles: Steps on the Road to Infinity. LNCS, vol. 3838, pp. 280–308. Springer, Heidelberg (2005). https://doi.org/10.1007/11601548_16
29. Plump, D.: Reasoning about graph programs. In: Proceedings 9th International Workshop on Computing with Terms and Graphs, TERMGRAPH 2016, Electronic Proceedings in Theoretical Computer Science, vol. 225, pp. 35–44. Open Publishing Association (2016). https://doi.org/10.4204/EPTCS.225.6

Computing Embeddings of Directed Bigraphs

Alessio Chiapperini[1], Marino Miculan[1(✉)], and Marco Peressotti[2]

[1] DMIF, University of Udine, Udine, Italy
marino.miculan@uniud.it
[2] IMADA, University of Southern Denmark, Odense, Denmark
peressotti@imada.sdu.dk

Abstract. *Directed bigraphs* are a meta-model which generalises Milner's bigraphs by taking into account the *request flow* between controls and names. A key problem about these bigraphs is that of *bigraph embedding*, i.e., finding the embeddings of a bigraph inside a larger one. We present an algorithm for computing embeddings of directed bigraphs, via a reduction to a *constraint satisfaction problem*. We prove soundness and completeness of this algorithm, and provide an implementation in jLibBig, a general Java library for manipulating bigraphical reactive systems, together with some experimental results.

1 Introduction

Bigraphical Reactive Systems (BRSs) are a family of graph-based formalisms introduced as a meta-model for distributed, mobile systems [17,22,25]. In this approach, system configurations are represented by *bigraphs*, graph-like data structures capable of describing at once both the locations and the logical connections of (possibly nested) components. The dynamics of a system is defined by means of a set of *graph rewriting rules*, which can replace and change components' positions and connections. BRSs have been successfully applied to the formalization of a wide spectrum of domain-specific models, including context-aware systems, web-service orchestration languages [4,5,20,28]. BRSs are appealing because they provide a range of general results and tools, which can be readily instantiated with the specific model under scrutiny: libraries for bigraph manipulation (e.g., DBtk [1] and jLibBig [23,24]), simulation tools [10,19,21], graphical editors [9], model checkers [27], modular composition [26], stochastic extensions [18], etc.

Along this line, [13,14] introduced *directed bigraphs*, a strict generalization of Milner's bigraphs where the link graph is directed (see Fig. 1). This variant is very suited for reasoning about *dependencies* and *request flows* between components,

M. Miculan—Supported by Italian MIUR project PRIN 2017FTXR7S *IT MATTERS* (Methods and Tools for Trustworthy Smart Systems).
M. Peressotti—Supported by the Independent Research Fund Denmark, Natural Sciences, grant DFF-7014-00041.

F. Gadducci and T. Kehrer (Eds.): ICGT 2020, LNCS 12150, pp. 38–56, 2020.
https://doi.org/10.1007/978-3-030-51372-6_3

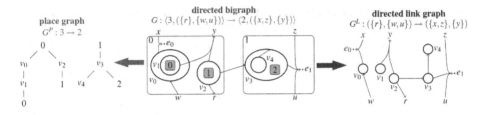

Fig. 1. An example of directed bigraph and its place and link graphs [15].

such as those found in client-server or producer-consumer scenarios. In fact, they have been used to design formal models of security protocols [12], molecular biology [2], access control [15], container-based systems [5], etc.

A key notion at the core of these results and tools is that of *bigraph embedding*. Informally, an embedding is a structure preserving map from a bigraph (called *guest*) to another one (called *host*), akin a subgraph isomorphism. Finding such embeddings is a difficult problem; in fact, the sole embedding of place graphs has been proved to be NP-complete [3]. Several algorithms have been proposed in literature for bigraphs with undirected links (see e.g. [7,11,23,29,30]), but there is no embedding algorithm for the more general case of directed bigraphs, yet.

In this work, we propose an algorithm for computing embedding of directed bigraphs (which subsume traditional ones), laying the theoretical and technical foundations for bringing directed bigraphs to tools like the ones listed above.

More precisely, in Sect. 2 we first introduce directed bigraphs and bigraphic reactive systems, generalizing [5,13]. Then, the notion of directed bigraph embedding is defined in Sect. 3. In Sect. 4 we present a reduction of the embedding problem for directed bigraphs to a constraint satisfaction problem (CSP) and show that it provides a sound and complete algorithm for computing embeddings. This reduction extends our previous (unpublished) work [23] on the embedding problem for undirected bigraphs. We have implemented this algorithm as an extension of jLibBig [24], a general Java library for BRSs; this implementation and some experimental results are reported in Sect. 5. Finally, some conclusions and directions for future work are drawn in Sect. 6.

2   Reactive Systems on Directed Bigraphs

In this section we introduce a conservative extension of the notions of *directed link graphs* and bigraphs, and *directed bigraphical reactive systems*, originally defined in [13,14].

2.1   Directed Bigraphs

Definition 1 (Polarized interface). *A polarized interface X is a pair (X^-, X^+), where X^+ and X^- are sets of names s.t. $X^- \cap X^+ = \varnothing$; the two sets are called* downward *and* upward *interfaces respectively.*

Definition 2 (Polarized signature). *A signature is a pair* $(\mathcal{K}, ar)$, *where* $\mathcal{K}$ *is the set of* controls, *and* $ar : \mathcal{K} \to \mathbb{N} \times \mathbb{N}$ *is a map assigning to each control its polarized arity, that is, a pair* $\langle n, m \rangle$ *where* n, m *are the numbers of* positive *and* negative *ports of the control, respectively.*

We define $ar^+, ar^- : \mathcal{K} \to \mathbb{N}$ *as shorthand for the* positive *and* negative *ports of controls:* $ar^+ \triangleq \pi_1 \circ ar$, $ar^- \triangleq \pi_2 \circ ar$.

The main difference between this definition and that from [13] is that we allow also for *inward* ports in controls, whereas in [13], like in [25], controls have only outward ports. This turns up also in the definition of *points* and *handles*. The addition of negative ports enables us to represent more faithfully the dependencies between processes, entities and components, according to the micro-services paradigm.

Definition 3 (Directed Link Graph). *A directed link graph* $A : X \to Y$ *is a quadruple* $A = (V, E, ctrl, link)$ *where* X, Y, V, E *and ctrl are defined as before, while the link map is defined as link* $: Pnt(A) \to Lnk(A)$ *where*

$$Prt^+(A) \triangleq \sum_{v \in V} ar^+(ctrl(v)) \qquad\qquad Prt^-(A) \triangleq \sum_{v \in V} ar^-(ctrl(v))$$
$$Pnt(A) \triangleq X^+ \uplus Y^- \uplus Prt^+(A) \qquad Lnk(A) \triangleq X^- \uplus Y^+ \uplus E \uplus Prt^-(A)$$

with the following additional constraints:

$$\forall x \in X^-, \forall y \in X^+ . link(y) = x \Rightarrow link^{-1}(x) = \{y\} \tag{1}$$
$$\forall y \in Y^+, \forall x \in Y^- . link(x) = y \Rightarrow link^{-1}(y) = \{x\}. \tag{2}$$

The elements of $Pnt(A)$ *are called the* points *of* A; *the elements of* $Lnk(A)$ *are called the* handles *of* A.

The constraint (1) means that if there is an upward inner name connected to a downward inner name, then nothing else can be connected to the latter; constraint (2) is similar, on the outer interface. Together, these requirements guarantee that composition of link graphs (along the correct interfaces) is well defined.

Direct link graphs are graphically depicted similarly to ordinary link graphs, with the difference that edges are represented as vertices of the graph and not as hyper-arcs connecting points and names.

Directed bigraphs are composed by a directed link graph and a place graph. Since the definition of place graph is the same as for pure bigraphs, we will omit it and refer the interested reader to [25].

Definition 4 (Directed Bigraph). *An interface* $I = \langle m, X \rangle$ *is composed by a finite ordinal* m, *called the* width, *and by a directed interface* $X = (X^-, X^+)$.

Let $I = \langle m, X \rangle$ *and* $O = \langle n, Y \rangle$ *be two interfaces; a directed bigraph with signature* $\mathcal{K}$ *from* I *to* O *is a tuple* $G = (V, E, ctrl, prnt, link) : I \to O$ *where*

- *I and O are the* inner *and* outer *interfaces;*

- *V and E are the sets of nodes and edges;*
- *ctrl, prnt, link are the* control, parent *and* link *maps;*

such that $G^L \triangleq (V, E, ctrl, link) : X \to Y$ is a directed link graph and $G^P \triangleq (V, ctrl, prnt) : m \to n$ is a place graph, that is, the map $prnt : m \uplus V \to n \uplus V$ is acyclic. The bigraph G is denoted also as $\langle G^P, G^L \rangle$.

Definition 5 (Composition and identities).

- *The composition of two place graphs $F : k \to m$ and $G : m \to n$, is defined in the same way as pure bigraphs (i.e., suitable grafting of forests);*
- *If $F : X \to Y$ and $G : Y \to Z$ are two link graphs, their composition is the link graph $G \circ F \triangleq (V, E, ctrl, link) : X \to Z$ such that $V = V_F \uplus V_G$, $E = E_F \uplus E_G$, $ctrl = ctrl_F \uplus ctrl_G$, and $link : Pnt(G \circ F) \to Lnk(G \circ F)$ is defined as follows:*

$$Pnt(G \circ F) = X^+ \uplus Z^- \uplus Prt^+(F) \uplus Prt^+(G)$$

$$Lnk(G \circ F) = X^- \uplus Z^+ \uplus Prt^-(F) \uplus Prt^-(G) \uplus E$$

$$link(p) \triangleq \begin{cases} prelink(p) & if\ prelink(p) \in Lnk(G \circ F) \\ link(prelink(p)) & otherwise \end{cases}$$

where $prelink : Pnt(G \circ F) \uplus Y^+ \uplus Y^- \to Lnk(G \circ F) \uplus Y^+$ is $link_F \uplus link_G$. The identity link graph at X is $id_X \triangleq (\varnothing, \varnothing, \varnothing_K, Id_{X^- \uplus X^+}) : X \to X$.
- *If $F : I \to J$ and $G : J \to K$ are two bigraphs, their composite is*

$$G \circ F \triangleq \langle G^P \circ F^P, G^L \circ F^L \rangle : I \to K$$

and the identity bigraph at $I = \langle m, X \rangle$ is $\langle id_m, id_{X^- \uplus X^+} \rangle$.

Definition 6 (Juxtaposition).

- *For place graphs, the juxtaposition of two interfaces m_0 and m_1 is $m_0 + m_1$; the unit is 0. If $F_i = (V_i, ctrl_i, prnt_i) : m_i \to n_i$ are disjoint place graphs (with $i = 0, 1$), their juxtaposition is defined as for pure bigraphs;*
- *For link graphs, the juxtaposition of two (directed) link graph interfaces X_0 and X_1 is $(X_0^- \uplus X_1^-, X_0^+ \uplus X_1^+)$. If $F_i = (V_i, E_i, ctrl_i, link_i) : X_i \to Y_i$ are two link graphs (with $i = 0, 1$), their juxtaposition is*

$$F_0 \otimes F_1 \triangleq (V_0 \uplus V_1, E_0 \uplus E_1, ctrl_0 \uplus ctrl_1, link_0 \uplus link_1) : X_0 \otimes X_1 \to Y_0 \otimes Y_1$$

- *For bigraphs, the juxtaposition of two interfaces $I_i = \langle m_i, X_i \rangle$ (with $i = 0, 1$) is $\langle m_0 + m_1, (X_0^- \uplus X_1^-, X_0^+ \uplus X_1^+) \rangle$ (the unit is $\epsilon = \langle 0, (\varnothing, \varnothing) \rangle$). If $F_i : I_i \to J_i$ are two bigraphs (with $i = 0, 1$), their juxtaposition is*

$$F_0 \otimes F_1 \triangleq \langle F_0^P \otimes F_1^P, F_0^L \otimes F_1^L \rangle : I_0 \otimes I_1 \to J_0 \otimes J_1.$$

Polarized interfaces and directed bigraphs over a given signature $\mathcal{K}$ form a monoidal category $\mathrm{DBIG}(\mathcal{K})$.

Milner's pure bigraphs [25] correspond precisely to directed bigraphs with positive interfaces only and over signatures with only positive ports. We observe also that the introduction of negative ports is more important than adding directions to interfaces: directed bigraphs as per [13] can be obtained as a traced category over the category of pure bigraphs, while we cannot properly represent controls with negative ports using those with positive ports only.

2.2   Reactive Systems over Directed Bigraphs

In order to define reactive systems over bigraphs, we need to define how a parametric reaction rule (i.e., a pair of "redex-reactum" bigraphs) can be instantiated. Essentially, in the application of the rule, the "sites" of the reactum must be filled with the parameters appearing in the redex. This relation can be expressed by specifying an *instantiation map* in the rule.

Definition 7 (Instantiation map). *An* instantiation map $\eta :: \langle m, X \rangle \to \langle m', X' \rangle$ *is a pair* $\eta = (\eta^P, \eta^L)$ *where*

- $\eta^P : m' \to m$ *is a function which maps sites of the reactum to sites of the redex; for each* $j \in m'$, *it determines that the* j-*th site of the reactum is filled with the* $\eta(j)$-*th parameter of the redex.*
- $\eta^L : \left(\sum_{i=0}^{m'-1} X \right) \to X'$ *is a wiring (i.e., a link graph without nodes nor edges), which is responsible for mapping names of the redex to names of the reactum. This can be described as a pair of functions* $\eta^L = (\eta^+, \eta^-)$ *where* $\eta^+ : \left(\sum_{i=0}^{m'-1} X^+ \right) \to X'^+$ *and* $\eta^- : X'^- \to \sum_{j=0}^{m'-1} X^-$.

We can now define the dynamics of directed bigraphs, starting with the formal definition of parametric reaction rules.

Definition 8 (Parametric reaction rule). *A* parametric reaction rule *for bigraphs is a triple of the form* $(R : I \to J, R' : I' \to J, \eta :: I \to I')$ *where* R *is the parametric redex,* R' *the parametric reactum and* η *is an instantiation map.*

We can now define the key notion of reactive systems over directed bigraphs, which is a generalization of that in [14,25]. Let $Ag(\mathcal{K})$ be the set of *agents* (i.e., bigraphs with no inner names nor sites) over a signature $\mathcal{K}$.

Definition 9 (DBRS). *A* directed bigraphical reactive system $DBG(\mathcal{K}, \mathcal{R})$ *is defined by a signature* $\mathcal{K}$ *and a set* $\mathcal{R}$ *of rewriting rules.*

A DBRS $DBG(\mathcal{K}, \mathcal{R})$ *induces a rewriting relation* $\twoheadrightarrow \subseteq Ag(\mathcal{K}) \times Ag(\mathcal{K})$ *according to the following rule:*

$$\frac{\begin{array}{c} (R_L, R_R, \eta) \in \mathcal{R} \\ A = C \circ (R_L \otimes Id_Z) \circ \omega \circ (D_0 \otimes \ldots \otimes D_{m-1}) \\ A' = C \circ (R_R \otimes Id_Z) \circ \omega' \circ (D_{\eta^P(0)} \otimes \ldots \otimes D_{\eta^P(m'-1)}) \end{array}}{A \twoheadrightarrow A'}$$

where ω and ω' are called wiring maps *and are defined as follows:*

$$\omega : \sum_{i=0}^{m-1} X_i \to X \oplus Z \qquad\qquad \omega' : \sum_{j=0}^{m'-1} X_{\eta^P(j)} \to X' \oplus Z$$

$$\omega^+ : \sum_{i=0}^{m-1} X_i^+ \to X^+ \uplus Z^+ \qquad\qquad \omega'^+ : \sum_{j=0}^{m'-1} X_{\eta^P(j)}^+ \to X'^+ \uplus Z^+$$

$$\omega^- : X^- \uplus Z^- \to \sum_{i=0}^{m-1} X_i^- \qquad\qquad \omega'^- : X'^- \uplus Z^- \to \sum_{j=0}^{m'-1} X_{\eta^P(j)}^-$$

$$\omega'^+(j,x) \triangleq \begin{cases} \eta^+(j, \omega^+(\eta(j), x)) & if \ \omega^+(\eta^P(j), x) \in X^+ \\ \omega^+(\eta^P(j), x) & if \ \omega^+(\eta(j), x) \in Z^+ \end{cases}$$

$$\omega'^-(x) \triangleq (j, y) \ for \ j \in \eta^{P^{-1}}(i) \ and \ (i, y) \in \eta^-(x)$$

The difference with respect to the previous versions of BRS is that now links can descend from the redex (and reactum) into the parameters, as it is evident from the fact that redexes and reactums in rules may have generic inner interfaces (I and I'). This is very useful for representing a request flow which goes "downwards", e.g. connecting a port of a control in the redex to a port of an inner component (think of, e.g., a linked library).

However, this poses some issues when the rules are not linear. If any of D_i's is cancelled by the rewriting, the controls in it disappear as well, and we may be not able to connect some name descending from R_L or Id_Z anymore. More formally, this means that the map ω^- can be defined only if for every $x \in (X'^- \uplus Z^-)$ there are j, y such that $(\eta^P(j), y) = \eta^-(x)$. We can have two cases:

1. for some x, there are no such j, y. This means that ω is not defined and hence the rule cannot be applied.
2. for each x, there are one or more pairs (j, y) such that $(\eta^P(j), y) = \eta^-(x)$. This means that for a given source agent decomposition, there can be several ways to define ω^-, each yielding a different application of the same rule.

Overall, the presence of downward names in parameters adds a new degree of non-determinism to Directed BRSs, with respect to previous versions of BRSs.

3   Directed Bigraph Embeddings

As we have seen in the previous section, to execute or simulate a BRS it is necessary to solve the *bigraph matching problem*, that is, finding the occurrences of a redex R within a given bigraph A. More formally, this translates to finding C, Z, ω and $D = (D_0 \otimes \ldots \otimes D_{m-1})$ such that $A = C \circ (R \otimes Id_Z) \circ \omega \circ D$. C and D are called *context* and *parameter*, respectively.

If we abstract from the decomposition of the agent A in context, redex and parameter we can see how the matching problem is related to the *subgraph isomorphism* problem. Therefore, in this section we define the notions of *directed bigraph embedding*. The following definitions are taken from [16], modified to suit the definition of directed bigraphs.

Directed Link Graph. Intuitively an embedding of link graphs is a structure preserving map from one link graph (the *guest*) to another (the *host*). As one would expect from a graph embedding, this map contains a pair of injections: one for the nodes and one for the edges (i.e., a support translation). The remaining of the embedding map specifies how names of the inner and outer interfaces should be mapped into the host link graph. Outer names can be mapped to any link; here injectivity is not required since a context can alias outer names. Dually, inner names can be mapped to hyper-edges linking sets of points in the host link graph and such that every point is contained in at most one of these sets.

Definition 10 (Directed link graph embedding). *Let $G : X_G \to Y_G$ and $H : X_H \to Y_H$ be two directed link graphs. A directed link graph embedding $\phi : G \hookrightarrow H$ is a map $\phi \triangleq \phi^v \uplus \phi^e \uplus \phi^i \uplus \phi^o$, assigning nodes, edges, inner and outer names with the following constraints:*

(L1) $\phi^v : V_G \rightarrowtail V_H$ and $\phi^e : E_G \rightarrowtail E_H$ *are injective;*

(L2) $ctrl_G = ctrl_H \circ \phi^v$;

(L3) $\phi^i : Y_H^- \uplus X_H^+ \uplus P_H^+ \rightharpoonup X_G^+ \uplus Y_G^- \uplus P_G^+$ *defined as follows*

$$\phi^i(x) \triangleq \begin{cases} \phi^{i^-}(x) & if\ x \in Y_H^- \uplus P_H^+ \\ \phi^{i^+}(x) & if\ x \in X_H^+ \uplus P_H^+ \end{cases} \quad where \quad \begin{matrix} \phi^{i^-} : Y_H^- \uplus P_H^+ \rightharpoonup Y_G^- \uplus P_G^+ \\ \phi^{i^+} : X_H^+ \uplus P_H^+ \rightharpoonup X_G^+ \uplus P_G^+ \\ dom(\phi^{i^+}) \cap dom(\phi^{i^-}) = \varnothing \end{matrix}$$

(L4) $\phi^o : X_G^- \uplus Y_G^+ \rightharpoonup E_H \uplus X_H^- \uplus Y_H^+ \uplus P_H^-$ *is a partial map s.t.:*

$$\phi^o(y) \triangleq \begin{cases} \phi^{o^-}(y) & if\ y \in X_G^- \\ \phi^{o^+}(y) & if\ y \in Y_G^+ \end{cases} \quad where \quad \begin{matrix} \phi^{o^-} : X_G^- \rightharpoonup E_H \uplus X_H^- \uplus P_H^- \\ \phi^{o^+} : Y_G^+ \rightharpoonup E_H \uplus Y_H^+ \uplus P_H^- \end{matrix}$$

(L5a) $img(\phi^e) \cap img(\phi^o) = \varnothing$;

(L5b) $\forall v \in V_G, \forall j \in ar(ctrl(v))\ .\ \phi^i((\phi^v(v), j)) = \bot$;

(L6a) $\phi^p \circ link_G^{-1}|_{E_G} = link_H^{-1} \circ \phi^e$;

(L6b) $\forall v \in V_G, \forall i \in ar(ctrl(v))\ .\ \phi^p \circ link_G^{-1}((v, i)) = link_H^{-1} \circ \phi^{port}((v, i))$;

(L7) $\forall p \in dom(\phi^i) : link_H(p) = (\phi^o \uplus \phi^e)(link_G \circ \phi^i(p))$.

where $\phi^p \triangleq \phi^{i^+} \uplus \phi^{o^-} \uplus \phi^{port}$ and $\phi^{port} : P_G \rightarrowtail P_H$ is $\phi^{port}(v, i) \triangleq (\phi^v(v), i)$.

The first three conditions are on the single sub-maps of the embedding. Conditions (L5a) and (L5b) ensures that no components (except for outer names) are identified; condition (L6a) imposes that points connected by the image of an edge are all covered. Finally, conditions (L2), (L6b) and (L7) ensure that the guest structure is preserved i.e. node controls and point linkings are preserved.

Place Graph. Like link graph embeddings, place graph embeddings are just a structure preserving injective map from nodes along with suitable maps for the inner and outer interfaces. In particular, a site is mapped to the set of sites and nodes that are "put under it" and a root is mapped to the host root or node that is "put over it" splitting the host place graphs in three parts: the guest image, the context and the parameter (which are above and below the guest image).

Definition 11 (Place graph embedding [16, Def 7.5.4]). *Let $G : n_G \to m_G$ and $H : n_H \to m_H$ be two place graphs. A place graph embedding $\phi : G \hookrightarrow H$ is a map $\phi \triangleq \phi^v \uplus \phi^s \uplus \phi^r$ (assigning nodes, sites and roots respectively) such that:*

(P1) $\phi^v : V_G \rightarrowtail V_H$ *is injective;*
(P2) $\phi^s : n_G \rightarrowtail \wp(n_H \uplus V_H)$ *is fully injective;*
(P3) $\phi^r : m_G \to V_H \uplus m_H$ *in an arbitrary map;*
(P4) $img(\phi^v) \cap img(\phi^r) = \emptyset$ *and* $img(\phi^v) \cap \bigcup img(\phi^s) = \emptyset$;
(P5) $\forall r \in m_G : \forall s \in n_G : prnt^*_H \circ \phi^r(r) \cap \phi^s(s) = \emptyset$;
(P6) $\phi^c \circ prnt^{-1}_G \big|_{V_G} = prnt^{-1}_H \circ \phi^v$;
(P7) $ctrl_G = ctrl_H \circ \phi^v$;
(P8) $\forall c \in n_G \uplus V_G : \forall c' \in \phi^c(c) : (\phi^f \circ prnt_G)(c) = prnt_H(c')$;

where $prnt^*_H(c) = \bigcup_{i<\omega} prnt^i(c)$, $\phi^f \triangleq \phi^v \uplus \phi^r$, and $\phi^c \triangleq \phi^v \uplus \phi^s$.

These conditions follow the structure of Definition 10, the main difference is (P5) which states that the image of a root cannot be the descendant of the image of another. Conditions (P1), (P2) and (P3) are on the three sub-maps composing the embedding; (P4) and (P5) ensure that no components are identified; (P6) imposes surjectivity on children and the last two conditions require the guest structure to be preserved by the embedding.

Directed Bigraph. Finally, a directed bigraph embedding can be defined as a pair composed by an directed link graph embedding and a place graph embedding, with a consistent interplay of these two structures. The interplay is captured by two additional conditions ensuring that points (resp. handles) in the image of guest upward (resp. downward) inner names reside in some parameter defined by the place graph embedding (i.e. descends from the image of a site).

Definition 12 (Directed bigraph embedding). *Let $G : \langle n_G, X_G \rangle \to \langle m_G, Y_G \rangle$ and $H : \langle n_H, X_H \rangle \to \langle m_H, Y_H \rangle$ be two directed bigraphs. A directed bigraph embedding is a map $\phi : G \hookrightarrow H$ given by a place graph embedding $\phi^P : G^P \hookrightarrow H^P$ and a link graph embedding $\phi^L : G^L \hookrightarrow H^L$ subject to the following constraints:*

(B1) $dom(\phi^{i^+}) \subseteq X^+_H \uplus \{(v,i) \in P^+_H \mid \exists s \in n_G, k \in \mathbb{N} : prnt^k_H(v) \in \phi^s(s)\}$;
(B2) $img(\phi^{o^-}) \subseteq X^-_H \uplus \{(v,i) \in P^-_H \mid \exists s \in n_G, k \in \mathbb{N} : prnt^k_H(v) \in \phi^s(s)\}$.

4   Implementing the Embedding Problem in CSP

In this Section we present a constraint satisfaction problem that models the directed bigraph embedding problem. The encoding is based solely on integer linear constraints and is proven to be sound and complete.

Initially, we present the encoding for the directed link graph embedding problem and for the place graph embedding problem. Then we combine them providing some additional "gluing constraints" to ensure the consistency of the two sub-problems. The resulting encoding contains 37 constraint families (reflecting the complexity of the problem definition, see Sect. 3); hence we take advantage of the orthogonality of link and place structures for the sake of both exposition and adequacy proofs. We observe that the overall number of variables and constraints produced by the encoding is polynomially bounded with respect to the size of the involved bigraphs, i.e., the number of nodes and edges.

4.1   Directed Link Graphs

Let us fix the guest and host bigraphs $G : X_G \to Y_G$ and $H : X_H \to Y_H$. We characterize the embeddings of G into H as the solutions of a suitable *multi-flux problem* which we denote as $\mathrm{DLGE}[G, H]$. The main idea is to see the host points (i.e. positive ports, upward inner names and downward outer names) as sources, and the handles (i.e. edges, negative ports, upward outer names and downward inner names) as sinks (see Fig. 2). Each point outputs a flux unit and each handle inputs one unit for each point it links. Units flow towards each point handle following H edges and optionally taking a "detour" along the linking structure of the guest G (provided that some conditions about structure preservation are met). The formal definition of the flux problem is in Fig. 3.

The flux network reflects the linking structure and contains an edge connecting each point to its handle; these edges have an integer capacity limited to 1 and

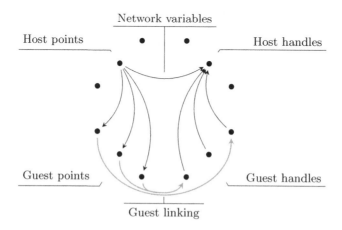

Fig. 2. Schema of the multi-flux network encoding.

$$N_{h,h'} \in \{0, \ldots, |link_H^{-1}(h')|\} \qquad \begin{aligned} &h \in E_G \uplus Y_G^+ \uplus X_G^- \uplus P_G^-,\\ &h' \in E_H \uplus Y_H^+ \uplus X_H^- \uplus P_H^- \end{aligned} \qquad (3)$$

$$N_{p,h'} \in \{0,1\} \qquad \begin{aligned} &h' \in E_H \uplus Y_H^+ \uplus X_H^- \uplus P_H^-,\\ &p \in link_H^{-1}(h') \end{aligned} \qquad (4)$$

$$N_{p,p'} \in \{0,1\} \qquad \begin{aligned} &p' \in X_G^+ \uplus P_G^+ \uplus Y_G^-,\\ &p \in X_H^+ \uplus P_H^+ \uplus Y_H^- \end{aligned} \qquad (5)$$

$$F_{h,h'} \in \{0,1\} \qquad \begin{aligned} &h \in E_G \uplus Y_G^+ \uplus X_G^- \uplus P_G^-,\\ &h' \in E_H \uplus Y_H^+ \uplus X_H^- \uplus P_H^- \end{aligned} \qquad (6)$$

$$\sum_k N_{p,k} = 1 \qquad p \in X_H^+ \uplus P_H^+ \uplus Y_H^- \qquad (7)$$

$$\sum_k N_{k,h} = |link_H^{-1}(h)| \qquad h \in E_H \uplus Y_H^+ \uplus X_H^- \uplus P_H^- \qquad (8)$$

$$\sum_k N_{h,k} = \sum_{p \in link_G^{-1}(h)} \sum_k N_{k,p} \qquad h \in E_G \uplus Y_G^+ \uplus X_G^- \uplus P_G^- \qquad (9)$$

$$\sum_k N_{k,p} \le 1 \qquad p \in X_G^+ \uplus P_G^+ \uplus Y_G^- \qquad (10)$$

$$N_{p,p'} = 0 \qquad p' \in P_G^+,\ p \in X_H^+ \uplus Y_H^- \qquad (11)$$

$$\frac{N_{h,h'}}{|link_H^{-1}(h')|} \le F_{h,h'} \le N_{h,h'} \qquad \begin{aligned} &h \in E_G \uplus Y_G^+ \uplus X_G^- \uplus P_G^-,\\ &h' \in E_H \uplus Y_H^+ \uplus X_H^- \uplus P_H^-,\\ &link_G^{-1}(h) \ne \varnothing,\ link_H^{-1}(h') \ne \varnothing \end{aligned} \qquad (12)$$

$$N_{p,p'} \le F_{h,h'} \qquad \begin{aligned} &h \in E_G \uplus Y_G^+ \uplus X_G^- \uplus P_G^-,\\ &h' \in E_H \uplus Y_H^+ \uplus X_H^- \uplus P_H^-,\\ &p \in link_G^{-1}(h),\ p' \in link_H^{-1}(h') \end{aligned} \qquad (13)$$

$$F_{h,h'} \le \sum_{\substack{p \in link_G^{-1}(h)\\ p' \in link_H^{-1}(h')}} N_{p,p'} \qquad \begin{aligned} &h \in E_G \uplus Y_G^+ \uplus X_G^- \uplus P_G^-,\\ &h' \in E_H \uplus Y_H^+ \uplus X_H^- \uplus P_H^-,\\ &link_G^{-1}(h) \ne \varnothing, link_H^{-1}(h') \ne \varnothing \end{aligned} \qquad (14)$$

$$\sum_k F_{h,k} = 1 \qquad h \in E_G \uplus Y_G^+ \uplus X_G^- \uplus P_G^- \qquad (15)$$

$$N_{p,h'} + F_{h,h'} \le 1 \qquad \begin{aligned} &h \in E_G,\ h' \in E_H \uplus Y_H^+ \uplus X_H^- \uplus P_H^-,\\ &p \in link_H^{-1}(h') \end{aligned} \qquad (16)$$

$$F_{h,h'} + F_{h'',h'} + F_{h''',h'} + F_{h^{iv},h'} \le 1 \qquad \begin{aligned} &h \in E_G,\ h' \in Y_H^+ \uplus X_H^- \uplus P_H^-,\\ &h'' \in Y_G^+,\ h''' \in X_G^-,\ h^{iv} \in P_G^- \end{aligned} \qquad (17)$$

$$F_{h,h'} = 0 \qquad h \in E_G,\ h' \in Y_H^+ \uplus X_H^- \uplus P_H^- \qquad (18)$$

$$F_{h,h'} \le 1 \qquad h \in E_G \uplus Y_G^+ \uplus X_G^- \uplus P_G^-,\ h' \in E_H \qquad (19)$$

$$N_{p,p'} = 0 \qquad \begin{aligned} &v \in V_G,\ v' \in V_H,\\ &ctrl_G(v) = ctrl_H(v) = c,\ i \ne i' \le c,\\ &p = (v,i) \in P_G^+ \uplus P_G^-,\\ &p' = (v',i') \in P_H^+ \uplus P_H^- \end{aligned} \qquad (20)$$

$$N_{p,p'} = 0 \qquad \begin{aligned} &v \in V_G,\ v' \in V_H,\\ &ctrl_G(v) \ne ctrl_H(v)\\ &p = (v,i) \in P_G^+ \uplus P_G^-,\\ &p' = (v',i') \in P_H^+ \uplus P_H^- \end{aligned} \qquad (21)$$

$$\sum_{j \le c} N_{(v,j),(v',j)} = c \cdot N_{p,p'} \qquad \begin{aligned} &v \in V_G,\ v' \in V_H,\\ &ctrl_G(v) = ctrl_H(v) = c,\ i \le c,\\ &p = (v,i) \in P_G^+ \uplus P_G^-,\\ &p' = (v',i') \in P_H^+ \uplus P_H^- \end{aligned} \qquad (22)$$

$$N_{p,p'} = 0 \qquad p \in P_H^+,\ p' \in X_G^+ \uplus Y_G^- \qquad (23)$$

Fig. 3. Constraints of DLGE$[G, H]$.

are represented by the variables defined in (4). The remaining edges of the network are organised in two complete biparted graphs: one between guest and host handles and one between guest and host points. Edges of the first sub-network are described by the variables in (3) and their capacity is bounded by the number of points linked by the host handle since this is the maximum acceptable flux and corresponds to the case where each point passes through the same hyperedge of the guest link graph. Edges of the second sub-network are described by the variables in (5) and, like the first group of links, have their capacity limited to 1; to be precise, some of these variables will never assume a value different from 0 because guest points can receive flux from anything but the host ports (as expressed by constraint (11)). Edges for the link structure of the guest are presented implicitly in the flux preservation constraints (see constraint (9)). In order to fulfil the injectivity conditions of link embeddings, some additional *flux variables* (whereas the previous are *network variables*) are defined by (6). These are used to keep track and separate each flux on the bases of the points handle.

The constraint families (7) and (8) define the outgoing and ingoing flux of host points and handles respectively. The former has to send exactly one unit considering every edge they are involved with and the latter receive one unit for each of their point regardless if this unit comes from the point directly or from a handle of the guest.

The linking structure of the guest graph is encoded by the constraint family (9) which states that flux is preserved while passing through the guest i.e. the output of each handle has to match the overall input of the points it connects.

Constraints (10), (11), (20), (21), (22) and (23) shape the flux in the sub-network linking guest and host points. Specifically, (10) requires that each point from the guest receives at most one unit; this is needed when we want to be able to embed a redex where some points (e.g. upward inner names) would not match with an entity of the agent and (those points) would be deleted anyway when composing the resulting agent back. Constraints (11), (20) and (21) disable edges between guest ports and host inner names, between mismatching ports of matching nodes and between ports of mismatching nodes. Constraint (23) ensures that ascending inner names or descending outer names of the redex are not matched with positive ports of the agent. Finally, the flux of ports of the same node has to act compactly, as expressed by (22): if there is flux between the i-th ports of two nodes, then there should be flux between every other ports.

Constraints (12), (13) and (14) relate flux and network variables ensuring that the formers assume a true value if, and only if, there is actual flux between the corresponding guest and host handles. In particular, (13) propagates the information about the absence of flux between handles disabling the sub-network linking handles points and, *vice versa*, (14) propagates the information in the other way disabling flux between handles if there is no flux between their points.

The remaining constraints prevent fluxes from mixing. Constraint (15) requires guest handles to send their output to exactly one destination thus rendering the sub-network between handles a function assigning guest handles to host handles. This mapping is subject to some additional conditions when edges are involved: (18) and (19) ensure that the edges are injectively mapped to edges only, (17) forbids host outer names to receive flux from an edge and an outer name at the same time. Finally, (16) states that the output of host points cannot bypass the guest if there is flux between its handle and an edge from the guest.

Adequacy. Let N be a solution of $DLGE[G, H]$. The corresponding link graph embedding $\phi : G \hookrightarrow H$ is defined as follows:

$$\phi^v(v) \triangleq v' \in V_H \text{ if } \exists i : N_{(v,i),(v',i)} = 1 \qquad \phi^e(e) \triangleq e' \in E_H \text{ if } F_{e,e'} = 1$$

$$\phi^i(x) \triangleq \begin{cases} \phi^{i^-}(x) & \text{if } x \in Y_H^- \uplus P_H^+ \\ \phi^{i^+}(x) & \text{if } x \in X_H^+ \uplus P_H^- \end{cases} \qquad \phi^o(y) \triangleq \begin{cases} \phi^{o^-}(y) & \text{if } y \in X_G^- \\ \phi^{o^+}(y) & \text{if } y \in Y_G^+ \end{cases}$$

where

$$\phi^{o^-}(y) \triangleq y' \in X_H^- \uplus P_H^- \text{ if } F_{y,y'} = 1 \qquad \phi^{o^+}(y) \triangleq y' \in Y_H^+ \uplus P_H^- \text{ if } F_{y,y'} = 1,$$

$$\phi^{i^-}(x) \triangleq x' \in Y_G^- \uplus P_G^+ \text{ if } N_{x,x'} = 1, \qquad \phi^{i^+}(x) \triangleq x' \in X_G^+ \uplus P_G^+ \text{ if } N_{x,x'} = 1$$

$$\text{and } dom(\phi^{i^+}) \cap dom(\phi^{i^-}) = \varnothing.$$

It is easy to check that these components of ϕ are well-defined and compliant with Definition 10.

On the other way around, let $\phi : G \hookrightarrow H$ be a link graph embedding. The corresponding solution N of $DLGE[G, H]$ is defined as follows:

$$N_{p,p'} \triangleq \begin{cases} 1 & \text{if } p \in X_H^+ \uplus Y_H^- \wedge p' = \phi^i(p) \\ 1 & \text{if } p' = (v, i) \in P_G^+ \wedge p = (\phi^v(v), i) \\ 0 & \text{otherwise} \end{cases}$$

$$N_{p,h'} \triangleq \begin{cases} 1 & \text{if } h' = link_H(p) \wedge \nexists p' : N_{p,p'} = 1 \\ 0 & \text{otherwise} \end{cases}$$

$$N_{h,h'} \triangleq \begin{cases} 1 & \text{if } h' \in E_H \wedge h \in E_G \wedge h' = \phi^e(h) \\ 1 & \text{if } h' \in Y_H^+ \uplus X_H^- \wedge h \in Y_G^+ \uplus X_G^- \wedge h' = \phi^o(h) \\ 1 & \text{if } h = (v, i) \in P_G^- \wedge h' = (\phi^v(v), i) \\ 0 & \text{otherwise} \end{cases}$$

$$F_{h,h'} = 1 \xLeftrightarrow{\triangle} N_{h,h'} \neq 0$$

Every constraint of $DLGE[G, H]$ is satisfied by the solution just defined.

The constraint satisfaction problem in Fig. 3 is sound and complete with respect to the directed link graph embedding problem given in Definition 10.

Proposition 1 (Adequacy of DLGE). *For any two concrete directed link graphs G and H, there is a bijective correspondence between the directed link graph embeddings of G into H and the solutions of DLGE[G, H].*

4.2 Place Graphs

Let us fix the guest and host place graphs: $G : n_G \to m_G$ and $H : n_H \to m_H$. We characterize the embeddings of G into H as the solutions of the constraint satisfaction problem in Fig. 4. The problem is a direct encoding of Definition 11 as a matching problem presented, as usual, as a bipartite graph. Sites, nodes and roots of the two place graphs are represented as nodes and partitioned into the guest and the host ones. For convenience of exposition, the graph is complete.

$$M_{h,g} \in \{0,1\} \qquad \begin{array}{l} g \in n_G \uplus V_G \uplus m_G, \\ h \in n_H \uplus V_H \uplus m_H \end{array} \qquad (24)$$

$$M_{h,g} = 0 \qquad g \in n_G \uplus V_G, \ h \in m_H \qquad (25)$$

$$M_{h,g} = 0 \qquad g \in V_G \uplus m_G, \ h \in n_H \qquad (26)$$

$$M_{h,g} = 0 \qquad \begin{array}{l} g \in V_G, \ h \in V_H, \\ ctrl_G(g) \neq ctrl_H(h) \end{array} \qquad (27)$$

$$M_{h,g} = 0 \qquad \begin{array}{l} g \in m_G, \ h \notin m_H, \\ v \in prnt_H^*(h) \cap V_G, \\ ctrl_G(v) \notin \Sigma_a \end{array} \qquad (28)$$

$$M_{h,g} \leq M_{h',g'} \qquad \begin{array}{l} g \notin m_G, \ g' \in prnt_G(g), \\ h \notin m_H, \ h' \in prnt_H(h) \end{array} \qquad (29)$$

$$\sum_{h \in V_H \uplus m_H} M_{h,g} = 1 \qquad g \in m_G \qquad (30)$$

$$\sum_{h \in n_H \uplus V_H} M_{h,g} = 1 \qquad g \in V_G \qquad (31)$$

$$m_G \cdot \sum_{g \in n_G \uplus V_G} M_{h,g} + \sum_{g \in m_G} M_{h,g} \leq m_G \qquad h \in V_H \qquad (32)$$

$$|prnt_H^{-1}(h)| \cdot M_{h,g} \leq \sum_{\substack{h' \in prnt_H^{-1}(h), \\ g' \in prnt_G^{-1}(g)}} M_{h',g'} \qquad g \in V_G, \ h \in V_H \qquad (33)$$

$$|prnt_G^{-1}(g) \setminus n_G| \cdot M_{h,g} \leq \sum_{\substack{h' \in prnt_H^{-1}(h) \setminus n_h, \\ g' \in prnt_G^{-1}(g) \setminus n_g}} M_{h',g'} \qquad g \in m_G, \ h \in V_H \qquad (34)$$

$$M_{h,g} + \sum_{h' \in prnt_H^*(h), g' \in m_G} M_{h',g'} \leq 1 \qquad g \in V_G, \ h \in V_H \qquad (35)$$

Fig. 4. Constraints of PGE$[G,H]$.

Edges are modelled by the boolean variables defined in (24); these are the only variables used by the problem. So far, a solution is nothing more than a relation between the components of guest and host containing only those pairs connected by an edge assigned a non-zero value. To capture exactly those assignments that are actual place graph embeddings some conditions have to be imposed.

Constraints (25) and (26) prevent roots and sites from the host to be matched with nodes or sites and nodes or roots respectively. (27) disables matching between nodes decorated with different controls. Constraint (28) prevents any matching for host nodes under a passive context (i.e. have an ancestor labelled with a passive control). (29) propagates the matching along the parent map from children to parents. Constraints (30) and (31) ensure that the matching is a function when restricted to guest nodes and roots (the codomain restriction follows by (25) and (26)). (32) says that if a node from the host cannot be matched with a root or a node/site from the guest at the same time; moreover, if the host node is matched with a node then it cannot be matched to anything else.

The remaining constraints are the counterpart of (29) and propagate matchings from parents to children. (33) applies to matchings between nodes and says that if parents are matched, then children from the host node are covered by children from the guest node. In particular, the matching is a perfect assignment when restricted to guest children that are nodes (because of (32)) and is a surjection on those that are sites. (34) imposes a similar condition on matchings between guest roots and host nodes. Specifically, it says that the matching has to cover child nodes from the guest (moreover, it is injective on them) leaving child sites to match whatever remains ranging from nothing to all unmatched children. Finally, (35) prevents matching from happening inside a parameter.

Adequacy. Let M be a solution of $\mathrm{PGE}[G, H]$. The corresponding place graph embedding $\phi : G \hookrightarrow H$ is defined as follows:

$$\phi^{\mathsf{v}}(g) \triangleq h \in V_H \text{ if } \exists i : M_{h,g} = 1 \qquad \phi^{\mathsf{s}}(g) \triangleq \{h \in n_h \uplus V_H \mid M_{h,g} = 1\}$$

$$\phi^{\mathsf{r}}(g) \triangleq h \in m_H \uplus V_H \text{ if } M_{h,g} = 1$$

These components of ϕ are well-defined and compliant with Definition 11. On the opposite direction, let $\phi : G \hookrightarrow H$ be a place graph embedding. The corresponding solution M of $\mathrm{PGE}[G, H]$ is defined as aside. It is easy to check that every constraint of $\mathrm{PGE}[G, H]$ is satisfied by this solution. Hence, the constraint satisfaction problem in Fig. 4 is sound and complete with respect to the place graph embedding problem (Definition 11).

$$M_{h,g} \triangleq \begin{cases} 1 & \text{if } g \in V_G \wedge h = \phi^{\mathsf{v}}(g) \\ 1 & \text{if } g \in m_G \wedge h = \phi^{\mathsf{r}}(g) \\ 1 & \text{if } g \in n_G \wedge h \in \phi^{\mathsf{s}}(g) \\ 0 & \text{otherwise} \end{cases}$$

Proposition 2 (Adequacy of PGE). *For any two concrete place graphs G and H, there is a bijective correspondence between the place graph embeddings of G into H and the solutions of $\mathrm{PGE}[G, H]$.*

4.3 Bigraphs

Let $G : \langle n_G, X_G \rangle \to \langle m_G, Y_G \rangle$ and $H : \langle n_H, X_H \rangle \to \langle m_H, Y_H \rangle$ be two bigraphs. By taking advantage of the orthogonality of the link and place structures we

$$M_{v,v'} = N_{p,p'} \qquad \begin{aligned} &v \in V_H,\ v' \in V_G, \\ &p = (v,k) \in P_H^+,\ p' = (v',k) \in P_G^+ \end{aligned} \qquad (36)$$

$$M_{v,v'} = F_{h,h'} \qquad v \in V_H,\ v' \in V_G,\ h \in P_G^-,\ h' \in P_H^- \qquad (37)$$

$$\sum_{p' \in X_G^+} N_{p,p'} \leq \sum_{h \in prnt_H^*(v),g \in n_G} M_{h,g} \qquad v \in V_H,\ p = (v,k) \in P_H^+ \qquad (38)$$

$$\sum_{h \in X_G^-} F_{h,h'} \leq \sum_{h \in prnt_H^*(v),g \in n_G} M_{h,g} \qquad v \in V_H,\ h' = (v,k) \in P_H^- \qquad (39)$$

Fig. 5. Constraints of DBGE[G, H].

can define the constraint satisfaction problem capturing bigraph embeddings by simply composing the constraints given above for the link and place graph embeddings and by adding four consistency constraints to relate the solutions of the two problems. These additional constraint families are reported in Fig. 5. The families (36) and (37) ensure that solutions for DLGE[G, H] and PGE[G, H] agree on nodes since the map ϕ^v has to be shared by the corresponding link and place embeddings. The families (38) and (39) respectively, ensure that positive ports (negative ports resp.) are in the same image as upward inner names (downward inner names resp.) only if their node is part of the parameter i.e. only if it is matched to a site from the guest or it descends from a node that is so.

Conditions (38) and (39) correspond exactly to (B1) and (B2). It thus follows from Propositions 1 and 2 that the CSP defined by Figs. 3 to 5 is sound and complete with respect to the bigraph embedding problem given in Definition 12.

Theorem 1 (Adequacy of BGE). *For any two concrete bigraphs G and H, there is a bijective correspondence between the bigraph embeddings of G into H and the solutions of DBGE[G, H].*

5   Experimental Results

The reduction algorithm presented in the previous section has been successfully integrated into jLibBig, an extensible Java library for manipulating bigraphs and bigraphical reactive systems which can be used for implementing a wide range of tools and it can be adapted to support several extensions of bigraphs [24].

The proposed algorithm is implemented by extending the data structures and the models for pure bigraphs to suit our definition of directed bigraphs.

In this section we test our implementation by simulating a system in which we want to track the position and the movements of a fleet of vehicles inside a territory divided in "zones", which are accessible via "roads". The rewriting rule in question and an example agent can be found in Fig. 6.

We evaluate the running time of the different components of our algorithm: model construction, CSP resolution, building of the actual embedding and execution of the rewriting rule. Moreover, we want to evaluate how these performances scale while increasing the size of the agent. The parameters used to build the

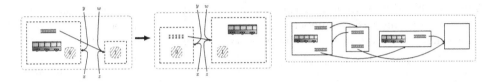

Fig. 6. Rewriting rule (left) and example of an agent (right) for the test cases.

tests are: number of zones, number of cars and "connectivity degree". The last parameter is a number between 1 and 100 representing the probability of the existence of a connection between two nodes; a value of 100 means that every node is connected to all its neighbours.

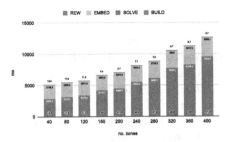

Fig. 7. Execution times vs. number of zones, 70 cars and 100% connectivity.

Fig. 8. Execution times vs. connectivity, 11×11 grid and 70 cars.

We consider the following kinds of tests:

1. varying number of zones, with fixed number of cars and connectivity degree;
2. varying connectivity degree, with fixed numbers of zones and cars.

Each test case is made up of four groups of instances, where for each group we choose an increasing value for their fixed parameters. For each group we choose ten values for its variable parameter. The instances generation works as follows: for each test case and for each group of that particular test case we generate ten random instances for each combination of the values of the fixed parameters and the variable one. We then take the average of the running times of those ten random instances. At the end of the process, for each group we have tested 100 instances, 10 for each value of the variable parameter, so 400 instances for each test case and 1200 in total.

All tests have been performed on an Intel Core i7-4710HQ (4 cores at 3.5 GHz), 8 GB of RAM running on ArchLinux with kernel 5.5.2 and using OpenJDK 12.

We briefly review the results obtained from these tests and refer to the companion technical report [6] for more details.

Time vs. Number of Zones. In this case we evaluate how our implementation scales with an increasing number of zones; see Fig. 7. We can see that the running time grows exponentially, especially the resolution time. Similarly to the previous test case, the time spent building the CSP and applying the reaction rule is negligible even though we can see that the time necessary to build the CSP increases linearly with the grid size. We can also observe that there is no correlation between the rewriting time and the number of zones.

Time vs. Connectivity Degree. In this case we evaluate how our implementation scales with an increasing connectivity degree; see Fig. 8. We can see that the running time scales exponentially, no matter the grid size or the number of cars. Once again, we see that although increasing, the time spent building the model and applying the rewriting rule is negligible.

6   Conclusions and Future Works

In this paper, we have presented a new version of *directed* bigraphs and bigraphic reactive systems, which subsume previous versions (such as Milner's bigraphs). For this kind of bigraphs we have provided a sound and complete algorithm for solving the embedding problem, based on a constraint satisfaction problem. The resulting model is compact and the a number of variables and linear constraints are polynomially bounded by the size of the guest and host bigraphs. Differently from existing solutions, this algorithm applies also to non-ground hosts.

The algorithm has been successfully integrated into jLibBig, an extensible library for manipulating bigraphical reactive systems. The empirical evaluation of the implementation of our algorithm in jLibBig looks promising. It cannot be considered a rigorous experimental validation yet, mainly because performance depends on the implementation and the solver and the model is not optimized for any specific solver. Moreover, up to now there are no "official" (or "widely recognized") benchmarks, nor any other algorithms or available tools that solve the directed bigraph embedding problem, to compare with.

The proposed approach offers great flexibility: it can be easily applied also to other extensions of bigraphs and directed bigraphs. An interesting direction for future work would be to extend the algorithm also to stochastic and probabilistic bigraphs [18]; this would offer useful modelling and verification tools for quantitative aspects, e.g. for systems biology [2,8]. Approximated and weighted embeddings are supported in jLibBig, but still as experimental feature. In fact, the theoretical foundations of these extensions have not been fully investigated yet, suggesting another line of research.

References

1. Bacci, G., Grohmann, D., Miculan, M.: DBtk: a toolkit for directed bigraphs. In: Kurz, A., Lenisa, M., Tarlecki, A. (eds.) CALCO 2009. LNCS, vol. 5728, pp. 413–422. Springer, Heidelberg (2009). https://doi.org/10.1007/978-3-642-03741-2_28

2. Bacci, G., Grohmann, D., Miculan, M.: A framework for protein and membrane interactions. arXiv preprint arXiv:0911.4513 (2009)
3. Bacci, G., Miculan, M., Rizzi, R.: Finding a forest in a tree. In: Maffei, M., Tuosto, E. (eds.) TGC 2014. LNCS, vol. 8902, pp. 17–33. Springer, Heidelberg (2014). https://doi.org/10.1007/978-3-662-45917-1_2
4. Bundgaard, M., Glenstrup, A.J., Hildebrandt, T., Højsgaard, E., Niss, H.: Formalizing higher-order mobile embedded business processes with binding bigraphs. In: Lea, D., Zavattaro, G. (eds.) COORDINATION 2008. LNCS, vol. 5052, pp. 83–99. Springer, Heidelberg (2008). https://doi.org/10.1007/978-3-540-68265-3_6
5. Burco, F., Miculan, M., Peressotti, M.: Towards a formal model for composable container systems. In: Proceedings of the SAC, pp. 173–175. ACM (2020)
6. Chiapperini, A., Miculan, M., Peressotti, M.: A CSP implementation of the directed bigraph embedding problem. CoRR abs/2003.10209 (2020)
7. Damgaard, T.C., Glenstrup, A.J., Birkedal, L., Milner, R.: An inductive characterization of matching in binding bigraphs. Form. Asp. Comp. 25(2), 257–288 (2013)
8. Damgaard, T.C., Højsgaard, E., Krivine, J.: Formal cellular machinery. Electr. Not. Theoret. Comput. Sci. 284, 55–74 (2012)
9. Faithfull, A.J., Perrone, G., Hildebrandt, T.T.: Big red: a development environment for bigraphs. Electr. Commun. EASST 61, 1–10 (2013)
10. Gassara, A., Rodriguez, I.B., Jmaiel, M., Drira, K.: Executing bigraphical reactive systems. Discr. Appl. Math. 253, 73–92 (2019)
11. Glenstrup, A.J., Damgaard, T.C., Birkedal, L., Højsgaard, E.: An implementation of bigraph matching. IT University of Copenhagen p. 22 (2007)
12. Grohmann, D.: Security, cryptography and directed bigraphs. In: Ehrig, H., Heckel, R., Rozenberg, G., Taentzer, G. (eds.) ICGT 2008. LNCS, vol. 5214, pp. 487–489. Springer, Heidelberg (2008). https://doi.org/10.1007/978-3-540-87405-8_41
13. Grohmann, D., Miculan, M.: Directed bigraphs. ENTCS 173, 121–137 (2007)
14. Grohmann, D., Miculan, M.: Reactive systems over directed bigraphs. In: Caires, L., Vasconcelos, V.T. (eds.) CONCUR 2007. LNCS, vol. 4703, pp. 380–394. Springer, Heidelberg (2007). https://doi.org/10.1007/978-3-540-74407-8_26
15. Grohmann, D., Miculan, M.: Controlling resource access in directed bigraphs. ECE-ASST 10, 1–21 (2008)
16. Højsgaard, E., et al.: Bigraphical languages and their simulation. IT University of Copenhagen, Programming, Logic, and Semantics (2012)
17. Jensen, O.H., Milner, R.: Bigraphs and transitions. In: ACM SIGPLAN Notices, vol. 38, pp. 38–49. ACM (2003)
18. Krivine, J., Milner, R., Troina, A.: Stochastic bigraphs. Electr. Not. Theor. Comput. Sci. 218, 73–96 (2008)
19. Mansutti, A., Miculan, M., Peressotti, M.: Distributed execution of bigraphical reactive systems. ECEASST 71, 1–21 (2014)
20. Mansutti, A., Miculan, M., Peressotti, M.: Multi-agent systems design and prototyping with bigraphical reactive systems. In: Magoutis, K., Pietzuch, P. (eds.) DAIS 2014. LNCS, vol. 8460, pp. 201–208. Springer, Heidelberg (2014). https://doi.org/10.1007/978-3-662-43352-2_16
21. Mansutti, A., Miculan, M., Peressotti, M.: Towards distributed bigraphical reactive systems. In: Echahed, R., Habel, A., Mosbah, M. (eds.) Proceedings of the GCM 2014, p. 45 (2014). workshop version
22. Miculan, M., Peressotti, M.: Bigraphs reloaded: a presheaf presentation. Technical report UDMI/01/2013, Dept. of Mathematics and Computer Science, Univ. of Udine (2013). http://www.dimi.uniud.it/miculan/Papers/UDMI012013.pdf

23. Miculan, M., Peressotti, M.: A CSP implementation of the bigraph embedding problem. CoRR abs/1412.1042 (2014)
24. Miculan, M., Peressotti, M.: jLibBig: a library for bigraphical reactive systems, November 2015. https://github.com/bigraphs/jlibbig
25. Milner, R.: The Space and Motion of Communicating Agents. Cambridge University Press, Cambridge (2009)
26. Perrone, G., Debois, S., Hildebrandt, T.T.: Bigraphical refinement. In: Refine@FM. EPTCS, vol. 55, pp. 20–36 (2011)
27. Perrone, G., Debois, S., Hildebrandt, T.T.: A model checker for bigraphs. In: Proceedings of the 27th Annual ACM Symposium on Applied Computing, pp. 1320–1325 (2012)
28. Sahli, H., Ledoux, T., Rutten, É.: Modeling self-adaptive fog systems using bigraphs. In: Proceedings of the FOCLASA, Oslo, Norway, pp. 1–16, September 2019
29. Sevegnani, M., Calder, M.: Bigraphs with sharing. Theoretical Computer Science **577**, 43–73 (2015)
30. Sevegnani, M., Unsworth, C., Calder, M.: A SAT based algorithm for the matching problem in bigraphs with sharing. University of Glasgow, Technical report (2010)

A Categorical Semantics for Guarded Petri Nets

Fabrizio Genovese[1]([⊠]) and David I. Spivak[2]

[1] Statebox, Amsterdam, The Netherlands
research@statebox.io
[2] MIT, Boston, USA
dspivak@mit.edu

Abstract. We build on the correspondence between Petri nets and free symmetric strict monoidal categories already investigated in the literature, and present a categorical semantics for Petri nets with guards. This comes in two flavors: Deterministic and with side-effects. Using the Grothendieck construction, we show how the guard semantics can be internalized in the net itself.

1 Introduction

Category theory has been used to study Petri nets at least since the beginning of the nineties [6]. Throughout this time, the main effort in this direction of research consisted in showing how Petri nets can be thought of as presenting various flavors of free monoidal categories [2,5,6,8] This idea has been very influential, successfully modeling the individual-token philosophy via process semantics.

On the other hand, shortly after Petri's first publications about the nets that carry his name [7] researchers started investigating what happens when nets are enriched with new features. One of the most successful extensions of Petri nets is *guarded (or coloured) nets* [3]. Modulo different flavors of modeling what boils down to be the same concept, a *guarded net* is a Petri net with the following extra properties:

- To each token is attached some "attribute". The kind of attributes we can attach to tokens depends on the place the token is in;
- Each arc is decorated with an expression, which modifies tokens' attributes as they flow through the net;
- Each transition is decorated with a predicate and only fires on tokens whose attributes satisfy the predicate.

At a fist glance, guarded nets allow for a more expressive form of modeling with respect to their unguarded counterparts, but as we will see shortly, this is not necessarily the case. Indeed, depending on the underlying theory from which properties, expressions, and predicates are drawn the gain in expressive power with respect to undecorated nets may be nil: With a wise choice of underlying

F. Gadducci and T. Kehrer (Eds.): ICGT 2020, LNCS 12150, pp. 57–74, 2020.
https://doi.org/10.1007/978-3-030-51372-6_4

theory, coloured nets amount to be nothing more than syntactic sugar for standard nets, though of course the availability of such syntactic sugar can greatly simplify the modeling of complex processes using the Petri net formalism.

Recently there has been renewed interest in employing Petri nets as the basis for a programming language [10]. In this setting, the categorical correspondence between nets and symmetric monoidal categories has been of the utmost importance, single-token philosophy being considered necessary to make the programming language usable [9]. Clearly, extending nets with new features such as guards or timings is desirable to make the language more expressive.

In this work we try to unify these two longstanding directions of research – the categorical approach to Petri nets and the study of guarded nets – by showing how guarded nets can be modeled as ordinary Petri nets with a particular flavor of semantics in the style of [1].

Importantly, we are able to define both a *deterministic semantics* and a *non-deterministic semantics* in our formalism. The first models the traditional notion of guards deterministically modifying data attached to tokens, while the second describes a setting where token data is modified depending on side effects.

Using the Grothendieck construction, we show how the guard semantics can be *internalized* in the net itself, providing a categorical proof that in our model, guarded nets do not increase expressivity, as compared to traditional nets. This is a desired feature, since it means that many nice properties of nets such as termination or decidability of the reachability relation are preserved. It also shows that the core mathematical abstraction in computer implementations of Petri nets need not be modified when offering users the flexibility of guarded nets.

We save all proofs for the appendix, which starts on page 14.

2   Guarded Nets

Having given an intuitive version of what a guarded net is, we now start modeling the concept formally. We will use the formalism developed in [1], of which we recall some core concepts.

We denote by $\mathfrak{F}(N)$ the free symmetric strict monoidal category associated to a Petri net N, and with $\mathfrak{U}(\mathcal{C})$ the Petri net associated to the free symmetric strict monoidal category $\mathcal{C}$. We denote composition in diagrammatic order; i.e. given $f: c \to d$ and $g: d \to e$, we denote their composite by $(f \mathbin{\mathring{,}} g): c \to e$.

Definition 1. *Given a strict monoidal category $\mathcal{S}$, a Petri net with $\mathcal{S}$-semantics is a pair $(N, N^\sharp)$, consisting of a Petri net N and a strict monoidal functor*

$$N^\sharp : \mathfrak{F}(N) \to \mathcal{S}.$$

A morphism $F : (M, M^\sharp) \to (N, N^\sharp)$ is just a strict monoidal functor $F : \mathfrak{F}(M) \to \mathfrak{F}(N)$ such that $M^\sharp = F \mathbin{\mathring{,}} N^\sharp$.

Nets equipped with $\mathcal{S}$-semantics and their morphisms form a monoidal category denoted **Petri**$^\mathcal{S}$*, with the monoidal structure arising from the product in* **Cat***.*

Definition 2. *We denote by $\mathbf{Set}_*$ the category of sets and partial functions, and by $\mathbf{Span}$ the 1-category of sets and spans, where isomorphic spans are identified. Both these categories are symmetric monoidal. From now on, we will work with the strictified version of $\mathbf{Set}_*$ and $\mathbf{Span}$, respectively.*

Example 1. Let $\mathbf{1}$ denote the terminal symmetric monoidal category. A Petri net with 1-semantics is just a Petri net. Petri nets are in bijective correspondence with free symmetric strict monoidal categories, so $\mathbf{Petri^1}$ denotes the usual category of free symmetric strict monoidal categories and strict monoidal functors between them.

Notation 1. *Recall that a morphism $A \to B$ in $\mathbf{Span}$ consists of a set S and a pair of functions $A \leftarrow S \to B$. When we need to notationally extract this data from f, we write*

$$A \xleftarrow{f_1} S_f \xrightarrow{f_2} B$$

We sometimes consider the span as a function $f \colon S_f \to A \times B$, thus we may write $f(s) = (a, b)$ for $s \in S_f$ with $f_1(s) = a$ and $f_2(s) = b$.

Perhaps unsurprisingly, $\mathbf{Set}_*$ and $\mathbf{Span}$ will be the target semantics corresponding to two different flavors for our guards, with $\mathbf{Span}$ allowing for some form of nondeterminism – expressed as the action of side-effects – whereas $\mathbf{Set}_*$ models a purely deterministic semantics. Expressing things formally:

Definition 3. *A* guarded net *is an object of $\mathbf{Petri^{Set_*}}$. A* guarded net with side effects *is an object of $\mathbf{Petri^{Span}}$. A* morphism of guarded nets (with side effects) *is a morphism in $\mathbf{Petri^{Set_*}}$ (resp. in $\mathbf{Petri^{Span}}$).*

Remark 1. Although it doesn't affect our formalism by any means, in practice the choice of semantics, both for $\mathbf{Set}_*$ and $\mathbf{Span}$, is limited by computational requirements: the places in a net are usually sent to *finite* sets, while transitions are usually sent to computable functions and spans[1], respectively. Such restrictions are necessary to make sure the net is executable and to keep model checking decidable.

Let us unroll the cryptic Definition 3, starting from the case $\mathbf{Petri^{Set_*}}$. An object in $\mathbf{Petri^{Set_*}}$ is a net N together with a strict monoidal functor $N^\sharp \colon \mathfrak{F}(N) \to \mathbf{Set}_*$. It assigns to each place p of N – corresponding to a generating object of $\mathfrak{F}(N)$ – a set $N^\sharp(p)$, representing all the possible colours a token in p can assume. A transition $f \colon p \to p'$ – corresponding to a generating morphism of $\mathfrak{F}(N)$ – gets sent to a partial function $N^\sharp(f) \colon N^\sharp(p) \to N^\sharp(p')$, representing how token colours are transformed during firing. Importantly, the fact that the functions in the semantics are *partial* means that a transition may not be defined for tokens of certain colors. An example of this is the net in Fig. 1a, which is shown together with its semantics. Although reachability in the base net seems quite straightforward, we see that a token in the leftmost place will never reach the rightmost place, since the rightmost transition is not defined on the tokens output by the leftmost one.

---

[1] A *computable span* is one for which both legs are computable functions.

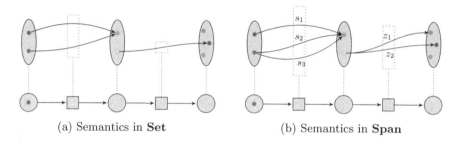

(a) Semantics in **Set**   (b) Semantics in **Span**

Fig. 1. The same net (below), equipped with a partial function and span semantics, respectively (above).

In the case of **Petri**$^{\mathbf{Span}}$ the intuition is similar. Objects are sent to sets, exactly as in **Petri**$^{\mathbf{Set}_*}$, but transitions are mapped to spans. Spans can be understood as *relations with witnesses*, provided by elements in the apex of the span. Practically, this means that each path from the span domain to its codomain is indexed by some element of the span apex, as it is shown in Fig. 1b. The presence of witnesses allows to consider different paths between the same elements. Moreover, an element in the domain can be sent to different elements in the codomain via different paths. We interpret this as *non-determinism*: The firing of the transition is not only a matter of the tokens input and output, it also includes the path chosen, which we interpret as having side-effects that are interpreted outside of our model. As one can see, in both Figs. 1a and 1b the composition of paths is the empty function (resp. span). Seeing things from a reachability point of view, the process given by firing the left transition and then the right will never occur. Let us make this intuition precise:

Definition 4. *Given a guarded Petri net (with side effects)* $(N, N^\sharp)$, *a marking for* $(N, N^\sharp)$ *is a pair* (X, x) *where* X *is an object of* $\mathfrak{F}(N)$ *and* $x \in N^\sharp X$. *We say that a marking* (Y, y) *is reachable from* (X, x) *if there is a morphism* $f : X \to Y$ *in* $\mathfrak{F}(N)$ *such that* $N^\sharp f(x) = y$.

The goal we will pursue in the next section will be to internalize the guard semantics in the free category $\mathfrak{F}(N)$ associated to a net.

3 Internalizing Guards

By "internalizing the semantics of a guarded net N in $\mathfrak{F}(N)$" we mean *obtaining an unguarded net M such that $\mathfrak{F}(M)$ represents all the possible runs of N.* For readers familiars with coloured Petri nets, this corresponds to the claim that reachability in a coloured net is equivalent to reachability in a suitably constructed "standard" net [3].

Since our point of view is process-theoretic, and we are working with symmetric strict monoidal categories and functors, such internalization must be built categorically. The main tool we will use is the *Grothendieck construction* [4], which in our context we will specialize to functors to **Set**$_*$ and **Span**, respectively.

Definition 5. *Let* $(M, M^\sharp) \in \textbf{Petri}^{Set_*}$ *be a guarded net. We define its* inter- *nalization, denoted* $\int M^\sharp$, *as the following category:*

- *The objects of* $\int M^\sharp$ *are pairs* (X, x), *where* X *is an object of* $\mathfrak{F}(M)$ *and* x *is an element of* $M^\sharp X$. *Concisely:*

$$\text{Obj} \int M^\sharp := \left\{ (X, x) \mid (X \in \text{Obj } \mathfrak{F}(M)) \wedge (x \in M^\sharp X) \right\}.$$

- *A morphism from* (X, x) *to* (Y, y) *in* $\int M^\sharp$ *is a morphism* $f \colon X \to Y$ *in* $\mathfrak{F}(M)$ *such that* x *is sent to* y *via* $M^\sharp f$. *Concisely:*

$$\text{Hom}_{\int M^\sharp} [(X, x), (Y, y)] := \left\{ f \mid (f \in \text{Hom}_{\mathfrak{F}(M)} [X, Y]) \wedge (M^\sharp f(x) = y) \right\}.$$

It is worth giving some intuition of what the Grothendieck construction does in our context. It basically makes a place for each element of the set we send a place to, and makes a transition for each path between these elements, as shown below:

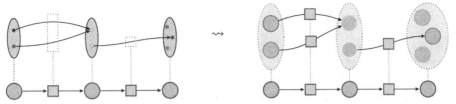

An equivalent definition exists when the semantics is taken to be in **Span**, which is the following:

Definition 6. *Let* $(M, M^\sharp) \in \textbf{Petri}^{Span}$ *be a guarded net with side effects. We define the* internalization *of* $(M, M^\sharp)$, *denoted with* $\int M^\sharp$, *as the following cate- gory:*

- *The objects of* $\int M^\sharp$ *are pairs* (X, x), *where* X *is an object of* $\mathfrak{F}(M)$ *and* x *is an element of* $M^\sharp X$. *Concisely:*

$$\text{Obj} \int M^\sharp := \left\{ (X, x) \mid (X \in \text{Obj } \mathfrak{F}(M)) \wedge (x \in M^\sharp X) \right\}.$$

- *A morphism from* (X, x) *to* (Y, y) *in* $\int M^\sharp$ *is a pair* (f, s) *where* $f \colon X \to Y$ *in* $\mathfrak{F}(M)$ *and* $s \in S_{M^\sharp f}$ *in the apex of the corresponding span connects* x *to* y. *Concisely:*

$$\text{Hom}_{\int M^\sharp} [(X, x), (Y, y)] :=$$
$$:= \left\{ (f, s) \mid (f \in \text{Hom}_{\mathfrak{F}(M)} [X, Y]) \wedge (s \in S_{M^\sharp f}) \wedge (M^\sharp f(s) = (x, y)) \right\}.$$

The intuition in the span case is exactly as for partial functions, and we don't deem it useful to draw the same picture again. Looking at the example, though, a couple of things become clear. The first is that to justify the idea of the Grothendieck construction turning an assignment of semantics into a net we have to prove that the resulting category is symmetric strict monoidal and free. The second is that the net thus built is fibered over the base net, and there should be an opposite construction sending $\int M^\sharp$ to M. Both of these claims are true, as we now prove:

Lemma 1. *In the case of both* **Set**$_*$ *and* **Span***, the category* $\int M^\sharp$ *has a strict symmetric monoidal structure.*

Theorem 1. *In both the case of* **Set**$_*$ *and of* **Span** *the strict symmetric monoidal category* $\int M^\sharp$ *is free.*

Counterexample 1 (Relations). *Theorem 1 does not hold – the Grothendieck construction does not yield a free symmetric strict monoidal category – if we replace* **Set**$_*$ *or* **Span** *with* **Rel***. To see this, consider* $\int M^\sharp$ *in the case that* $M^\sharp \colon \mathfrak{F}(M) \to$ **Rel***. Let* M *be the Petri net consisting of three places* X, Y, Z *and two transitions* $f \colon X \to Y$ *and* $g \colon Y \to Z$*. Let* $M^\sharp$ *send* X *to* $\{x\}$*,* Y *to* $\{y_1, y_2\}$*, and* Z *to* $\{z\}$*. On morphisms, let* $M^\sharp$ *send* f *to the maximal relation on* $\{x\} \times \{y_1, y_2\}$ *and* g *to the maximal relation on* $\{y_1, y_2\} \times \{z\}$*. Then we have the following four generating morphisms in* $\int M^\sharp$*:*

$$f_1 \colon (X, x) \to (Y, y_1) \qquad f_2 \colon (X, x) \to (Y, y_2)$$
$$g_1 \colon (Y, y_1) \to (Z, z) \qquad g_2 \colon (Y, y_2) \to (Z, z)$$

There is an equality $f_1 \,\mathring{\,}\, g_1 = f_2 \,\mathring{\,}\, g_2$ *as morphisms* $(X, x) \to (Z, z)$ *in* $\int M^\sharp$*, proving* $\int M^\sharp$ *is not free.*

The reason that Theorem 1 holds in the span case is that spans keep track of different paths between elements, whereas relations do not. To see this, *consider the span composition:*

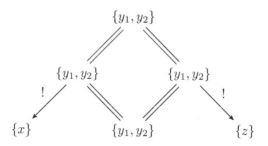

It is clear that in this composition the two paths from x to z *are considered as separated in the* **Span** *case, and witnessed by* y_1, y_2 *respectively, while in the case of* **Rel** *they would have been conflated to one. The result is that these paths correspond to the same morphism in the relational case of* $\int M^\sharp$*, introducing new equations and breaking freeness, while they stay separated in the span case.*

Lemma 2. *In the case of both* **Set**$_*$ *and* **Span***, there is a strict monoidal functor* $\pi_M \colon \int M^\sharp \to \mathfrak{F}(M)$ *sending* (X, x) *to* X *and* $f \colon (X, x) \to (Y, y)$ *to* $f \colon X \to Y$ *(resp.* $(f, s) \colon (X, x) \to (Y, y)$ *to* $f \colon X \to Y$*).*

Remark 2. In general, π_M is not an opfibration. This is because our target categories **Set**$_*$ and **Span** allow for partial functions. Indeed, if $f \colon X \to Y$ in M is sent by $M^\sharp$ to a partial function that is not defined on $x \in M^\sharp X$, then there is no coCartesian lift emanating from (X, x) for the morphism f.

We conclude this section by proving that the reachability semantics of a guarded net coincides with the reachability semantics of its internalization.

Theorem 2. *Let $(N, N^\sharp)$ be a guarded Petri net (with side effects). (Y, y) is reachable from (X, x) if and only if (Y, y) is reachable from (X, x) in the net $\mathfrak{U}\left(\int N^\sharp\right)$.*

4   Properties of Internalizations

The Grothendieck construction provides a way to internalize partial function and span semantics to nets. As such, it acts on objects of the categories **Petri**^{Set_*} and **Petri**^{Span}, respectively. It is thus worth asking what happens to morphisms in these categories. The answer is, luckily, easy to find:

Lemma 3. *Let $F\colon (M, M^\sharp) \to (N, N^\sharp)$ be a morphism in **Petri**^{Set_*} (resp. in **Petri**^{Span}). Then it lifts to strict monoidal functor $\vec{F}\colon \int M^\sharp \to \int N^\sharp$ (resp. $\widehat{F}\colon \int M^\sharp \to \int N^\sharp$), such that the following diagram on the left (resp. on the right) commutes:*

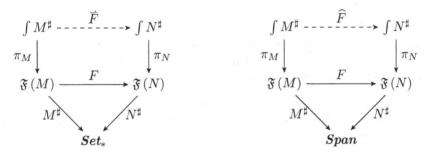

Notation 2. *The notation for the liftings in Lemma 3 is easy to remember: The arrow over $\vec{F}$ looks like a stylized function, while the hat over $\widehat{F}$ looks like a stylized span.*

The lifting of Lemma 3 is quite well-behaved. First of all, it is worth stressing how it preserves some relevant categorical properties:

Lemma 4. *For any map $F\colon (M, M^\sharp) \to (N, N^\sharp)$ in **Petri**^{Set_*} (respectively in **Petri**^{Span}), the functor F is faithful if and only if $\vec{F}$ is faithful (resp. $\widehat{F}$ is faithful). If F is full, then so is $\vec{F}$ (resp. $\widehat{F}$).*

Having ascertained that "basic" categorical properties are preserved, it is worth asking what happens to particular classes of functors in **Petri**^{Set_*} and **Petri**^{Span}, respectively.

Following [1], there are three relevant kinds of morphisms in a category of Petri nets with semantics. On one hand there are transition-preserving functors, which represent morphisms of free monoidal categories arising purely from the topological structure of their underlying net. On the other there are functors representing glueings of nets, which are themselves divided into synchronizations (defined in terms of addition and erasing of generators, that is, double pushouts) and identifications (defined in terms of pushouts). Let us investigate which ones of these properties are preserved.

Definition 7. *A strict symmetric monoidal functor F between FSSMCs $\mathcal{C}, \mathcal{D}$ is said to be* transition-preserving *when each generating morphism f of $\mathcal{C}$ is mapped to $\sigma \, \mathring{,} \, g \, \mathring{,} \, \sigma'$ for some generating morphism g of $\mathcal{D}$ and symmetries σ, σ'.*

Lemma 5. *If F is transition-preserving, so are $\vec{F}$ and $\widehat{F}$.*

Lemma 6. *If F is injective on objects, so are $\vec{F}$ and $\widehat{F}$.*

Another interesting class of morphisms is *identifications*:

Definition 8. *A Petri net $\left(N, N^{\sharp}\right)$ is said to be an* identification *of $\left(M, M^{\sharp}\right)$ if there is a morphism $F : \left(M, M^{\sharp}\right) \to \left(N, N^{\sharp}\right)$ such that:*

– *There is a Petri net O, and a pair of transition-preserving functors $l, r :$ $\mathfrak{F}(O) \to \mathfrak{F}(M)$;*
– *$l \, \mathring{,} \, M^{\sharp} = r \, \mathring{,} \, M^{\sharp}$; and*
– *F is the coequalizer of l and r.*

Identifications are also preserved. The ultimate reason for this is that identifications are defined purely in terms of coequalizers of transition-preserving functors, which are preserved by the Grothendieck construction.

Lemma 7. *If $\left(N, N^{\sharp}\right)$ is an identification of $\left(M, M^{\sharp}\right)$ via F and witnesses O, l, r, then $\int N^{\sharp}$ is an identification of $\int M^{\sharp}$ via $\vec{F}$ and witnesses $\mathfrak{U}\left(\int \left(l \, \mathring{,} \, M^{\sharp}\right)\right), \vec{l}, \vec{r}$. The span case is analogous.*

Preservation of identifications also entails that addition of generators for a net are preserved by internalizations.

Definition 9. *A net $\left(M, M^{\sharp}\right)$ is an* addition of generating morphisms *to $\left(K, K^{\sharp}\right)$ via W, w if:*

– *There is a net W together with a strict monoidal functor $w : \mathfrak{F}(W) \to \mathfrak{F}(K)$ which sends generating objects to generating objects, is injective on objects and faithful;*
– *$\mathfrak{F}(M)$ is the pushout of $\mathfrak{F}(\overline{W}) \hookrightarrow \mathfrak{F}(W) \xrightarrow{w} \mathfrak{F}(K)$ and $\mathfrak{F}(\overline{W}) \hookrightarrow \mathfrak{F}(W)$, where $\overline{W}$ denotes the net with the same places of W and no transitions; and*
– *$M^{\sharp}$ arises from the universal property of the pushout.*

Lemma 8. *Let* $(M, M^\sharp)$ *be an addition of generating morphisms to* $(K, K^\sharp)$ *via* W, w. *Then* $\int M^\sharp$ *is an addition of generating morphisms to* $\int K^\sharp$ *via* $\mathfrak{U}\left(\int (w \,\r{;}\, K^\sharp)\right), \bar{w}$. *The span case is analogous.*

Finally, we investigate what happens when considering erasings of generators from a net. To do this, we first follow [1] to define subnets:

Definition 10. *Given Petri nets* N, N_w, *we say that* N_w *is a subnet of* N *if its places and transitions are a subset of places and transitions of* N, *and input and output functions on* N_w *are restrictions of the input and output functions on* N. *If* N_w *is a subnet of* N, *then there is an obvious identity on objects, identity on morphisms strict monoidal functor,* $\iota : \mathfrak{F}(N_w) \hookrightarrow \mathfrak{F}(N)$ *between their associated free symmetric strict monoidal categories. From this, we say that a net* $(N_w, N_w{}^\sharp)$ *is a subnet of* $(N, N^\sharp)$ *if* N_w *is a subnet of* N *and* $N_w{}^\sharp = \iota \,\r{;}\, N^\sharp$.

This enables us to define erasings of generators:

Definition 11. *Let* $(N_w, (N_w)^\sharp)$ *be a subnet of* $(N, N^\sharp)$. *An* erasing of generators *of* $(N, N^\sharp)$ *via* N_w *is a net* $(K, K^\sharp)$ *such that:*

- $(K, K^\sharp)$ *is a subnet of* $(N, N^\sharp)$;
- $\left(\overline{N_w}, \overline{N_w}^\sharp\right)$, *where* $\overline{N_w}$ *denotes the net with the same places of* N_w *and no transitions, is a subnet of* $(K, K^\sharp)$; *and*
- $\mathfrak{F}(N)$ *is the pushout of* $\mathfrak{F}\left(\overline{N_w}\right) \hookrightarrow \mathfrak{F}(N_w)$; *and* $\mathfrak{F}\left(\overline{N_w}\right) \hookrightarrow \mathfrak{F}(K)$.

Indeed, erasings of generators are preserved as well by our internalization:

Lemma 9. *Let* $(K, K^\sharp)$ *be an erasing of generating morphisms from* $(N, N^\sharp)$ *via a subnet* N_w. *Then* $\int K^\sharp$ *is an erasing of generators from* $\int N^\sharp$ *via* $\int sub_{N_w} \,\r{;}\, N^\sharp$. *The span case is analogous.*

Surprisingly, even if erasing and addition of generators are preserved by internalizations, synchronizations are not. Indeed, following [1], $(M, M^\sharp)$ is a synchronization of $(N, N^\sharp)$ via W, w when $\mathfrak{F}(M)$ is defined to be the result of applying the following double pushout rewrite rule to $\mathfrak{F}(N)$:

$$\mathfrak{F}(N_w) \xleftarrow{w'} \mathfrak{F}(W) \xleftarrow{in_W} \mathfrak{F}(\overline{W}) \xrightarrow{in_W} \mathfrak{F}(W)$$

Here, we require that w factorizes through w'. In internalizing this construction the pushouts are preserved, but the rewrite rule is not! This becomes evident by lifting the definition of synchronization altogether, where in the following diagram we are sticking to the notation developed in [1]:

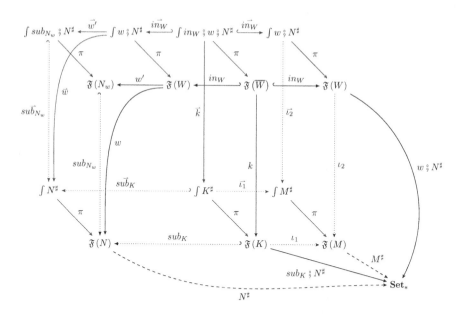

The black arrows are just the definition of synchronization. The dotted arrows denote the pushout arrows, while the dashed arrows arise from the universal property of the pushout. The maroon arrows and objects represent the Grothendieck construction and the lifting of the functors obtained from Lemma 3, while the πs stand for the functors obtained in Lemma 2, where we omitted subscripts to avoid clutter.

As one can see the pushout squares are both preserved, but $\int M^\sharp$ is not a synchronization of $\int N^\sharp$ via $\int in_W \,\raisebox{0.2ex}{\scriptsize$\circ$}\, w \,\raisebox{0.2ex}{\scriptsize$\circ$}\, N^\sharp$ since

$$\int sub_{N_w} \,\raisebox{0.2ex}{\scriptsize$\circ$}\, N^\sharp \neq \left(\int N^\sharp \right)_{\int w \,\raisebox{0.2ex}{\scriptsize$\circ$}\, N^\sharp}$$

In other words, $\int sub_{N_w} \,\raisebox{0.2ex}{\scriptsize$\circ$}\, N^\sharp$ is too big of a subcategory of $\int N^\sharp$ to make $\int M^\sharp$ into a synchronization. An analogous observation holds for spans.

Counterexample 2 (Synchronizations not preserved)**.** *We provide a practical counterexample of why synchronizations are not preserved by internalizations. Consider the following nets, where we are borrowing the graphical notation developed in [1], decorating net elements with their images in* **Set**$_*$*.*

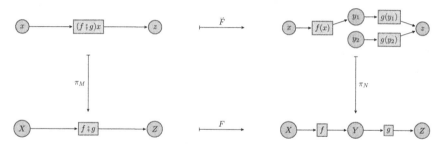

At the base level we have two nets, M on the left and N on the right. Eliding the functor $N^\sharp$, the places of N are mapped to sets:

$$X := \{x\} \qquad Y := \{y_1, y_2\} \qquad Z := \{z\}$$

While transitions are mapped to partial functions $f\colon X \to Y$ and $g\colon Y \to Z$, defined as follows:

$$f(x) = y_1 \qquad g(y_1) = g(y_2) = z$$

M is clearly a synchronization of N via F: The generators f and g have been erased and a generator corresponding to $f \,\mathring{,}\, g$ has been added. Taking the Grothendieck construction on M and N (top left and top right in the figure, respectively), we see how the erasing of generators is problematic: The morphism g in N branches into $g(y_1)$ and $g(y_2)$ in $\int N^\sharp$, of which only $g(y_1)$ forms a path with $f(x)$. In lifting the synchronization M to $\int M^\sharp$, we would expect $g(y_1)$ and $f(x)$ to be erased and conflated into $f(x) \,\mathring{,}\, g(y_1)$, whereas $g(y_2)$ stays. But this is not the case, since in M the generator g has already been erased "before being allowed to branch", taking $g(y_2)$ with it when we take $\int M^\sharp$!

As we said before, this ultimately depends on the fact that the internalization of the subnet provided by the synchronization witness contains too many morphisms, and ends up erasing more generators than we would like it to.

5   Internalization as a Functor

In this final section, we put together some of the properties we have proved so far about internalizations, and prove that internalization is a functor. The intuitive argument behind the results that are about to follow is this: If $\int N^\sharp$ internalizes the semantics of $(N, N^\sharp)$, in either the case $N^\sharp\colon N \to \mathbf{Set}_*$ or $N^\sharp\colon N \to \mathbf{Span}$, then $\int N^\sharp$ should be considered as "just a ne", that is, an object of $\mathbf{Petri^1}$; see Example 1.

Putting together results about lifting of functors obtained in the previous section, we are indeed able to prove this.

Theorem 3. *Denote with $\mathbf{1}$ the terminal category, together with the trivial symmetric monoidal structure on it. There is a faithful, strong monoidal functor $emb_{\mathbf{Set}_*}\colon \mathbf{Petri^{Set_*}} \to \mathbf{Petri^1}$ defined as follows:*

- *On objects, it sends $(M, M^\sharp)$ to $\left(\mathfrak{U}\left(\int M^\sharp\right), \mathfrak{U}\left(\int M^\sharp\right)^\sharp\right)$.*
- *On morphisms, we send the functor $F\colon (M, M^\sharp) \to (N, N^\sharp)$ to the functor[2]*

$$\hat{F}\colon \left(\mathfrak{U}\left(\int M^\sharp\right), \mathfrak{U}\left(\int M^\sharp\right)^\sharp\right) \to \left(\mathfrak{U}\left(\int N^\sharp\right), \mathfrak{U}\left(\int N^\sharp\right)^\sharp\right)$$

Similarly, there is a faithful, strong monoidal functor $emb_{\mathbf{Span}}\colon \mathbf{Petri^{Span}} \to \mathbf{Petri^1}$ defined as follows:

---

[2] To be absolutely precise, we are referring to the functor $\mathfrak{F}\left(\mathfrak{U}\left(\int M^\sharp\right)\right) \simeq \int M^\sharp \xrightarrow{\hat{F}} \int N^\sharp \simeq \mathfrak{F}\left(\mathfrak{U}\left(\int N^\sharp\right)\right)$.

- *On objects, it sends* $(M, M^\sharp)$ *to* $\left(\mathfrak{U}\left(\int M^\sharp\right), \mathfrak{U}\left(\int M^\sharp\right)^\sharp\right)$.
- *On morphisms, we send the functor* $F\colon (M, M^\sharp) \to (N, N^\sharp)$ *to the functor*[3]

$$\hat{F}\colon \left(\mathfrak{U}\left(\int M^\sharp\right), \mathfrak{U}\left(\int M^\sharp\right)^\sharp\right) \to \left(\mathfrak{U}\left(\int N^\sharp\right), \mathfrak{U}\left(\int N^\sharp\right)^\sharp\right)$$

Finally, it is worth nothing that for each choice of semantics $\mathcal{S}$ there is another obvious functor from **Petri**$^\mathcal{S}$ to **Petri**^{1}, which just forgets the semantics altogether. It is worth asking how this functor and the ones provided in Theorem 3 are related.

Proposition 1. *Denote with*

$$for_{Set_*}\colon Petri^{Set_*} \to Petri^1 \qquad for_{Span}\colon Petri^{Span} \to Petri^1$$

the "forgetful" functors defined by sending each Petri net $(M, M^\sharp)$ *to* $(M, M^\sharp)$. *Then there are natural transformations:*

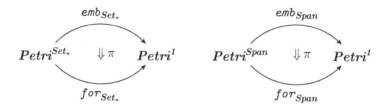

6   Conclusion and Future Work

In this work, we described guarded Petri nets as Petri nets endowed with a functorial semantics. We provided two different styles of semantics: a deterministic one, realized using the category of sets and partial functions, and a non-deterministic one that allows for side effects, realized using the category of partial functions and that of spans.

We moreover showed how, using the Grothendieck construction, the guards can be internalized, obtaining a Petri net whose reachability relation is equivalent to the one of the guarded one. We proved that internalizations have nice properties, and the internalization construction is functorial in the choice of the guarded net we start from.

Regarding directions of future work, a pretty straightforward thing to do would be to figure out which semantics, other than **Set**$_*$ and **Span**, are internalizable. That is, if $F\colon \mathfrak{F}(N) \to \mathcal{S}$ is a symmetric monoidal functor, which properties do $\mathcal{S}$ and F need to have so that $\int F$ is a free symmetric strict monoidal category.

---

[3] To be absolutely precise, we are referring to the functor $\mathfrak{F}\left(\mathfrak{U}\left(\int M^\sharp\right)\right) \simeq \int M^\sharp \xrightarrow{\hat{F}} \int N^\sharp \simeq \mathfrak{F}\left(\mathfrak{U}\left(\int N^\sharp\right)\right)$.

Acknowledgements. David Spivak acknowledges support from Honeywell Inc. as well as from AFOSR grants FA9550-17-1-0058 and FA9550-19-1-0113. Fabrizio Genovese wants to thank his fellow team members at Statebox for useful discussion and support.

Appendix – Proofs

Lemma 1. *In the case of both $\mathbf{Set}_*$ and $\mathbf{Span}$, the category $\int M^\sharp$ has a strict symmetric monoidal structure.*

Proof. We start with the case of $\mathbf{Set}_*$. Since $M^\sharp$ is strict monoidal, $M^\sharp(X \otimes Y) = M^\sharp X \times M^\sharp Y$. Thus on objects, we can set $(X, x) \otimes (Y, y) := (X \otimes Y, (x, y))$. On morphisms, we just use the monoidal product $f \otimes g$ from $\mathfrak{F}(M)$. The monoidal unit is $(I, *)$, where I is the monoidal unit of $\mathfrak{F}(M)$ and $*$ is the unique element of the monoidal unit $\{*\}$ of $\mathbf{Set}_*$. The rest of the proof is a straightforward check.

Now we consider the case of $\mathbf{Span}$. On objects, we set again $(X, x) \otimes (Y, y) := (X \otimes Y, (x, y))$. On morphisms, we set $(f, s) \otimes (g, t) := (f \otimes g, (s, t))$, where $f \otimes g$ is as in $\mathfrak{F}(M)$ and (s, t) is the pair of span-apex elements:

$$
\begin{array}{ccccc}
M^\sharp X & \xleftarrow{\;f_1\;} & S & \xrightarrow{\;f_2\;} & M^\sharp X' \\
(x, y) & \xleftarrow{(f_1, g_1)} & (s, t) & \xrightarrow{(f_2, g_2)} & (x', y') \\
M^\sharp Y & \xleftarrow{\;g_1\;} & T & \xrightarrow{\;g_2\;} & M^\sharp Y'
\end{array}
$$

The monoidal unit is $(I, *)$, where I is the monoidal unit of $\mathfrak{F}(M)$ and $*$ is the unique element of the monoidal unit $\{*\}$ of $\mathbf{Span}$.

The remainder of the proof is as in the previous case.   □

Theorem 1. *In both the case of $\mathbf{Set}_*$ and of $\mathbf{Span}$ the strict symmetric monoidal category $\int M^\sharp$ is free.*

Proof. We start with the case of $\mathbf{Set}_*$. Consider the free symmetric strict monoidal category $\hat{M}$ generated as follows:

- Object generators are pairs (X, x), with X object generator in $\mathfrak{F}(M)$ and $x \in M^\sharp X$;
- A morphism generator $(X, x) \to (Y, y)$ is a morphism generator $f: X \to Y$ of $\mathfrak{F}(M)$ such that $M^\sharp f(x) = y$.

We want to prove that $\hat{M}$ and $\int M^\sharp$ are isomorphic.

First, let (X, x) be an object in $\int M^\sharp$. Then X is an object of $\mathfrak{F}(M)$, which is free, and hence we have $X = X_1 \otimes \ldots \otimes X_n$ for generating objects $X_1 \ldots X_n$ in $\mathfrak{F}(M)$. By definition, we have $x \in M^\sharp X$. Being $M^\sharp$ strict, this means:

$$
\begin{aligned}
x \in M^\sharp X &\Leftrightarrow x \in M^\sharp(X_1 \otimes \ldots \otimes X_n) \\
&\Leftrightarrow x \in M^\sharp X_1 \times \cdots \times M^\sharp X_n \\
&\Leftrightarrow \exists!(x_1 \in M^\sharp X_1), \ldots, \exists!(x_n \in M^\sharp X_n).(x = (x_1, \ldots, x_n))
\end{aligned}
$$

Hence $(X, x) = (X_1, x_1) \otimes \ldots \otimes (X_n, x_n)$, and the object generators of $\int M^\sharp$ are the pairs (X, x) with X object generator in $\mathfrak{F}(M)$ and $x \in M^\sharp X$. This means that there is a bijection on objects of $\hat{M}$ and $\int M^\sharp$:

$$(X_1, x_1) \otimes \ldots \otimes (X_n, x_n) \mapsto (X_1 \otimes \ldots \otimes X_n, (x_1, \ldots, x_n))$$

We can then define a symmetric monoidal functor $T \colon \hat{M} \to \int M^\sharp$ that is bijective on objects, and sends a generating morphism $f \colon (X, x) \to (Y, y)$ of $\hat{M}$ to the morphism $f \colon (X, x) \to (Y, y)$ in $\int M^\sharp$.

We want to prove that T is full and faithful. Faithfulness is obvious; given $f_1, f_2 \colon (X, x) \to (Y, y)$ in $\hat{M}$, if $T(f_1) = T(f_2)$ then in particular $f_1 = f_2$ in $\mathfrak{F}(M)$. It follows from the fact that $M^\sharp$ is at most single-valued – i.e. $M^\sharp g(x) = x'$ and $M^\sharp g(x) = x''$ imply $x' = x''$ – that $f_1 = f_2$ also in $\hat{M}$. For fullness, take a morphism $f \colon (X, x) \to (Z, z)$ in $\int M^\sharp$, and notice the following:

- If $f \colon (X, x) \to (Z, z)$ is equal to $f_1 \, \mathbin{\raisebox{0.5ex}{\scriptsize$\circ$}} \, f_2$, where $f_1 \colon X \to Y$ and $f_2 \colon Y \to Z$, then we have:

$$z = M^\sharp f(x) = M^\sharp (f_1 \, \mathbin{\raisebox{0.5ex}{\scriptsize$\circ$}} \, f_2)(x) = (M^\sharp f_1 \, \mathbin{\raisebox{0.5ex}{\scriptsize$\circ$}} \, M^\sharp f_2)(x)$$

So there is a $y \in M^\sharp Y$ such that $M^\sharp f_1(x) = y$ and $M^\sharp f_2(y) = z$. This means that $f_1 \colon (X, x) \to (Y, y)$ and $f_2 \colon (Y, y) \to (Z, z)$ are morphisms in $\int M^\sharp$;
- If $f \colon (X_1 \otimes X_2, (x_1, x_2)) \to (Y_1 \otimes Y_2, (y_1, y_2))$ is equal to $f_1 \otimes f_2$, where $f_1 \colon X_1 \to Y_1$ and $f_2 \colon X_2 \to Y_2$, then we have:

$$(y_1, y_2) = M^\sharp f(x_1, x_2) = M^\sharp (f_1 \otimes f_2)(x_1, x_2) = (M^\sharp f_1 \times M^\sharp f_2)(x_1, x_2)$$

This means that $M^\sharp f_1(x_1) = y_1$ and $M^\sharp f_2(x_2) = y_2$, and hence that $f_1 \colon (X_1, x_1) \to (Y_1, y_1)$ and $f_2 \colon (X_2, x_2) \to (Y_2, y_2)$ are morphisms in $\int M^\sharp$.

By definition, since $\mathfrak{F}(M)$ is free, any morphism $f \colon X \to Z$ can be decomposed into a composition of monoidal products of morphism generators, symmetries and identities. The points above prove that $f \colon (X, x) \to (Z, z)$ can be decomposed in the same way, and hence is in the image of T; thus it is full.

Our correspondence is bijective on objects and fully faithful, proving that $\hat{M}$ and $\int M^\sharp$ are isomorphic as categories. Since $\hat{M}$ is free so is $\int M^\sharp$, completing the proof.

We now consider the case of **Span**. The structure of the proof is similar. Consider the free symmetric strict monoidal category $\hat{M}$ generated as follows:

- Object generators are pairs (X, x), with X object generator in $\mathfrak{F}(M)$ and $x \in M^\sharp X$;
- For each morphism generator $f \colon X \to Y$ and $s \in S$ such that $M^\sharp f(s) = (x, y)$, there is a morphism generator $(f, s) \colon (X, x) \to (Y, y)$.

We want to prove that $\hat{M}$ and $\int M^\sharp$ are isomorphic.

On objects, the proof of bijectivity is as in the previous case. We can then define a symmetric monoidal functor $\hat{M} \to \int M^\sharp$ that is bijective on objects, and sends a generating morphism $(f, s): (X, x) \to (Y, y)$ of $\hat{M}$ to the morphism $(f, s): (X, x) \to (Y, y)$ in $\int M^\sharp$.

We want to prove that this functor is full and faithful. Faithfulness is again straightforward. Suppose given $(f_1, s_1), (f_2, s_2): (X, x) \to (Y, y)$ in $\hat{M}$. By construction, $(f_1, s_1) = (f_2, s_2)$ in $\int M^\sharp$ if and only if $f_1 = f_2$ in $\mathfrak{F}(M)$ and $s_1 = s_2$. But this means that $(f_1, s_1) = (f_2, s_2)$ also in $\hat{M}$.

For fullness, take a morphism $(f, s): (X, x) \to (Z, z)$ in $\int M^\sharp$, and notice the following:

- Each morphism (f, s) such that f is a generator, an identity or a symmetry in $\mathfrak{F}(M)$ is also in $\hat{M}$;
- If $(f, s): (X, x) \to (Z, z)$ is such that $f = g \, \mathring{,} \, h$, where $g: X \to Y$ and $h: Y \to Z$, then by definition of composition in **Span**, we have $s = (t, u)$ for some t, u with $M^\sharp g(t) = (x, y)$ and $M^\sharp h(u) = (y, z)$. This means that $(g, t): (X, x) \to (Y, y)$ and $(h, u): (Y, y) \to (Z, z)$ are morphisms in $\int M^\sharp$;
- If $(f, s): (X_1 \otimes X_2, (x_1, x_2)) \to (Y_1 \otimes Y_2, (y_1, y_2))$ is such that $f = f_1 \otimes f_2$, where $f_1: X_1 \to Y_1$ and $f_2: X_2 \to Y_2$, then $s = (s_1, s_2)$ for some s_1, s_2, and we have:

$$((x_1, x_2), (y_1, y_2)) = M^\sharp f(s) = M^\sharp (f_1 \otimes f_2)(s_1, s_2) = (M^\sharp f_1 \times M^\sharp f_2)(s_1, s_2)$$

This means that $M^\sharp f_1(s_1) = (x_1, y_1)$ and $M^\sharp f_2(s_2) = (x_2, y_2)$, and hence that $(f_1, s_1): (X_1, x_1) \to (Y_1, y_1)$ and $(f_2, s_2): (X_2, x_2) \to (Y_2, y_2)$ are morphisms in $\int M^\sharp$.

By definition, since $\mathfrak{F}(M)$ is free, any morphism f can be decomposed into a composition of monoidal products of morphism generators, symmetries and identities. The points above prove that each of such morphisms is also in $\int M^\sharp$, and hence in $\hat{M}$. So f in $\int M^\sharp$ is the image of f in $\hat{M}$, and the functor is full.

Since our correspondence is bijective on objects and fully faithful, this proves that $\hat{M}$ and $\int M^\sharp$ are isomorphic as categories. Since $\hat{M}$ is free so is $\int M^\sharp$, completing the proof. □

Theorem 2. Let $(N, N^\sharp)$ be a guarded Petri net (with side effects). (Y, y) is reachable from (X, x) if and only if (Y, y) is reachable from (X, x) in the net $\mathfrak{U}(\int N^\sharp)$.

Proof. By definition (Y, y) is reachable from (X, x) if and only if there is a morphism $f: X \to Y$ in $\mathfrak{F}(N)$ such that $N^\sharp f(x) = y$ (resp. $N^\sharp f(s) = (x, y)$ for some $s \in S_f$). Again by definition, this means that $f: (X, x) \to (Y, y)$ (resp. $f_s: (X, x) \to (Y, y)$) is a morphism of $\int N^\sharp$. Since $\int N^\sharp$ is free, f (resp. f_s) can be decomposed as a composition of monoidal products of generating morphisms. But every generating morphism of $\int N^\sharp$ corresponds to a transition of $\mathfrak{U}(\int N^\sharp)$, from which the thesis follows. □

Lemma 5. *If F is transition-preserving, so are $\vec{F}$ and $\hat{F}$.*

Proof. The proof is obvious considering that, by construction, $f : (X, x) \to (Y, y)$ is a generator (resp. a symmetry) in $\int N^{\sharp}$ if and only if it is a generator (resp. a symmetry) in $\mathfrak{F}(N)$.

An analogous argument holds for $\widehat{F}$.   □

Theorem 3. *Denote with* **1** *the terminal category, together with the trivial symmetric monoidal structure on it. There is a faithful, strong monoidal functor* $emb_{Set_*} : \textbf{Petri}^{Set_*} \to \textbf{Petri}^{1}$ *defined as follows:*

- *On objects, it sends* $(M, M^{\sharp})$ *to* $\left(\mathfrak{U}(\int M^{\sharp}), \mathfrak{U}(\int M^{\sharp})^{\sharp} \right)$.
- *On morphisms, we send the functor* $F : (M, M^{\sharp}) \to (N, N^{\sharp})$ *to the functor[4]*

$$\vec{F} : \left(\mathfrak{U}(\textstyle\int M^{\sharp}), \mathfrak{U}(\textstyle\int M^{\sharp})^{\sharp} \right) \to \left(\mathfrak{U}(\textstyle\int N^{\sharp}), \mathfrak{U}(\textstyle\int N^{\sharp})^{\sharp} \right)$$

Similarly, there is a faithful, strong monoidal functor $emb_{Span} : \textbf{Petri}^{Span} \to \textbf{Petri}^{1}$ *defined as follows:*

- *On objects, it sends* $(M, M^{\sharp})$ *to* $\left(\mathfrak{U}(\int M^{\sharp}), \mathfrak{U}(\int M^{\sharp})^{\sharp} \right)$.
- *On morphisms, we send the functor* $F : (M, M^{\sharp}) \to (N, N^{\sharp})$ *to the functor[5]*

$$\widehat{F} : \left(\mathfrak{U}(\textstyle\int M^{\sharp}), \mathfrak{U}(\textstyle\int M^{\sharp})^{\sharp} \right) \to \left(\mathfrak{U}(\textstyle\int N^{\sharp}), \mathfrak{U}(\textstyle\int N^{\sharp})^{\sharp} \right).$$

Proof. The proofs for **Set$_*$** and **Span** are very similar, so we just provide the one for **Set$_*$**. Clearly if F is the identity functor $(M, M^{\sharp}) \to (M, M^{\sharp})$ then so is $\vec{F}$. For composition, consider $F : (M, M^{\sharp}) \to (M, M^{\sharp}) N^{\sharp}$ and $G : (N, N^{\sharp}) \to (P, P^{\sharp})$. We have to prove that $\overrightarrow{F \mathbin{\fatsemi} G} = \vec{F} \mathbin{\fatsemi} \vec{G}$. On objects, $\overrightarrow{F \mathbin{\fatsemi} G}$ sends (X, x) in $\int M^{\sharp}$ to $((F \mathbin{\fatsemi} G)X, x)$ in $\int P^{\sharp}$, so it coincides with $\vec{F} \mathbin{\fatsemi} LiftSetSG$. Now consider a morphism $f : (X, x) \to (Y, y)$ in $\int M^{\sharp}$. This is sent by $\vec{F}$ to $Ff : (FX, x) \to (FY, y)$, and applying $\vec{G}$ to it one gets $G(Ff) : (G(FX), x) \to (G(FY), y)$. Since $G(F(\_))$ is $(F \mathbin{\fatsemi} G)(\_)$, we are done. This proves that emb_{Set_*} is a functor. Faithfulness is trivial.

Now we focus on monoidality. First of all we have to prove that

$$emb_{Set_*} \left((M, M^{\sharp}) \otimes (N, N^{\sharp}) \right) \simeq emb_{Set_*} (M, M^{\sharp}) \otimes emb_{Set_*} (N, N^{\sharp})$$

Remembering from Definition 1 that for each choice of semantics $\mathcal{S}$ the monoidal structure on **Petri**$^{\mathcal{S}}$ is defined in terms of coproduct of symmetric monoidal

---

[4] To be absolutely precise, we are referring to the functor $\mathfrak{F}(\mathfrak{U}(\int M^{\sharp})) \simeq \int M^{\sharp} \xrightarrow{\vec{F}} \int N^{\sharp} \simeq \mathfrak{F}(\mathfrak{U}(\int N^{\sharp}))$.

[5] To be absolutely precise, we are referring to the functor $\mathfrak{F}(\mathfrak{U}(\int M^{\sharp})) \simeq \int M^{\sharp} \xrightarrow{\widehat{F}} \int N^{\sharp} \simeq \mathfrak{F}(\mathfrak{U}(\int N^{\sharp}))$.

categories, and hence from the coproduct of the underlying nets, this means that:

$$\left(\mathfrak{U}\left(\int [M^\sharp, N^\sharp]\right), \mathfrak{U}\left(\int [M^\sharp, N^\sharp]\right)^\sharp\right) =$$

$$= \mathsf{emb}_{\mathbf{Set}_*}(M + N, [M^\sharp, N^\sharp])$$

$$= \mathsf{emb}_{\mathbf{Set}_*}((M, M^\sharp) \otimes (N, N^\sharp))$$

$$\simeq \mathsf{emb}_{\mathbf{Set}_*}(M, M^\sharp) \otimes \mathsf{emb}_{\mathbf{Set}_*}(N, N^\sharp)$$

$$= \left(\mathfrak{U}\left(\int M^\sharp\right), \mathfrak{U}\left(\int M^\sharp\right)^\sharp\right) \otimes \left(\mathfrak{U}\left(\int N^\sharp\right), \mathfrak{U}\left(\int N^\sharp\right)^\sharp\right)$$

$$= \left(\mathfrak{U}\left(\int M^\sharp\right) + \mathfrak{U}\left(\int N^\sharp\right), \left[\mathfrak{U}\left(\int M^\sharp\right)^\sharp, \mathfrak{U}\left(\int N^\sharp\right)^\sharp\right]\right)$$

Since $\mathfrak{U}(\_)$ preserves isomorphisms and coproducts, it is sufficient to prove:

$$\int [M^\sharp, N^\sharp] \simeq \int M^\sharp + \int N^\sharp$$

– By definition, objects of $\int [M^\sharp, N^\sharp]$ are pairs (X, x) with $X \in \mathfrak{F}(M + N) \simeq \mathfrak{F}(M) + \mathfrak{F}(N)$ and $x \in [M^\sharp, N^\sharp]X$. This is clearly isomorphic to $\mathsf{Obj} \int M^\sharp \sqcup \mathsf{Obj} \int N^\sharp$.
– Again by definition, we have

$$\mathsf{Hom}_{\int [M^\sharp, N^\sharp]}[(X, x), (Y, y)] :=$$

$$:= \{f \in \mathsf{Hom}_{\mathfrak{F}(M) + \mathfrak{F}(N)}[X, Y] \mid | [M^\sharp, N^\sharp]f(x) = y\}$$

This follows noting that by definition the set of morphisms of $\mathfrak{F}(M) + \mathfrak{F}(N)$ is the disjoint union of the sets of morphisms of $\mathfrak{F}(M)$ and $\mathfrak{F}(N)$.

Then we have to prove that $\mathsf{emb}_{\mathbf{Set}_*}(F \otimes G) = \mathsf{emb}_{\mathbf{Set}_*} F \otimes \mathsf{emb}_{\mathbf{Set}_*} G$. Unrolling definitions this amounts to prove that $F \vec{+} G = \vec{F} + \vec{G}$, which is obvious.

Finally, we need to prove that $\mathsf{emb}_{\mathbf{Set}_*}$ preserves the monoidal unit. Notice that for each choice of semantics S the monoidal unit in $\mathbf{Petri}^S$ is taken to be $(\emptyset, \emptyset^\sharp S)$. $\mathfrak{F}(\emptyset)$ is the free category consisting of only the monoidal unit I, and $\emptyset^\sharp S : \mathfrak{F}(\emptyset) \to S$ sends the monoidal unit to the monoidal unit and its identity to itself. In our case, the monoidal unit of $\mathbf{Petri}^{\mathbf{Set}_*}$ is $(\emptyset, \emptyset^\sharp)$, with $\emptyset^\sharp$ sending the monoidal unit of $\mathfrak{F}(\emptyset)$ to the singleton set $\{*\}$ in $\mathbf{Set}_*$. In particular, this means that $\int \emptyset^\sharp \simeq \mathfrak{F}(\emptyset)$, proving that

$$\mathsf{emb}_{\mathbf{Set}_*}(\emptyset, \emptyset^\sharp) = \left(\mathfrak{U}\left(\int \emptyset^\sharp\right), \mathfrak{U}\left(\int \emptyset^\sharp\right)^\sharp\right) \simeq \left(\mathfrak{U}(\mathfrak{F}(\emptyset)), \mathfrak{U}(\mathfrak{F}(\emptyset))^\sharp\right) \simeq (\emptyset, \emptyset^\sharp)$$

□

Proposition 1. *Denote with*

$$for_{\mathbf{Set}_*} : \mathbf{Petri}^{\mathbf{Set}_*} \to \mathbf{Petri}^1 \qquad for_{\mathbf{Span}} : \mathbf{Petri}^{\mathbf{Span}} \to \mathbf{Petri}^1$$

the functors defined by sending each Petri net $(M, M^\sharp)$ to $(M, M^\sharp)$. Then there are natural transformations:

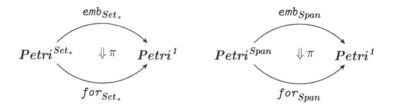

Proof. For each object $(M, M^\sharp)$ in **Petri**$^{\mathbf{Set}_*}$, we set

$$\pi_{(M,M^\sharp)} : \left(\mathfrak{U} \left(\int M^\sharp \right), \mathfrak{U} \left(\int M^\sharp \right)^\sharp \right) \to (M, M^\sharp)$$

to be the functor $\mathfrak{F} \left(\mathfrak{U} \left(\int M^\sharp \right) \right) \simeq \int M^\sharp \xrightarrow{\pi_M} \mathfrak{F} (M)$, where π_M is defined as in Lemma 3. The naturality condition follows from Lemma 3 as well. The span case is analogous.   □

References

1. Genovese, F.: The essence of petri net gluings. http://arxiv.org/abs/1909.03518
2. Genovese, F., Herold, J.: Executions in (semi-)integer petri nets are compact closed categories, vol. 287, pp. 127–144. https://doi.org/10.4204/EPTCS.287.7
3. Jensen, K., Kristensen, L.M.: Coloured Petri Nets. Springer, Berlin and Heidelberg (2009). https://doi.org/10.1007/b95112. http://link.springer.com/10.1007/b95112
4. MacLane, S., Moerdijk, I.: Sheaves in Geometry and Logic, a First Introduction to Topos Theory, Universitext, vol. 13. Springer, Heidelberg (1994). https://doi.org/10.1007/978-1-4612-0927-0
5. Master, J.: Generalized petri nets. http://arxiv.org/abs/1904.09091
6. Meseguer, J., Montanari, U.: Petri nets are monoids. **88**(2), 105–155. https://doi.org/10.1016/0890-5401(90)90013-8
7. Petri, C., Reisig, W.: Petri net. http://www.scholarpedia.org/article/Petri_net
8. Sassone, V.: On the category of Petri net computations. In: Mosses, P.D., Nielsen, M., Schwartzbach, M.I. (eds.) CAAP 1995. LNCS, vol. 915, pp. 334–348. Springer, Heidelberg (1995). https://doi.org/10.1007/3-540-59293-8_205
9. Statebox Team: The mathematical specification of the statebox language. http://arxiv.org/abs/1906.07629
10. Statebox Team: Statebox github page. https://github.com/statebox/

Unfolding Symbolic Attributed Graph Grammars

Maryam Ghaffari Saadat[1]([✉]), Reiko Heckel[1], and Fernando Orejas[2]

[1] Department of Informatics, Leicester University, Leicester, UK
{mgs17,rh122}@leicester.ac.uk
[2] Department of Computer Science, Polytechnic University of Catalonia,
Barcelona, Spain
orejas@cs.upc.edu

Abstract. Attributed graph grammars can specify the transformation of complex data and object structures within a natural rule-based model of concurrency. This is crucial to their use in modelling interfaces of services and components as well as the evolution of complex systems and networks. However, the established concurrent semantics of graph grammars by unfolding does not cover attributed grammars. We develop a theory of unfolding for attributed graph grammars where attribution is represented symbolically, via logical constraints. We establish a functorial representation (a coreflection) of unfolding which guarantees it to be correct, complete and fully abstract.

As a case study and running example we demonstrate the use of visual contracts to specify an escrow smart contract.

Keywords: Symbolic attribute graph transformation · Unfolding semantics · Visual smart contracts

1 Introduction

The majority of software developed today is distributed and concurrent, including mobile, service-oriented, cloud- and component-based applications, smart contracts and decentralised applications (Dapps) on blockchain platforms such a Ethereum or Neo, P2P applications, etc. Technical and organisational challenges, such as the lack of a central authority both at runtime and during development are being addressed by model-based software engineering methods. In order to model such applications using graph transformation systems, we rely on a natural rule-based model of concurrency. In its most comprehensive form, the concurrent behaviour of graph grammars is formalised by their *unfolding* [1,3,4]. This represents in one structure the branching computations of the grammar in (what could be called) a partial-order variant of its derivation tree.

Many practical applications of graph transformation, including to model-based software engineering, require graphs attributed by numerical or textual data and transformations combining structural with attribute updates, i.e.,

© Springer Nature Switzerland AG 2020
F. Gadducci and T. Kehrer (Eds.): ICGT 2020, LNCS 12150, pp. 75–90, 2020.
https://doi.org/10.1007/978-3-030-51372-6_5

attributed graph grammars [5]. However, the theory of unfolding has so far only been developed for the un-attributed case. Indeed, in the presence of attributes, the concurrent and non-deterministic behaviour of a graph grammar becomes significantly more complex. While, traditionally, dependencies and conflicts between transformations are based on how their left- and right-hand sides overlap in given or derived graphs, and how these overlaps include elements created or deleted by one rule and required or preserved by the other, the presence of data requires a deeper analysis of conflicts and dependencies. For example, two rules updating the same attribute, one after the other, may still be exchangeable if the attribute operations performed are commutative. Also, the choice of different assignments for variables may lead to infinitely branching systems even if the branching structure of their underlying structural grammar is finite.

In this paper we lift the theory of unfolding to attributed graph grammars, providing the semantic foundations for concurrent and distributed system models with data. By treating attribution purely logically, based on the notion of symbolic attributed graphs, we are able to separate attribute computations from structural unfolding. In particular, the use of *lazy symbolic graph transformations* [9] allows to abstract from specific attribute values while keeping track of all constraints such values have to satisfy. The resulting theory follows the un-attributed case in establishing unfolding as a coreflection, i.e., a right adjoint to the inclusion of the subcategory of occurrence grammars into the category of attributed graph grammars. This shows the construction to be correct and complete in representing only and all steps of the given grammar, and to be fully abstract (i.e. represented in a minimal way).

We introduce the main concepts of symbolic attributed graph grammars, including the notion of lazy symbolic transformations, and present the escrow contract model as running example. Section 3 recalls the unfolding of classical DPO graph grammars, which is extended to attributed grammars in Sects. 4 and 5. In Sect. 6 we draw conclusions and discuss related and future work.

2   Symbolic Attributed Graph Transformation

Given a data algebra D with non-empty carrier sets, an *attributed graph* [5] over D is a graph whose *graph objects* (nodes and edges) are labelled over D. Elements of D are represented as *data nodes* and connected to graph objects by *attribute links*. Formally this structure is known as an *E-graph* [5]. A *symbolic attributed graph* [8] specifies a class of attributed graphs by means of constraints. Formally $\langle G, \Phi \rangle$ is an E-graph whose graph objects are labelled by a set of *variables* $x \in X_G$, together with a set of *constraints* Φ over X_G.

In our case, constraints are equations over X_G using values in D as constants. Each substitution $\sigma : X_G \rightarrow D$ extends canonically to such equations. If they hold in D, denoted $D \models \sigma(\Phi)$, this defines an *attributed graph* $\sigma(G)$ replacing variables x with data values $\sigma(x)$. Hence, the semantics of $\langle G, \Phi \rangle$ is $Sem(\langle G, \Phi \rangle) = \{\sigma(G) \mid D \models \sigma(\Phi)\}$. To simplify notation we will identify G with the symbolic attributed graph, denoting its constraints Φ_G.

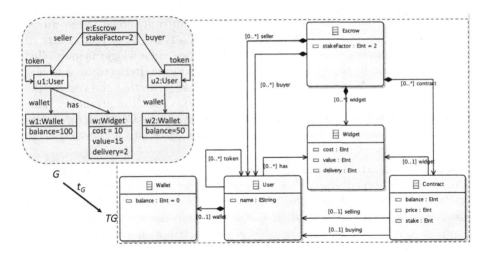

Fig. 1. Symbolic attributed type and instance graphs

An *attributed graph morphism* $h : \langle G_1, \Phi_1 \rangle \rightarrow \langle G_2, \Phi_2 \rangle$ is a graph homomorphism $h : G_1 \rightarrow G_2$ such that $D \models (\Phi_2 \Rightarrow h(\Phi_1))$, where $h(\Phi_1)$ are the formulas obtained by replacing in Φ_1 every variable x_1 by $h(x_1)$. Symbolic graphs and morphisms over D form the category $\mathbf{SGraph}_D$, or just $\mathbf{SGraph}$ if D is understood.[1,2]

In the rest of the paper, we will work with typed graphs. Given a dedicated *type graph TG* defining the types of nodes, edges and attributes, a *TG-typed (symbolic attributed) instance graph* is a graph G with a morphism $t_G : G \rightarrow TG$ mapping elements in G to their types. The category $\mathbf{SGraph}_{TG}$ has TG-typed instance graphs $\langle G_i, t_{G_i} \rangle$ as objects and as arrows $h : \langle G_1, t_{G_1} \rangle \rightarrow \langle G_2, t_{G_2} \rangle$ attributed graph morphisms $h : G_1 \rightarrow G_2$ such that $t_{G_2} \circ h = t_{G_1}$.

Example 1 (type and instance graphs of escrow model). Figure 1 shows a type graph in the bottom right corner and one of its instances in the top left. Note that, according to our definitions, this is an E-graph whose data nodes are variables related to their values by equations. For example, node $w1$ is a node of type Wallet and has an attribute link of type balance pointing to an (invisible) variable node (say b) whose value is given by an equation $b = 100$ in the constraints Φ_G.

Modern blockchain technologies allow developers to offer smart contracts as services based on the blockchain's distributed data model and other basic features, such as wallets and payments. A popular example is an escrow service,

---

[1] We use D to denote both the many-sorted algebra and the disjoint union of its carrier sets.

[2] We write $h(\Phi_1)$ in a slight abuse of notation. To be more rigorous (and less readable) we should write $h_X^\#(\Phi_1)$, where h_X is the restriction of h to the set of variables of G_1, mapping them to variables in G_2, and $h_X^\#$ is the (unique) extension of h_X to terms over D's signature.

to act as intermediary between sellers and buyers in commercial transactions. The type graph defines the data model for such a service. Type escrow acts as container and defines a stake factor, by which the price of a widget is multiplied to define the stake to be raised by both seller and buyer. Items to be sold and bought are of type Widget, with attributes for cost (reflecting the cost to the seller at which a widget was obtained or produced), value (the presumed value to a prospective buyer), and delivery (shipping cost). Type Contract represents the smart contract, created when buyer and seller enter into a transaction, to maintain the relevant data.

In general, an attribute can have a set of values, i.e., link to several variable nodes. For a TG-typed graph G and a graph object (node or edge) o in G we write $o.at$ for the set of variables x such that there is an attribute link al in G from o to x with $t_G(al) = at$. If $o.at$ is singleton, we write $o.at = x \in G$. Generally, in rules, input and reachable graphs, attribution is unique, i.e., $o.at$ is singleton for all graph objects o and attributes at.

As illustrated in the example above, an attributed graph can be presented as a symbolic one by replacing all its attribute values in the data algebra D by variables, and for each value v, where x_v is the variable replacing it, including an equation $x_v = v$ in the set of constraints. We call such symbolic graphs $grounded$.

A symbolic attributed graph G is $linear$ if each variable $x \in X_G$ occurs as an attribute value of at most one node or edge in G. A TG-$typed$ $symbolic$ $attributed$ $graph$ $transformation$ $rule$ over D is a triple $r = \langle \Phi_L, L \hookleftarrow K \hookrightarrow R, \Phi_R \rangle$, where $L \hookleftarrow K \hookrightarrow R$ is a span of TG-typed graph inclusions such that L is linear, $X_L = X_K \subseteq X_R$, and where Φ_L, Φ_R are sets of constraints over $X_L \cup D$ and $X_R \cup D$, respectively, satisfying $D \models (\Phi_R \Rightarrow \Phi_L)$.

$Typed$ $symbolic$ $attributed$ $graph$ $transformation$ is defined following the double-pushout (DPO) approach [5] by a transformation of the underlying E-graphs satisfying the constraints: Given a rule r as above and a morphism $m : L \rightarrow G$, a transformation $\langle G, \Phi_G \rangle \xrightarrow{r,m} \langle H, \Phi_G \cup m'(\Phi_R) \rangle$ exists iff there is a DPO step $G \xrightarrow{r,m} H$ of the E-graphs such that $D \models (\Phi_G \Rightarrow m(\Phi_L))$ and $\Phi_G \cup m'(\Phi_R)$ is satisfiable in D. That means, if the left-hand side constraints are implied by the constraints in the given graph, the right-hand side constraints are added to the derived graph assuming that the result is consistent.

The use of the DPO approach to define the structural transformation means that rule r is only applicable at match m if the $dangling$ and $dangling$ $identification$ $conditions$ are satisfied. Match m satisfies the dangling condition if no node in G about to be deleted by r is the source or target of an edge outside $m(L)$ (because such an edge would be left "dangling" when deleting the node). The identification condition requires that a node or edge x in $L \setminus R$ (i.e., $m(x)$ is deleted by the rule) cannot be identified with any other object y in L, i.e., $m(x) = m(y)$ implies $x = y$. This ensures resource consciousness for deleted graph objects while allowing non-injective matching in general.

Example 2 (sell rule and transformation). The top half of Fig. 2 shows an attributed graph transformation rule from our escrow example specifying an operation sell(u1:User, w:Widget, p:Int): Contract. We only show graphs L and R, with the interface left K implicit as their intersection. The rule assumes objects and data on the left of the arrow, introducing variables such as $e : Escrow$ and sf to capture them for use in the right-hand side to create links to new objects such as $c : Contract$ and in expressions of attribute assignments and updates. The match m is controlled by input parameters $u1, w, p$ subject to the condition in the yellow box restricting the choice of price p.

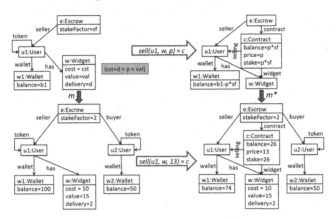

Fig. 2. Attributed rule and transformation (Color figure online)

The variables in this rule are $X_L = \{p, sf, b1, cst, val, d\}$ and $X_R = X_L \cup \{b1', b, s\}$, the latter referring to the new values of the $w.balance, c.balance$, and $c.stake$, respectively. The input parameter p is part of the left-hand side and therefore chosen by the match. This leads to branching over all values of p, a degree of non-determinism that makes the direct computation of all derivations impossible. $\Phi_L = \{cst + d < p < val\}$ and $\Phi_R = \Phi_L \cup \{balance = p \times sf, price = p, stake = p \times sf, balance = b1 - p \times sf\}$. In the bottom half we show an application of the rule, transforming an instance graph which, in addition to the objects in the rule, has a second user. Note that attributes here have actual values, and how new values are computed in the derived graph on the right.

Constraints added by r can be satisfied by several non-isomorphic graphs. Rule r is *ground-preserving* if for any grounded graph G and transformation $G \xrightarrow{r,m} H$, graph H is grounded. All rules in this paper are ground-preserving.

To separate the handling of constraints, we may want to apply a rule $r = \langle \Phi_L, L \hookleftarrow K \hookrightarrow R, \Phi_R \rangle$ without checking entailment of $m(\Phi_L)$. This is equivalent to applying the corresponding *lazy rule* $r^\emptyset = \langle \emptyset, L \hookleftarrow K \hookrightarrow R, \Phi_R \rangle$, in a *lazy transformation*.

Proposition 1 (lazy transformation of grounded graphs). *A symbolic graph transformation rule* $r = \langle \Phi_L, L \hookleftarrow K \hookrightarrow R, \Phi_R \rangle$ *and its lazy counterpart* $r^\emptyset = \langle \emptyset, L \hookleftarrow K \hookrightarrow R, \Phi_R \rangle$ *are equivalent on grounded graphs. That means, given a symbolic transformation* $G \xrightarrow{r,m} H$ *with G grounded, this also represents a lazy transformation using* $r^\emptyset$, *and vice versa.*

Sometimes we want to translate graphs, rules and transformations from one type graph to another. This is possible because a morphism $f : TG_1 \to TG_2$ between type graphs induces a translation of instances. Formally, we denote by $f_{TG}^> : \mathbf{SGraphs}_{TG_1} \to \mathbf{SGraphs}_{TG_2}$ the covariant retyping functor from TG_1-typed to TG_2-typed instance graphs, defined by composition of f_{TG} with the typing morphisms as $f_{TG}^>(\langle G_1, t_{G_1} \rangle) = \langle G_1, f_{TG} \circ t_{G_1} \rangle$. Note that, the functor only affects the typing, i.e., the attributed graphs with their variables and constraints remain unchanged. Hence $f_{TG}^>$ also acts as identity on morphisms.

Definition 1 (attributed graph grammar). *A (typed symbolic) attributed graph grammar (AGG) over a data algebra D is a 4-tuple $\mathcal{AG} = \langle TG, G_{in}, P, \pi \rangle$, where TG is a type graph over D, G_{in} is the TG-typed initial graph, P is a set of rule names, and π is a mapping associating to any $p \in P$ a TG-typed rule $\pi(p) = \langle \Phi_{L_p}, L_p \hookleftarrow K_p \hookrightarrow R_p, \Phi_{R_p} \rangle$ over D.*

An AGG morphism $f : \langle TG, G_{in}, P, \pi \rangle \to \langle TG', G'_{in}, P', \pi' \rangle$ between grammars over the same data algebra D is a pair $\langle f_{TG}, f_P \rangle$ where $f_{TG} : TG \to TG'$ is a retyping morphism and $f_P : P \to P'$ is a mapping of rule names such that

- $f_{TG}^>(G_{in}) = G'_{in}$
- *For all $p \in P$, $\pi'(f_P(p)) = \langle \Phi_{L_p}, f_{TG}^>(L_p) \leftarrow f_{TG}^>(K_p) \to f_{TG}^>(R_p), \Phi_{R_p} \rangle$.*

The category of attributed graph grammars and their morphisms is **AGG**.

A *derivation* in $\mathcal{AG}$ is a finite sequence of transformations $s = (G_{in} = G_0 \xrightarrow{p_1, m_1} \cdots \xrightarrow{p_n, m_n} G_n)$ with $p_i \in P$. Since $f_{TG}^>$ preserves pushouts and rules spans and constraints remain the same but for typing, morphisms of attributed graph grammars preserve derivations. That is, a derivation s in $\mathcal{AG}$ is translated by f to a derivation $f(s) = (f_{TG}^>(G_{in}) = f_{TG}^>(G_0) \xrightarrow{f_P(p_1), f_{TG}^>(m_1)} \cdots \xrightarrow{f_P(p_n), f_{TG}^>(m_n)} f_{TG}^>(G_n))$.

Given an attributed graph grammar $\mathcal{AG} = \langle TG, G_{in}, P, \pi \rangle$, its *underlying (unattributed) grammar* is $\mathcal{AG}_G = \langle TG_G, G_{in,G}, P, \pi_G \rangle$ where TG_G and $G_{in,G}$ are the restrictions of TG and G_{in} to their graph components (excluding variables and attribution edges) and for all $p \in P$ with $\pi(p) = \langle \Phi_{L_p}, L_p \hookleftarrow K_p \hookrightarrow R_p, \Phi_{R_p} \rangle$ their rules reduce to $\pi_G(p) = \langle \emptyset, (L_p)_G \hookleftarrow (K_p)_G \hookrightarrow (R_p)_G, \emptyset \rangle$. It is easy to see that a derivation s in $\mathcal{AG}$ reduces to a derivation s_G in the underlying grammar $\mathcal{G}_G$. Similarly, by $\mathcal{AG}_{EG}$ we denote the underlying E-graph grammar, retaining attribute links and variables, but dropping all constraints.

Example 3 (attributed graph grammar for escrow smart contract). The rules specifying the service are shown in Fig. 3 using an integrated notation merging left- and right hand side graphs into a single rule graph $L \cup R$. We indicate by colours and labels which elements are required but preserved (grey), required and deleted (red), or newly created (green). That means, the left-hand side L is given by all grey and red elements, the right-hand side by all grey and green ones, and the interface K by the grey elements only.

The service works as follows. The price is set by the seller using input parameter $p > 0$, which is recorded in the contract created as part of the sell operation.

This step also incurs the stake of $stakeFactor \times p$, which is transferred from the seller's wallet to the balance of the contract. If the seller changes their mind and withdraws the widget, the stake is refunded.

The buyer uses the buy operation to indicate their intend to buy at the price $c.price$ as listed in contract c while transferring the same stake as the seller. When entering into a contract, seller and buyer use up a token, modelled by a loop, used to control the number of contracts each user can engage in. This can either be returned when they leave the contract, allowing them to enter into a new one (in the current version) or it can be a single-use token (in a later variant). If the seller withdraws after the buyer entered into the contract, the buyer can leave with the stake returned, and then the contract is deleted. Note that, due to the dangling condition, leave is only applicable if the contract does

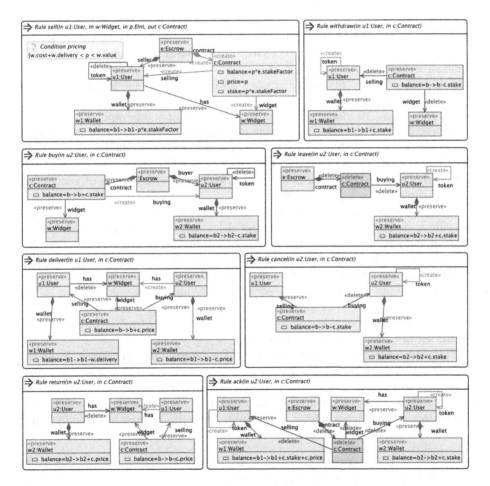

Fig. 3. Rules of the escrow model, in Henshin notation (Color figure online)

not have any further connections, in particular no selling link. That means it is only applicable after withdraw.

Once buyer and seller have entered the contract, the widget can be delivered. This leads to the transfer of the price from the buyer to the contract. The buyer can also change their mind and cancel either before delivery, in which case the buyer's and seller's stakes are refunded, or after delivery, e.g., if they are unsatisfied with the widget received. In the latter case they need to return the widget to get the price refunded. If the widget is to the buyer's satisfaction, they can acknowledge receipt leading to the resolution of the contract and the ultimate payment to the seller's wallet of the price and the refund of both stakes.

3   DPO Occurrence Graph Grammars and Unfolding

As a prerequisite to considering attributed grammars, in this section we review the theory of unfolding for typed DPO graph grammars [4]. In the rest of the paper we assume that all rules are *consuming*, i.e., their left-hand side morphism is not an isomorphism.

A grammar $\mathcal{G} = \langle TG, G_{in}, P, \pi \rangle$ is *(strongly) safe* if, for all H such that $G_{in} \Rightarrow^* H$, H has an injective typing morphism. Instance graphs with injective typing can be safely identified with the corresponding subgraphs of the type graph (thinking of injective morphisms as inclusions). Therefore, in particular, each graph $\langle G, t_G \rangle$ reachable in a safe grammar can be identified with the subgraph $t_G(G)$ of the type graph TG. For a safe grammar $\mathcal{G}$, the set of its *elements* is defined as $Elem(\mathcal{G}) = TG_E \cup TG_N \cup P$, assuming without loss of generality that the three sets are mutually disjoint.

Using a net-like language, we speak of *pre-set* $\bullet q$, *context* $\underline{q}$ and *post-set* $q\bullet$ of a rule q, defined as the sets of element deleted, preserved, or created by the rule. Similarly for a node or edge x in TG we write $\bullet x$, $\underline{x}$ and $x\bullet$ to denote the sets of rules which produce, preserve and consume x. The *causal relation* of a grammar $\mathcal{G}$ is the binary relation $<$ over $Elem(\mathcal{G})$ defined as the least transitive relation satisfying, for any node or edge x in TG and $q_1, q_2 \in P$: (1) if $x \in \bullet q_1$ then $x < q_1$; (2) if $x \in q_1\bullet$ then $q_1 < x$; and (3) if $q_1\bullet \cap q_2 \neq \emptyset$ then $q_1 < q_2$. As usual $\leq$ is the reflexive closure of $<$. Moreover, for $x \in Elem(\mathcal{G})$ we denote by $\lfloor x \rfloor$ the set of causes of x in P, namely $\{q \in P : q \leq x\}$.

The *asymmetric conflict relation* of $\mathcal{G}$ is the binary relation $\nearrow$ over P, given by (1) if $\underline{q_1} \cap \bullet q_2 \neq \emptyset$ then $q_1 \nearrow q_2$; (2) if $\bullet q_1 \cap \bullet q_2 \neq \emptyset$ and $q_1 \neq q_2$ then $q_1 \nearrow q_2$; and if $q_1 < q_2$ then $q_1 \nearrow q_2$.

A (nondeterministic) occurrence grammar is an acyclic grammar which represents, in a branching structure, several possible computations starting from its initial graph and using each rule at most once.

Definition 2 (occurrence grammar). *An occurrence grammar is a safe graph grammar $\mathcal{O} = \langle TG, G_{in}, P, \pi \rangle$ where*

1. *for each rule $q : \langle L, t_L \rangle \xleftarrow{l} \langle K, t_K \rangle \xrightarrow{r} \langle R, t_R \rangle$, typing morphisms t_L, t_K, t_R are injective;*

2. *its causal relation $\leq$ is a partial order, and for any $q \in P$, the set $\lfloor q \rfloor$ is finite and asymmetric conflict $\nearrow$ is acyclic on $\lfloor q \rfloor$;*
3. *the initial graph G_{in} coincides with the set $Min(\mathcal{O})$ of minimal elements of $\langle Elem(\mathcal{O}), \leq \rangle$ (with the graphical structure inherited from TG and typed by the inclusion);*
4. *each arc or node x in TG is created by at most one rule in P: $| {}^\bullet x | \leq 1$.*

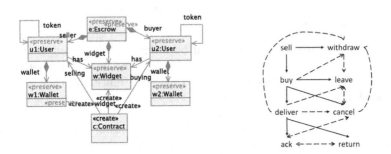

Fig. 4. Type graph, causality (solid) and asymmetric conflicts (dashed) of attributed occurrence grammar (Color figure online)

Since the initial graph of an occurrence grammar $\mathcal{O}$ is determined by $Min(\mathcal{O})$, we often do not mention it explicitly.

One can show that each occurrence grammar is safe [4].

Example 4 (occurrence graph grammar). An occurrence grammar based on the underlying grammar $\mathcal{AG}_G$ of the escrow model is shown in Fig. 4 with its type graph on the left and relations on the right. In the graph, all preserved (grey) elements are minimal, making up the input graph. The rules are the same as in Fig. 3. Matches and co-matches are defined by name, e.g., $c : Contract$ in any of the rules maps to $c : Contract$ in the type graph. This is possible here because we apply each rule exactly once. The causality and asymmetric conflict relations are justified below, with relations covered by transitive closure omitted.

- *sell < withdraw, buy* because *sell* creates the contract used by *withdraw*, *buy*;
- *buy < deliver, cancel, leave* because *buy* creates the buying link used by *deliver* and consumed by both *leave, cancel; buy $\nearrow$ withdraw* because *buy* uses the widget link deleted by *withdraw*;
- *withdraw $\nearrow$ leave* because *leave* deletes the contract used by *withdraw*;
- *deliver < ack, return* because *deliver* creates the has link used by both *ack, return; deliver $\nearrow$ withdraw, leave, cancel* because *withdraw* and *cancel* both consume the selling link and *leave* the contract used by *deliver*;
- *cancel $\nearrow$ withdraw* because *withdraw* deletes the selling link used by *cancel*
- *cancel, leave, ack* are in mutual conflict because all delete the buying link;

– *ack* ↗ *return* because *return* deletes the has link used by *ack*;
– *return* ↗ *ack* because *ack* deletes the contract used by *return*;

Note the lack of a causal dependency *withdraw* < *leave*; however due to the dangling condition *leave* can only be applied after *withdraw*. For a similar reason, *leave* cannot occur after *deliver* because *sell, buy, deliver, leave* violates the dangling condition due to the lack of *withdraw* while *sell, buy, withdraw, deliver, leave* is disallowed by *deliver* ↗ *withdraw*. Such subsets of rules, that are not fully executable in any order, do not form *configurations*. Subsets of rules that do, represent reachable graphs.

Occurrence grammars can be created by unfolding consuming graph grammars. The idea is to start from the initial graph of the grammar, then apply in all possible ways its rules while recording each occurrence and each new graph item generated with their causal histories. The basic ingredient here is the gluing operation, that we borrow literally from [4].

By $glue_*(q, m, G)$ we represent the additive application of rule q at match m to graph G, generating new items as specified by the rule and labelling them by $*$, but not removing items that should have been deleted. This is because such items may still be used by another rule in the nondeterministic unfolding.

4   Attributed Occurrence Grammars

In this and the following section we lift the theory of occurrence grammars and unfolding from typed graph grammars to lazy typed symbolic attributed ones. On grounded graphs, lazy rules are equivalent to general symbolic rules. Limiting ourselves to ground-preserving rules and a grounded start graph, all derivations will be grounded, which will allow us to transfer the results to the general case, including non-lazy rules.

Definition 3 (safe attributed graph grammar). *An attributed graph grammar* $\mathcal{AG} = \langle TG, P, G_{in}, \pi \rangle$ *is safe if* TG *is linear and its underlying E-graph grammar* $\mathcal{AG}_{EG}$ *is safe (reachable graphs* H *have injective typing* $t_H : H \to TG$*).*

A reachable graphs H will be seen as subgraphs of TG, and hence $X_H \subseteq X_{TG}$ and $D \models (\Phi_{TG} \implies \Phi_H)$. That means, X_{TG} is a global set of variables for $\mathcal{AG}$. The set of elements of a safe grammar $\mathcal{AG}$ is $Elem(\mathcal{AG}) = P \cup TG_G \cup X_{TG}$, including the rules, graph objects and variables in its type graph, but not its attribute links. With this, the causal and asymmetric conflict relations < and ↗ carry over to the rules and graph objects of $\mathcal{AG}$ via its underlying E-graph grammar $\mathcal{AG}_{EG}$. Applied to E-graphs, these relations include variable nodes (although variables are never deleted, so $^\bullet q \cap X_{TG} = \emptyset$) but no attribution links. In particular, $x \in q^\bullet$ iff $x \in X_{R_q} \setminus X_{L_q}$ and $x \in \underline{q}$ iff $x \in X_{L_q} \cap X_{R_q}$.

Let's examine the impact on the causal and asymmetric conflict relations of omitting attribute links from context, pre- and post-sets. When an attribute is updated, its attribute link is deleted and a new one is created pointing to a new variable. If two rules q_1, q_2 update the same attribute one after the other, q_1 creates the attribute link resulting in the first update and q_2 deletes the link. That means, in a causal relation including attribute links $q_1 < q_2$ based on clauses 1 and 2 in the definition of $<$. Now, without considering attribute links $q_1 < q_2$ by clause 3 because $v \in q_1^\bullet \cap \underline{q_2}$. However, disregarding attribute links impacts on asymmetric conflicts because variable nodes are never deleted, so neither clause 1 nor 2 of $\nearrow$ apply. That means, in a structural sense, attribute updates never cause conflicts. This is significant because it allows to separate attribute computations from structural unfolding.

For $e \in Elem(\mathcal{AG})$ we let $\lfloor e \rfloor = \{q \in P \mid q \leq e\}$ and $\lfloor E \rfloor$ the extension to sets $E \subseteq Elem(\mathcal{AG})$. The set of constraints of $R \subseteq P$ is $\Phi(R) = \bigcup_{p \in R} \left(t_{L_p}(\Phi_{L_p}) \cup t_{R_p}(\Phi_{R_p}) \right)$. Given $G \subseteq TG$, variable $x \in X_G$ is *current in G* if there is no rule $p \in \lfloor G \rfloor$ with $o.a = x \in L_p$ and $o.a = y \in R_p \setminus L_p$. Graph G *is current* if all variables in X_G are and every attribute has at most one variable as a label (i.e. cannot have both $o.a = y$ and $o.a = z$ with $y \neq z$ in the same graph G).

Definition 4 (attributed occurrence grammar). *An attributed occurrence grammar is a safe attributed graph grammar $\mathcal{AG} = \langle TG, G_{in}, P, \pi \rangle$ where*

1. *for each rule q the typing morphisms t_{L_p}, t_{K_q} and t_{R_q} are injective*
2. *for all $q \in P$, $(\nearrow)_{\lfloor q \rfloor}$ is acyclic, L_q current, $\lfloor q \rfloor$ finite, and $\Phi(\lfloor q \rfloor \cup \{q\}) \cup \Phi_{G_{in}}$ satisfiable*
3. *the initial graph G_{in} consists of the minimal elements $Min(\mathcal{AG})$ of $\langle Elem(\mathcal{AG}), \leq \rangle^3$ with attribution and graph structure inherited from TG and typed by the inclusion*
4. *each element (arc, node, variable or attribution edge) e in TG is created by at most one rule in P: $\mid {}^\bullet e \mid \leq 1$*

The category of attributed occurrence grammars and AGG morphisms is **AOG**.

Given attributed occurrence grammar $\mathcal{AO}$, its underlying grammar $\mathcal{AO}_G$ is an occurrence grammar. In particular, $\mathcal{AO}_G$ is obtained from $\mathcal{AO}$ by dropping its variables, attribute links and constraints. Hence conditions (2–4) in Definition 4 reduce to (2–4) in Definition 2.

Example 5 (attributed occurrence grammar). Figure 5 shows the rules and causal relation of an attributed occurrence grammar extending that of Example 4. Dependencies caused by attribute updates (of the contract's balance attribute) are shown as red dotted lines. Structurally, each such update reads the variable node representing the old value, creates a new variable node and changes the attribute link from the old to the new variable node. That means, subsequent rule applications access the new version of the variable node, which makes them different from applications of the same rules to the same graph objects accessing earlier versions of the attribute. Hence the duplication of rule occurrences in the

---

[3] Notice that $Min(\mathcal{AG}) \subseteq N_{TG} \cup E_{TG}$, i.e., it does not contain rules, since the grammar is consuming.

dependency structure. All asymmetric conflicts carry over from the unattributed case, i.e., if $p \nearrow q$ in Example 4, then $p? \nearrow q?$ for all copies $p?, q?$ of p, q. As discussed earlier, there are no new conflicts due to attribute updates.

Furthermore, in any current graph every attribute is labelled by at most one variable. To clarify the importance of this condition, consider the branching from $cancel2$ to $withdraw6$ and $return2$ both of which update attribute $c.balance$. The former substitutes it by $c.balance - c.stake$ and the latter by $c.balance - c.price$. Applying both rules in parallel yields $\{b_2 = b_1 - c.stake, b_3 = b_1 - c.price\} \subset \Phi(\lfloor G \rfloor)$ which is always satisfiable. However this result does not correspond to any sequence of rule applications since both updates of $c.balance$ use b_1 as the previous value. By preventing an attribute from having more than one label in a current graph, we disallow multiple versions of the same attribute.

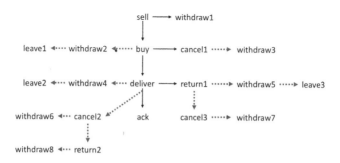

Fig. 5. Causality relation of attributed occurrence grammar (Color figure online)

5   Unfolding Attributed Graph Grammars

This section introduces the unfolding construction which, applied to an attributed graph grammar $\mathcal{AG}$, produces an attributed occurrence grammar $\mathcal{U}_{\mathcal{AG}}$ describing the behaviour of $\mathcal{AG}$. The unfolding is equipped with a morphism $u_{\mathcal{AG}}$ to the original grammar $\mathcal{AG}$ which allows to see rules in $\mathcal{U}_{\mathcal{AG}}$ as rule applications in $\mathcal{AG}$, and items of the type graph of $\mathcal{U}_{\mathcal{AG}}$ as instances of items of the type graph of $\mathcal{AG}$. Starting from the initial graph of the grammar, we apply in all possible ways its rules, and record in the unfolding each redex and new graph item generated with their causal histories. In order for these rule applications to reflect reachable transformations, we introduce attribute concurrent graphs representing reachable matches.

Definition 5 (attribute concurrent graph). *Let* $\mathcal{AO} = \langle TG, G_{in}, P, \pi \rangle$ *be an attributed occurrence grammar. A subgraph* G *of* TG *is called* attribute concurrent *if*

1. $\neg(x < y)$ *for all* $x, y \in G$;

2. G is current and $\lfloor G \rfloor$ finite;
3. $\nearrow$ is acyclic on $\lfloor G \rfloor$;
4. for all $e \in TG$ and $n \in \{s(e), t(e)\}$, if $n^\bullet \cap \lfloor G \rfloor \neq \emptyset$ and $^\bullet e \subseteq \lfloor G \rfloor$ then $e^\bullet \cap \lfloor G \rfloor \neq \emptyset$;
5. $\Phi_{G_{in}} \cup \Phi(\lfloor G \rfloor)$ is satisfiable.

Proposition 2 (attribute concurrent graphs are coverable). *A graph* $G \subseteq TG$ *is attribute concurrent if and only if it is coverable in* $\mathcal{AO}$, *i.e. there is a derivation* $Min(\mathcal{AO}) \Rightarrow^* H$ *with* $G \subseteq H$.

For every concurrent graph G one can find a derivation $Min(\mathcal{O}) \Rightarrow^* H$ which applies exactly once every rule in $\lfloor G \rfloor$, in any order consistent with $(\nearrow_{\lfloor G \rfloor})^*$. Vice versa for each derivation $Min(\mathcal{O}) \Rightarrow^* G$ in $\mathcal{O}$, the set of rules it applies contains $\lfloor G \rfloor$ and their order is compatible with $\nearrow^*$. Therefore reachable graphs are concurrent. Furthermore, each subgraph of a concurrent graph is concurrent as well, thus so are all coverable graphs.

The proof is based on an analogous result for the un-attributed case, analysing the additional dependencies arising from attribute updates, and deriving the satisfiability of constraints added throughout derivations from that of $\Phi(\lfloor q \rfloor \cup \{q\}) \cup \Phi_{G_{in}}$ in Definition 4.2.

The unfolding of an attributed grammar $\mathcal{AG} = \langle TG, G_{in}, P, \pi \rangle$ is defined as follows. For each n, we construct a partial unfolding $\mathcal{U}(\mathcal{AG})^{(n)} = \langle \mathcal{U}^{(n)}, u^{(n)} \rangle$, where $\mathcal{U}^{(n)} = \langle TG^{(n)}, G_{in}, P^{(n)}, \pi^{(n)} \rangle$ is an attributed occurrence grammar and $u^{(n)} = \langle u_{TG}^{(n)}, u_P^{(n)} \rangle : \mathcal{U}^{(n)} \to \mathcal{AG}$ an AGG morphism. Intuitively, the occurrence grammar generated at level n contains all possible computations of the grammar with "causal depth" at most n.

- $(\mathbf{n = 0})$ $\langle \mathcal{U}^{(0)}, u^{(0)} \rangle$ is defined as $\mathcal{U}^{(0)} = \langle G_{in}, G_{in}, \emptyset, \emptyset \rangle$ with $u_{TG}^{(0)} = t_{G_{in}}$.

- $(\mathbf{n \to n+1})$ Given $\mathcal{U}(\mathcal{AG})^{(n)}$, the partial unfolding $\mathcal{U}(\mathcal{AG})^{(n+1)}$ is obtained by applying all enabled redexes $m : L_q \to \langle TG^{(n)}, u_{TG}^{(n)} \rangle$ of $q \in P$ in $TG^{(n)}$. Let $P_{(n+1)}$ be the set of all triples $q_i^* = \langle q_i, m_{i,G}, m_{i,X} \rangle$ such that $m(L_q) \subseteq TG^{(n)}$ is attribute concurrent and $m_i(\Phi_p)$ is satisfiable. Then, $\mathcal{U}(\mathcal{AG})^{(n+1)}$ is given by
  - $TG^{(n+1)} = glue_{q_k^*}(q_k, m_k, \ldots glue_{q_1^*}(q_1, m_1, TG^{(n)}) \ldots)$, the consecutive gluing of $TG^{(n)}$ with $R_{q_1} \ldots R_{q_k}$ along $K_{q_1} \ldots K_{q_k}$, respectively, with constraints $\Phi_{TG^{(n+1)}} = \Phi_{TG^{(n)}} \cup m_1(\Phi_{R_{q_1}}) \cup \cdots \cup m_k(\Phi_{R_{q_k}})$;
  - morphism $u_{TG}^{(n)}$ extended canonically to $u_{TG}^{(n+1)} : TG^{(n+1)} \to TG$;
  - $P^{(n+1)} = P^{(n)} \cup \{\langle q, m_G, m_X \rangle \mid \langle q, m \rangle \in P_{(n+1)}\}$;
  - $u_P^{(n+1)} = u_P^{(n)} \cup u_{P,(n+1)}$ where $u_{P,(n+1)}(\langle q, m_G, m_X \rangle) = q$;
  - $\pi^{(n+1)}(\langle q, m_G, m_X \rangle)$ coinciding with $\pi(q)$ except for retyping.

The result of the gluing operation is independent of the order of redexes applied.

The match of a rule, which is part of its name in the unfolding, records the causal history by way of the graph and data elements it maps to, which in turn are labelled by the name of the rule that created them. This allows to distinguish rule occurrences with different histories. Note that if a rule is applied twice (also in different steps) with the same graphical match and variable mapping, the generated items are the same and thus they appear only once in the unfolding. By construction it is evident that $\mathcal{U}(\mathcal{AG})^{(n)} \subseteq \mathcal{U}(\mathcal{AG})^{(n+1)}$, component-wise.

Definition 6 (attributed unfolding). *The unfolding* $\mathcal{U}(\mathcal{AG}) = \langle \mathcal{U}_{\mathcal{AG}}, u_{\mathcal{AG}} \rangle$ *is defined as* $\bigcup_n \mathcal{U}(\mathcal{AG})^{(n)}$, *where union is applied componentwise.*

The folding morphism u is defined by mapping rules by their names, while on types $u_{TG}(x : X) = X$.

Example 6 (unfolding). In Example 5 we presented a partial unfolding of the grammar in Example 3. The set of rules $\{sell, buy, deliver, cancel, return\}$ admits two possible linearisation *sell, buy, deliver, cancel, return* and *sell, buy, deliver, return, cancel* that only differ in the order of applying *cancel* and *return*. While the structural unfolding does not distinguish between the sequences, in the attributed unfolding they both update attributes *c.balance* and *w2.balance*. Hence, whichever occurs second will depend on the first. The shared part of both linearisation is shown below, the two distinct parts in the following tables.

i	Rule	Rule application	$X^{(i)} \setminus X^{(i-1)}$	Constraint
1	*sell*	*sell1*	$\langle w1.balance, sell1 \rangle$	$\langle w1.balance, in \rangle - \langle p, in \rangle \times \langle e.stakeFactor, in \rangle$
			$\langle c.balance, sell1 \rangle$	$\langle p, in \rangle \times \langle e.stakeFactor, in \rangle$
			$\langle c.price, sell1 \rangle$	$\langle p, in \rangle$
			$\langle c.stake, sell1 \rangle$	$\langle p, in \rangle \times \langle e.stakeFactor, in \rangle$
2	*buy*	*buy1*	$\langle w2.balance, buy1 \rangle$	$\langle w2.balance, in \rangle - \langle c.stake, sell1 \rangle$
			$\langle c.balance, buy1 \rangle$	$\langle c.balance, sell1 \rangle + \langle c.stake, sell1 \rangle$
3	*deliver*	*deliver1*	$\langle c.balance, deliver1 \rangle$	$\langle c.balance, buy1 \rangle + \langle c.price, sell1 \rangle$
			$\langle w1.balance, deliver1 \rangle$	$\langle w1.balance, sell1 \rangle - \langle w.delivery, in \rangle$
			$\langle w2.balance, deliver1 \rangle$	$\langle w2.balance, buy1 \rangle - \langle c.price, sell1 \rangle$

i	Rule	Rule application	$X^{(i)} \setminus X^{(i-1)}$	Constraint
4	*cancel*	*cancel2*	$\langle c.balance, cancel2 \rangle$	$\langle c.balance, deliver1 \rangle - \langle c.stake, sell1 \rangle$
			$\langle w2.balance, cancel2 \rangle$	$\langle w2.balance, deliver1 \rangle + \langle c.stake, sell1 \rangle$
5	*return*	*return2*	$\langle c.balance, return2 \rangle$	$\langle c.balance, cancel2 \rangle - \langle c.price, sell1 \rangle$
			$\langle w2.balance, return2 \rangle$	$\langle w2.balance, cancel2 \rangle + \langle c.price, sell1 \rangle$

Through substitution we find that in both linearisation the end result is the same, i.e., both *c.balance* and *w2.balance* are equal to their initial values. However, the causal dependencies are different due to different data flow:

The two paths are confluent (produce the same result) but not equivalent (do not represent the same concurrent computation). For applications where we are

i	Rule	Rule application	$X^{(i)} \setminus X^{(i-1)}$	Formula
4	return	return1	$\langle c.balance,\ return1 \rangle$	$\langle c.balance,\ deliver1 \rangle - \langle c.stake,\ sell1 \rangle$
			$\langle w2.balance,\ return1 \rangle$	$\langle w2.balance, deliver1 \rangle + \langle c.price,\ sell1 \rangle$
5	cancel	cancel3	$\langle c.balance,\ cancel3 \rangle$	$\langle c.balance,\ return1 \rangle - \langle c.stake,\ sell1 \rangle$
			$\langle w2.balance, cancel3 \rangle$	$\langle w2.balance, return1 \rangle + \langle c.stake,\ sell1 \rangle$

cancel before return		return before cancel	
cancel2	$<$ return2	return1	$<$ cancel3
$\langle c.balance,\ cancel2 \rangle$	$< \langle c.balance,\ return2 \rangle$	$\langle c.balance,\ return1 \rangle$	$< \langle c.balance,\ cancel3 \rangle$
$\langle w2.balance, cancel2 \rangle$	$< \langle w2.balance,\ return2 \rangle$	$\langle w2.balance,\ return1 \rangle$	$< \langle w2.balance, cancel3 \rangle$

only interested in the end results of non-deterministic concurrent computations, it makes sense to explore whether confluence can be used to restrict the unfolding construction, avoiding exploration of confluent alternatives.

Proposition 3 (attributed unfolding). $\mathcal{U}_{A\mathcal{G}}$ *is an attributed occurrence grammar and* $u_{A\mathcal{G}} : \mathcal{U}_{A\mathcal{G}} \to A\mathcal{G}$ *is an attributed graph grammar morphism.*

Since AGG morphisms preserve derivations, this implies the correctness of the unfolding, i.e., $u_{\mathcal{G}}$ maps all derivations in $\mathcal{U}_{A\mathcal{G}}$ to derivations in $A\mathcal{G}$. The construction can be lifted to a functor between attributed grammars and occurrence grammars which is a right adjoint to the inclusion functor.

Theorem 1 (coreflection). *Unfolding is a coreflection* $\mathcal{U} : \mathbf{AGG} \to \mathbf{AOG}$, *that is, right-adjoint to the inclusion of* $\mathbf{AOG}$ *in* $\mathbf{AGG}$.

The proof established the unfolding as a co-free construction from attributed grammars to attributed occurrence grammars. That means, the unfolding is maximal among all attributed occurrence grammars equipped with a morphism back to the original grammar, in the sense that there exists a unique commuting morphisms from every other candidate. The existence of this morphism shows that the unfolding is complete in representing all possible behaviours present in the given grammar. Uniqueness shows that they are represented in a minimal way, so there is only one choice as to how the morphism can be defined.

6   Related Work and Conclusion

We presented occurrence grammars and unfoldings for attributed graph grammars. Since unfolding is a coreflection, the semantics is correct (only representing derivations specified by the grammar), complete (representing all such derivations) and fully abstract (represented in a minimal way). Our approach uses symbolic graph transformation [8,9] and the theory of unfolding of graph grammars [1,2], in turn generalising results for Petri Nets [6]. [7] uses approximated unfolding and abstract interpretation to analyse attributed graph grammars.

While their focus is on scalable analysis with numerical data, rather than comprehensive semantics, we share the objective of separating structural from attribute computations.

We will use the relation of occurrence grammars and event structures to develop, based on Rideau and Winskel's concurrent games as event structures [10], a notion of concurrent game over attributed graph grammars, with applications to analysing smart contracts. We also plan to integrate the theories of unfolding for attributed and conditional grammars.

Acknowledgement. We would like to thank the authors of [2] for their support and advice.

References

1. Baldan, P., Corradini, A., Heindel, T., König, B., Sobociński, P.: Unfolding grammars in adhesive categories. In: Kurz, A., Lenisa, M., Tarlecki, A. (eds.) CALCO 2009. LNCS, vol. 5728, pp. 350–366. Springer, Heidelberg (2009). https://doi.org/10.1007/978-3-642-03741-2_24
2. Baldan, P., Corradini, A., König, B.: Unfolding graph transformation systems: theory and applications to verification. In: Degano, P., De Nicola, R., Meseguer, J. (eds.) Concurrency, Graphs and Models. LNCS, vol. 5065, pp. 16–36. Springer, Heidelberg (2008). https://doi.org/10.1007/978-3-540-68679-8_3
3. Baldan, P., Corradini, A., Montanari, U., Ribeiro, L.: Coreflective concurrent semantics for single-pushout graph grammars. In: Wirsing, M., Pattinson, D., Hennicker, R. (eds.) WADT 2002. LNCS, vol. 2755, pp. 165–184. Springer, Heidelberg (2003). https://doi.org/10.1007/978-3-540-40020-2_9
4. Baldan, P., Corradini, A., Montanari, U., Ribeiro, L.: Unfolding semantics of graph transformation. Inf. Comput. **205**(5), 733–782 (2007). https://doi.org/10.1016/j.ic.2006.11.004
5. Ehrig, H., Ehrig, K., Prange, U., Taentzer, G.: Fundamentals of Algebraic Graph Transformation. Springer, Heidelberg (2006). https://doi.org/10.1007/3-540-31188-2
6. Esparza, J., Heljanko, K.: Unfoldings: A Partial-Order Approach to Model Checking. Monographs in Theoretical Computer Science. An EATCS Series. Springer, Heidelberg (2008). https://doi.org/10.1007/978-3-540-77426-6
7. König, B., Kozioura, V.: Towards the verification of attributed graph transformation systems. In: Ehrig, H., Heckel, R., Rozenberg, G., Taentzer, G. (eds.) ICGT 2008. LNCS, vol. 5214, pp. 305–320. Springer, Heidelberg (2008). https://doi.org/10.1007/978-3-540-87405-8_21
8. Orejas, F., Lambers, L.: Symbolic attributed graphs for attributed graph transformation. ECEASST **30** (2010). http://journal.ub.tu-berlin.de/index.php/eceasst/article/view/405
9. Orejas, F., Lambers, L.: Lazy graph transformation. Fundam. Inform. **118**(1–2), 65–96 (2012). https://doi.org/10.3233/FI-2012-706
10. Rideau, S., Winskel, G.: Concurrent strategies. In: 2011 IEEE 26th Annual Symposium on Logic in Computer Science, pp. 409–418. IEEE (2011)

Single Pushout Rewriting in Comprehensive Systems

Harald König[1,2](✉) and Patrick Stünkel[2]

[1] University of Applied Sciences, FHDW, Hannover, Germany
harald.koenig@fhdw.de
[2] Høgskulen på Vestlandet, Bergen, Norway
past@hvl.no

Abstract. The elegance of the *single-pushout* (SPO) approach to graph transformations arises from substituting total morphisms by partial ones in the underlying category. Thus, SPO's applicability depends on the durability of pushouts after this transition. There is a wide range of work on the question when pushouts exist in categories with partial morphisms starting with the pioneering work of Löwe and Kennaway and ending with an essential characterisation in terms of an exactness property (for the interplay between pullbacks and pushouts) and an adjointness condition (w.r.t. inverse image functions) by Hayman and Heindel.

Triple graphs and graph diagrams are frameworks to synchronize two or more updatable data sources by means of internal mappings, which identify common sub-structures. *Comprehensive systems* generalise these frameworks, treating the network of data sources and their structural inter-relations as a homogeneous comprehensive artifact, in which partial maps identify commonalities. Although this inherent partiality produces amplified complexity, Heindel's characterisation still yields cocompleteness of the category of comprehensive systems equipped with closed partial morphisms and thus enables computing by SPO graph transformation.

Keywords: Single Pushout Rewriting · Partial morphism · Category theory · Hereditary pushout · Upper adjoint · Comprehensive system

1 Introduction and Motivation

We want to dedicate this paper to Michael Löwe, the founder of the single-pushout approach [12] and simultaneously a pioneer in the investigation of categories of partial algebras with partial morphisms between them [13].

In this paper, we want to combine these two theories. We introduce the category of *comprehensive systems*, formally a category in which the inner structure of the objects can be described with partial maps, and will show that SPO rewriting is applicable in this category.

In Honour of Michael Löwe, 1956 - 2019.

© Springer Nature Switzerland AG 2020
F. Gadducci and T. Kehrer (Eds.): ICGT 2020, LNCS 12150, pp. 91–108, 2020.
https://doi.org/10.1007/978-3-030-51372-6_6

Comprehensive Systems have been introduced in [21] (see also [22]) as a means for global consistency management, representing a collection of inter-related systems. To provide an intuition of a comprehensive system (Definition 3 in Sect. 3), take a look at Fig. 1. There are three conceptual models A_1, A_2, A_3, which depict persons (👤) with certain features: In A_1 a phone number (☎) is assigned to the person, in A_2 and A_3 the person possesses a home address (🏠), and in A_3 persons additionally may have a business address

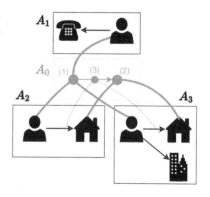

Fig. 1. Comprehensive system

(🏢). You may think of $A_{1/2/3}$ representing (excerpts of) the contents of three databases possibly in three distributed IT-systems, the first stores records of persons and phone numbers while the second and third store records of persons and addresses.

In many cases it is necessary to maintain global consistency of the databases' contents, especially in the presence of inter-model constraints [20]. Let us assume the following constraint:

IMC Every person with a business address must either provide a phone number or a home address or both.

Violations of this constraint can only be discovered, if common elements in the models are identified. Thus one has to specify that a recorded person in A_1 is actually *the same* as some person record in A_2 and/or A_3. Such commonality specifications extend the modelling language and are commonly used in practice, e.g. [5]. In Fig. 1 we employ grey-coloured "tentacles" to visualise three commonalities: (1) The three person records, (2) the two home addresses, and also (3) the two assignments of the home address to the person.

Note that $A_{1/2/3}$ formally represent directed graphs. But the junctions of each commonality (•) – called *commonality representatives* – form another graph A_0 in Fig. 1. Elements of A_0 witness common elements among $A_{1/2/3}$ and these commonalities must obviously respect node-edge-incidences (see the person to home address assignment), such that their respective outgoing grey lines are in fact graph morphisms $a_i : A_0 \rightarrow A_i$, $i \in I = \{1, 2, 3\}$.

For $|I| = 2$ the underlying star-shape of comprehensive systems (finite collections of arrows $(a_i)_{i \in \{1,...,n\}}$ with common source) reduces to the span shape $\bullet \leftarrow \bullet \rightarrow \bullet$, which is the underlying setting for triple graphs [18], the common source in the middle specifying the commonality graph. An extension of triple graphs are graph diagrams [23], a framework for *multi-ary* model synchronisation. Since multi-ary commonality relations such as the ternary tentacles of identical person records in Fig. 1 can not be encoded with several binary relations [19], one must distinguish relations of different arity in the underlying shape for graph diagrams: E.g. in Fig. 1, a shape with two nodes is required: One node specifies the existence of a graph containing binary home address commonalities and one node is used for ternary commonalities of person records. In larger system landscapes ($n > 3$), there may be many more commonality relations of arbitrary arity $k \leq n$, which would cause a considerable amount of heterogeneity of commonality nodes in the underlying shape for graph diagrams. Moreover, this schema and hence the basic setting for implementations must be altered, whenever new commonality relations are added.

We showed in [21,22] that comprehensive systems are a homogeneous generalisation of graph diagrams. They are *homogeneous*, because we need only one node to cope with commonality relations of arbitrary arity (the center of the star-shape specifying commonality graph A_0) and must not alter the base setting, if new relations are added. It is a *generalisation*, because we can implement each graph diagram as a comprehensive system, i.e. we can jointly collect different commonalities into one graph A_0.

An important distinction, however, is that graph morphisms $a_i : A_0 \rightarrow A_i$ in comprehensive systems are allowed to be *partial*. E.g. $a_1 : A_0 \rightarrow A_1$ in Fig. 1 is undefined on (2) and (3) in A_0. Nevertheless can we show that comprehensive systems form a category $\mathbb{CS}$, in which graph rewriting, especially SPO rewriting, is possible. For this, we will also consider the category $\mathrm{Par}(\mathbb{CS})$, i.e. $\mathbb{CS}$ equipped with *partial* morphisms, cf. Sect. 2. Although this requires handling both *intrinsic* and *extrinsic* partiality, we can prove existence of all pushouts in this category (Theorem 1 in Sect. 4) and hence demonstrate applicability of SPO rewriting.

We expect the reader to have basic knowledge in category theory. For categorical artefacts, we employ the following notations: *Categories* like $\mathbb{C}$ will be denoted in a double-struck font. When distinguishing between members of $\mathbb{C}$, we write $|\mathbb{C}|$ (or just $\mathbb{C}$) for its objects and $Arr(\mathbb{C})$ for its morphisms. Moreover, there are identities $id_A : A \rightarrow A$ and composition, e.g. $g \circ f$, for $f : A \rightarrow B$ and $g : B \rightarrow C$. $\mathbb{SET}$ is the category of sets and total mappings and we will use the letter $\mathbb{G}$ for categories that are based on a signature with unary operation

Fig. 2. Pullback along f

symbols only, see Sect. 3 for more details. Monomorphisms ($\rightarrowtail$), epimorphisms ($\twoheadrightarrow$) and – if applicable – inclusions ($\hookrightarrow$) have special notations. We furthermore expect the reader to be familiar with basic *universal constructions* like pullbacks, coproducts, and pushouts. When describing a pullback as in Fig. 2, we either say that the span (g', f') is the pullback of co-span (f, g) or we call g' the pullback of g along f. Moreover, the pullback object D in this construction is highlighted with a small adjacent right angle.

2   SPO Rewriting

2.1   Graph-Like Structures

A category $\mathbb{G}$ is called *based on a signature with unary operation symbols only*, if it is isomorphic to a category of total algebras w.r.t. a signature, which only contains sorts and unary operation symbols. The simplest example in our context –

and the rationale behind using letter $\mathbb{G}$ – are directed graphs, which are based on a signature with sorts E and V and two unary operation symbols $s, t : E \rightarrow V$. We do not endorse directed graphs in particular and could likewise choose $\mathbb{G}$ to be given by E-Graphs [4] or bipartite artefacts like condition-event-nets. It is well-known that all these categories are topoi and thus possess all limits (e.g. pullbacks) and colimits (coproducts, pushouts) [6].

Objects of such categories $\mathbb{G}$ are sometimes called "graph-like structures", as e.g. in [14], and thus we will simply call $\mathbb{G}$-objects and morphisms "graphs" and "graph morphisms" bearing in mind the above mentioned more general setting.

Remark: $\mathbb{G}$ will serve as the base category (or base structure) for assembling comprehensive systems, cf. Fig. 1. Actually, we could have traded $\mathbb{G}$ for more general (weak) adhesive (HLR) categories w.r.t. an admissible subclass $\mathcal{M}$ of all monomorphisms.[1] Adhesive HLR categories have mainly been introduced to model *attributes*, which poses some challenges regarding adhesiveness, in turn requiring to work with special subclasses of morphisms, which are isomorphic on the "data part". However, we restrict ourselves to graph-like structures because we are not focusing on attributes here and we want to stay in the tradition of Michael Löwe, who originally investigated graph-like structures only. Since graph-like structures are sufficiently concrete, we can actually refer to an element $x \in A$ for some $A \in \mathbb{G}$, i.e. an element of some carrier set of graph A. Likewise, "$\forall x \in A$" means "for all x of any sort s in the carrier set of A".

2.2 Partial Map Categories

Consider an arbitrary category $\mathbb{C}$ with pullbacks. Michael had the courage to leave the comfortable world of *total* $\mathbb{C}$-morphisms and utilised *partial* morphisms [17] for the SPO approach. While other researchers adhered to total morphisms, he forcefully followed through with partiality and proved that it is worthwhile [12]. He used the following definitions: A $\mathbb{C}$-span $X \xleftarrow{m} dom(f) \xrightarrow{f} Y$, is equivalent to a second span $X \xleftarrow{m'} dom(f') \xrightarrow{f'} Y$, if and only if there is an isomorphism $i : dom(f) \rightarrow dom(f')$ such that $m' \circ i = m$ and $f' \circ i = f$. A *partial morphism* is an equivalence class w.r.t. this relation, denoted by $\langle m, f \rangle : X \rightharpoonup Y$, i.e. the pair (m, f) is a representative of its equivalence class. $\langle m, f \rangle$ is called total, if m is an isomorphism, and we use the usual arrow tip in this case: $\langle m, f \rangle : X \rightarrow Y$.

It is then easy to see that the objects of $\mathbb{C}$ together with partial morphisms constitute a new category $\mathtt{Par}(\mathbb{C})$: An identity is the equivalence class of the identity span and composition of $\langle m_1, f_1 \rangle : G_1 \rightharpoonup G_2$ and $\langle m_2, f_2 \rangle : G_2 \rightharpoonup G_3$ is given by constructing the pullback span (m_2', f_1') of co-span (f_1, m_2) yielding the composed partial morphism $\langle m_1 \circ m_2', f_2 \circ f_1' \rangle$. This is in fact independent of the choice of pullbacks and independent of the choice of representatives.

---

[1] It seems that the subsequent proofs can still be carried out, if $\mathbb{G}$ is such a more general structure. We will provide respective facts from time to time.

Furthermore we obtain an identity-on-objects functor

$$\Gamma : \begin{cases} \mathbb{C} \to \mathbf{Par}(\mathbb{C}) \\ (A \xrightarrow{f} B) \mapsto (A \xrightarrow{\langle id_A, f \rangle} B) \end{cases}$$

called the *graphing functor* [8], i.e. a canonical embedding of totality into partiality.

Definition 1 (Hereditary Pushout). *[8] Any pushout in $\mathbb{C}$ is called* hereditary, *if its Γ-image is a pushout in $\mathbf{Par}(\mathbb{C})$. If all pushouts exist in $\mathbb{C}$ and they are all hereditary, we say that $\mathbb{C}$ is a* hereditary pushout category.

The following result can be found in [10]:

Proposition 1. *If the Γ-image of a $\mathbb{C}$-span has a pushout in $\mathbf{Par}(\mathbb{C})$, then this cocone consists of two total morphisms, which are the Γ-image of the pushout of this span in $\mathbb{C}$.* □

For $\mathbb{C} := \mathbb{G}$, we can refer to elements inside objects of $\mathbb{G}$, such that we will work with representatives $G \xhookleftarrow{m} dom(f) \xrightarrow{f} H$ of a partial morphism, in which the left leg m is chosen as the effective inclusion of $dom(f)$, the domain of definition of the partial morphism, into G. Since the name m is of minor importance, we may as well write $f : G \rightharpoonup H$. In this setting, we will call f "total", if the inclusion m is the identity. For the remainder of this paper we will use $\mathbb{G}$-inclusions when there is a choice for monomorphisms (replacing $\rightarrowtail$ with $\hookrightarrow$).

The following result was stated in [7] but fully worked out already in [12]:

Proposition 2. $\mathbb{G}$ *is a hereditary pushout category.* □

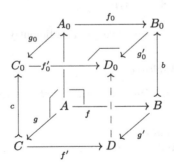

Fig. 3. Commutative cube

Finally, hereditariness can equivalently be characterised as follows:

Proposition 3 (Equivalent Characterisation of Hereditariness). *[9] A pushout like the top face in the cube in Fig. 3 is hereditary, if and only if in any commutative cube as in Fig. 3 with rear faces being pullbacks and vertical* *front left and back right arrows (c and b in Fig. 3) being monomorphisms, the following equivalence holds: The bottom face is a pushout if and only if (1) the two front faces are pullbacks and (2) the vertical front right arrow (the dashed arrow in Fig. 3) is a monomorphism.* □

2.3   Rewriting Rule and Derivation

While the *double-pushout approach* (DPO) [4] requires the construction of a pushout complement and another pushout in the underlying category $\mathbb{C}$, the introduction of partial morphisms enables graph transformations to be expressed by a *single pushout* in $\mathsf{Par}(\mathbb{C})$. This elegant alternative to DPO was initiated by Raoult [16] and then fully worked out in Michael's PhD Thesis [12].

Definition 2 (Rule, Match, Derivation, Conflict-Freeness). *An SPO rule is a morphism $L \overset{\rho}{\rightharpoonup} R$ of $\mathsf{Par}(\mathbb{C})$. A match for ρ at (host) $G \in \mathbb{C}$ is a total morphism $\mu : L \to G$. A pushout of ρ and μ in $\mathsf{Par}(\mathbb{C})$ generates the (SPO-) derivation $G \overset{\rho,\mu}{\Rightarrow} H$ with trace ρ' and co-match μ', see Fig. 4. The match μ is* called *conflict-free, if μ' is a total morphism.*

Computing by SPO graph transformation requires the existence of pushouts in $\mathsf{Par}(\mathbb{C})$ and it requires conflict freeness of matches, cf. Definition 2, if one wants to avoid partial matches. We can prove existence of pushouts for $\mathsf{Par}(\mathbb{CS})$ in Sect. 4, but - in contrast to a simple criterion for conflict freeness in terms of injectivity of μ on delete-preserve pairs [12] - we have to leave the question of a criterion for conflict freeness in $\mathsf{Par}(\mathbb{CS})$ for future research.

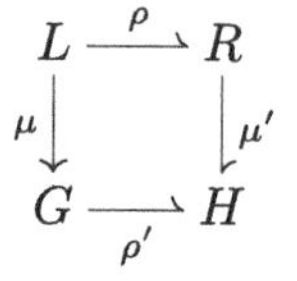

$$\begin{array}{ccc} L & \overset{\rho}{\rightharpoonup} & R \\ \mu\downarrow & & \downarrow\mu' \\ G & \underset{\rho'}{\rightharpoonup} & H \end{array}$$

Fig. 4. SPO derivation

3   Comprehensive Systems

For now and the rest of the paper, we fix a sufficiently large number $n \in \mathbb{N}$.[2]

3.1   Definitions and Background

Definition 3 (Comprehensive System). *Let $(A_i)_{0\leq i\leq n}$ be an $n+1$-tuple of $\mathbb{G}$-objects. We call*

- *$(A_j)_{1\leq j\leq n}$ the* Components *and*
- *A_0 the* Commonality Representatives

of a Comprehensive System

$$\mathbf{A} := (a_j : A_0 \rightharpoonup A_j)_{j\in\{1,...,n\}}$$

i.e. of an n-tuple of partial graph morphisms $(a_j)_{1\leq j\leq n}$, which we call projections[3].

---

[2] Usually the number of distributed systems under consideration.
[3] One might consider elements of A_0 to be vectors (of arbitrary arity), in which common elements are listed, hence the term "projection" for the a_j.

In order to make reading easier, we always use letter i, if indexing comprises graphs $A_0, A_1, \ldots, A_n$ and we use letter j, if indexing is only over the components $A_1, \ldots, A_n$. Moreover, we denote comprehensive systems with bold face letters.

Comprehensive systems admit an all-embracing view on a system of possibly heterogeneously typed components, in which all necessary informational overlaps are coded, cf. Fig. 1. They have been treated on the level of graphs in [20] and – on a more abstract level – in [3].

Definition 4 (Morphism of Comprehensive Systems). *Let* $\mathbf{A} := (a_j : A_0 \rightharpoonup A_j)_{j \in \{1,\ldots,n\}}$ *and* $\mathbf{A}' := (a'_j : A'_0 \rightharpoonup A'_j)_{j \in \{1,\ldots,n\}}$ *be two comprehensive systems. A morphism* $\mathbf{f} : \mathbf{A} \to \mathbf{A}'$ *is a family* $(f_i : A_i \to A'_i)_{0 \leq i \leq n}$ *of total* $\mathbb{G}$-*morphisms, such that for all* $1 \leq j \leq n$

$$a'_j \circ f_0 = f_j \circ a_j \tag{1}$$

holds in $Par(\mathbb{G})$, *the category of graphs and partial morphisms.*

Whenever we mention morphisms $\mathbf{f} : \mathbf{A} \to \mathbf{A}'$ between comprehensive systems, we implicitly assume the components of $\mathbf{A}$ and $\mathbf{A}'$ be denoted as in Definition 3 and we assume the constituents of $\mathbf{f}$ be denoted as in Definition 4.

There is the obvious identical morphism $\mathbf{id_A}$ for each comprehensive system $\mathbf{A}$ and composition can be defined componentwise. Hence we obtain

Proposition 4 (Category $\mathbb{CS}$ and Component Functors). *Let* $\mathbb{G}$ *be a category as described above.*

- *Comprehensive Systems and morphisms between them constitute a category, denoted* $\mathbb{CS}_{\mathbb{G}}$ *or often just* $\mathbb{CS}$, *if the base category is clear from the context.*
- *For each* $i \in \{0, \ldots, n\}$ *there is the component functor* $\mathcal{C}_i : \mathbb{CS} \to \mathbb{G}$ *defined by* $\mathcal{C}_i(\mathbf{f} : \mathbf{A} \to \mathbf{A}') = A_i \xrightarrow{f_i} A'_i$ *for any* $\mathbf{f}$ *defined as in Definition 4.* □

It is important to note that we claim (1) to hold in $Par(\mathbb{G})$ and not in $\mathbb{G}$! Let's investigate the consequences: Recall that the definition of composition of partial morphisms involves pullbacks. In the situation in Fig. 5 this enforces that a pullback of $\subseteq'_j$ along f_0 (to express the composition of $\langle \subseteq'_j, a'_j \rangle$ and $\langle id_{A_0}, f_0 \rangle$) can be chosen to be equal to $\subseteq_j$, the inclusion of the domain of definition of $f_j \circ a_j$ into A_0, see the upper square in Fig. 5, and it enforces that the lower square commutes. If $x \in A_0$, this observation is equivalent to the statement $x \in dom(a_j) \iff f_0(x) \in dom(a'_j)$, because "$\Rightarrow$" corresponds to commutatitivity of the upper square and "$\Leftarrow$" corresponds to the pullback property (in $\mathbb{SET}$ and hence in $\mathbb{G}$, in which pullbacks are constructed sortwise). This yields a more handy admissibility characterisation for $\mathbb{CS}$-morphisms:

Proposition 5 (Preservation and Reflection of Definedness). $\mathbf{f} : \mathbf{A} \to \mathbf{A}'$ *as defined in Definition 4 is a morphism of comprehensive systems if and only if for all* $j \in \{1, \ldots, n\}$ *and for all* $x \in A_0$:

$$a_j(x) \text{ is defined} \iff a'_j(f_0(x)) \text{ is defined} \tag{2}$$

and the lower square in Fig. 5 commutes. □

Usually a morphism between two partial algebras A and B requires only preservation of definedness, i.e. "$\Rightarrow$" in (2). In the next section, we justify why we additionally need reflection of definedness.

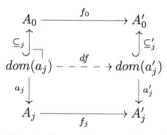

Fig. 5. Composing partial and total morphisms

3.2 Why Must Definedness Be Reflected?

Since our goal is to show that SPO rewriting is applicable for comprehensive systems, we must show that $\mathtt{Par}(\mathbb{CS})$ possesses all pushouts.

Assume we would not have claimed reflection of definedness for $\mathbb{CS}$-morphisms, but only commutativity of the two squares in Fig. 5, which is equivalent to claiming the properties of Proposition 5 except for the implication direction "$\Leftarrow$" in (2). In this case let's consider for $n = 1$ two simple comprehensive systems. Let $\mathbb{G} = \mathbb{SET}$ and $A_0 = \{*\}$ and $A_1 = \{\bullet\}$ be two one-element sets and let $\mathbf{A} = (a_1 : A_0 \rightharpoonup A_1)$ with a_1 the totally undefined map depicted with $(* \quad \bullet)$ and $\mathbf{A}' = (a'_1 : A_0 \to A_1)$ with a'_1 the unique total map from A_0 to A_1 depicted $(* \mapsto \bullet)$. If we only work with preservation of definedness, then morphism $\langle id_A, f \rangle : \mathbf{A} \to \mathbf{A}'$, in which f_0 maps $*$ to $*$ and f_1 maps $\bullet$ to $\bullet$, is an admissible morphism. We claim that the span $\mathbf{A}' \xleftarrow{\langle id_A, f \rangle} \mathbf{A} \xrightarrow{\langle id_A, f \rangle} \mathbf{A}'$ does not possess a pushout in $\mathtt{Par}(\mathbb{CS})$.

If there would be a pushout of this span of two total morphisms in $\mathtt{Par}(\mathbb{CS})$, then, by Proposition 1, it must coincide with the pushout of them in $\mathbb{CS}$. Since f is an epimorphism in $\mathbb{CS}$ (because all f_j are epimorphims in $\mathbb{G}$), the pushout in $\mathbb{CS}$ must have $p_1 = p_2 = id_{A'}$ as cocone, see the left top square in Fig. 6. The two partial morphisms $\langle m, h \rangle$ and $\langle id_{A'}, id_{A'} \rangle$ let the outer rectangle of partial morphisms commute, i.e.

$$\langle m, h \rangle \circ \langle id_A, f \rangle = \langle id_{A'}, id_{A'} \rangle \circ \langle id_A, f \rangle$$

in $\mathtt{Par}(\mathbb{CS})$, because the pullback object of m and f equals the pullback object of $id_{A'}$ and f in $\mathbb{CS}$. If there would be a unique mediator u, see the dashed line in the diagram, we must have $u = \langle id_{A'}, id_{A'} \rangle$, because the lower rhombus must be commutative. However, for this u the right rhombus fails to be commutative, because $u \circ \langle id_{A'}, p_2 \rangle = \langle id_{A'}, id_{A'} \rangle \neq \langle m, h \rangle$.

This example shows that we cannot expect to have all pushouts in $\mathtt{Par}(\mathbb{CS})$, if we would not require reflection of definedness. And this is true, even if the two morphisms, for which the pushout shall be constructed, are total monomorphisms.

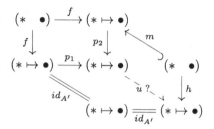

Fig. 6. Counterexample

3.3 Important Properties

Let's now assume all $\mathbb{CS}$-morphisms to reflect definedness.

In the sequel, we will use formulations like "a property is valid *component-wise*" in $\mathbb{CS}$ or some construction "is carried out *componentwise*". Since many of the following considerations are based on this methodology, we give a formalisation: "Pushout", "Pullback", "Monomorphism", "Commutativity" impose truth of a predicate (a certain property) on a diagram in a category $\mathbb{C}$. For pushouts and pullbacks the underlying diagram is a square, for the predicate "Monomorphism" it is a single arrow, for "Commutativity" it is an appropriate triangle of arrows. E.g. $\mathbb{CS}$-morphism $\mathbf{f} : \mathbf{A} \to \mathbf{B}$ is a componentwise monomorphism means that each f_i is a $\mathbb{G}$-monomorphism. More precisely: Given a diagram $\mathcal{D}$ of any of the above mentioned shapes in $\mathbb{CS}$, let $\mathcal{D}_i := \mathcal{C}_i(\mathcal{D})$ with component functor $\mathcal{C}_i$ from Proposition 4, then the predicate p is true componentwise if and only if it is true for $\mathcal{D}_i$ in $\mathbb{G}$ for all $i \in \{0, \ldots, n\}$.

Another formulation is *"componentwise construction* of predicate p", where p is based on a certain universal property. If e.g. p is the predicate for pushouts, componentwise construction of a $\mathbb{CS}$-cospan $\mathbf{C} \xrightarrow{\mathbf{f}'} \mathbf{D} \xleftarrow{\mathbf{g}'} \mathbf{B}$ from a $\mathbb{CS}$-span $\mathbf{C} \xleftarrow{\mathbf{g}} \mathbf{A} \xrightarrow{\mathbf{f}} \mathbf{B}$ consists of two steps: In a first step one constructs pushout cospans $C_i \xrightarrow{f_i'} D_i \xleftarrow{g_i'} B_i$ of spans $\mathcal{C}_i(\mathbf{C} \xleftarrow{\mathbf{g}} \mathbf{A} \xrightarrow{\mathbf{f}} \mathbf{B})$ for each $i \in \{0, \ldots, n\}$. In a second step one tries to define the projections d_j in $\mathbf{D} := (d_j : D_0 \rightharpoonup D_j)_{1 \le j \le n}$, cf. Definition 3, with the help of the pushouts' unique mediators. The cospan morphisms $\mathbf{f}'$ and $\mathbf{g}'$ consist of the respective components $(f_i')_{0 \le i \le n}$ and $(g_i')_{0 \le i \le n}$. The phrase *"p can be constructed componentwise"* then means that the newly constructed object $\mathbf{D}$ is an admissible object according to Definition 3, that the newly created morphisms $\mathbf{f}'$ and $\mathbf{g}'$ are admissible according to Definition 4, and that predicate p holds on the resulting diagram in $\mathbb{CS}$, i.e. the square that arises from enhancing the above $\mathbb{CS}$-span by the $\mathbb{CS}$-cospan yields a pushout in $\mathbb{CS}$. Of course, this procedure applies to other universal constructions in a similar way and after such a construction, we know that property p is valid componentwise.

"Commutatitivity" is valid componentwise by definition, but we also obtain.

Proposition 6 (Componentwise Properties of $\mathbb{CS}$). *Morphism* $\mathbf{f} : \mathbf{A} \to \mathbf{B}$ *is a monomorphism if and only if it is such componentwise.*

Moreover, $\mathbb{CS}$ *has*

1. *all pullbacks*
2. *all pushouts*
3. *all coproducts*

(and is thus cocomplete) and they are constructed componentwise, resp.

Proof. Componentwise validity of monomorphy and componentwise construction of pullbacks have been proved in [11] for so-called S-cartesian functor categories. We showed in [21] (see also [22]) that - for a certain schema category - this functor category is equivalent to $\mathbb{CS}$.

Thus, it remains to prove 2 and 3. For the proof of 2, let a span $(\mathbf{f} : \mathbf{A} \to \mathbf{B}, \mathbf{g} : \mathbf{A} \to \mathbf{C})$ of $\mathbb{CS}$-morphisms be given where $\mathbf{f} = (f_i : A_i \to B_i)_{0 \le i \le n}$ and

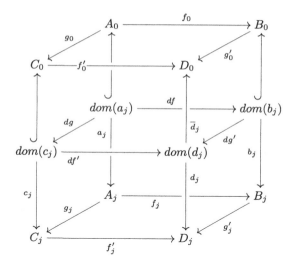

Fig. 7. Two commutative cubes

$\mathbf{g} = (g_i : A_i \to B_i)_{0 \le i \le n}$. Resolving these two morphisms into a triple of $\mathbb{G}$-morphisms for each $j \in \{1, \dots, n\}$ as in Fig. 5 and constructing pushouts componentwise in $\mathbb{G}$, i.e. for f_0 and g_0, f_j and g_j, and for the span of resulting domain mappings df and dg, see the dashed arrow in Fig. 5, yields two cubes on top of each other for each $j \in \{1, \dots, n\}$, cf. Fig. 7, in which the vertical front right arrows d_j and $\overline{d}_j$ are unique mediators w.r.t. the middle pushout. Because $\mathbb{G}$ is a hereditary pushout category by Proposition 2 and because the top face in the upper cube in Fig. 7 is a $\mathbb{G}$-pushout and the two back faces are pullbacks (cf. Definition 4 and the definition of composition of partial arrows on page 4), the prerequisite of the equivalent characterization of hereditaryness in Proposition 3 are fulfilled. Hence the fact that the middle layer in Fig. 7 is also a pushout (by construction) implies that the two upper front faces become pullbacks and the vertical upward arrow $\overline{d}_j$ in the front right can be chosen to be an inclusion arrow. This shows that the componentwise construction indeed yields an admissible comprehensive system

$$\mathbf{D} := (D_0 \xrightarrow{d_j} D_j)_{1 \le j \le n}$$

and a commutative square $(\mathbf{f} : \mathbf{A} \to \mathbf{B}, \mathbf{g}' : \mathbf{B} \to \mathbf{D}, \mathbf{g} : \mathbf{A} \to \mathbf{C}, \mathbf{f}' : \mathbf{C} \to \mathbf{D})$ in $\mathbb{CS}$. It remains to show that it is also a pushout.

Let for this a $\mathbb{CS}$-object $\mathbf{Z} := (Z_0 \xrightarrow{z_j} Z_j)_{1 \le j \le n}$ and two $\mathbb{CS}$-morphisms $\mathbf{h} : \mathbf{B} \to \mathbf{Z}$ and $\mathbf{k} : \mathbf{C} \to \mathbf{Z}$ be given such that $\mathbf{h} \circ \mathbf{f} = \mathbf{k} \circ \mathbf{g}$. Then componentwise considerations easily yield unique $\mathbf{u} := (u_i : D_i \to Z_i)_{0 \le i \le n}$ factoring through the components of $\mathbf{h}$ and $\mathbf{k}$, resp., see Fig. 8, which shows the situation involving $\mathbf{h}$ only. It is easy to see that universality of d_j and $\overline{d}_j$ yield commutativity of all squares in Fig. 8, such that it remains to show that $\mathbf{u}$ is a $\mathbb{CS}$-morphism. For this we use the equivalent characterization in Proposition 5, in particular we have to show (2) for $\mathbf{u}$. Let for this $x \in D_0$ be given. It is well known that pushouts in $\mathbb{G}$ yield jointly surjective cospans, i.e. x has a preimage y in C_0 or in B_0, cf. again Fig. 7. Assume w.l.o.g. that there is $y \in B_0$ and $g_0'(y) = x$ (the case, where there is a preimage in C_0, is similar). Then again using Proposition 5 several times yields

$$
\begin{aligned}
d_j(x) \text{ is defined} &\iff b_j(y) \text{ is defined} && (\text{because } \mathbf{g'} : \mathbf{B} \to \mathbf{D} \in Arr(\mathbb{CS})) \\
&\iff z_j(h_0(y)) \text{ is defined} && (\mathbf{h} : \mathbf{B} \to \mathbf{Z} \in Arr(\mathbb{CS}), \text{ see Fig. 8}) \\
&\iff z_j(u_0(x)) \text{ is defined} && (h_0 = u_0 \circ g_0' \text{ and } x = g_0'(y)),
\end{aligned}
$$

which shows that $\mathbf{u}$ is a $\mathbb{CS}$-morphism.[4]

The proof of the existence of coproducts is similar: Let $(\mathbf{A}^k := (A_0^k \xrightarrow{a_j^k} A_0^k)_{1 \leq j \leq n})_{k \in I}$ be a family of comprehensive systems indexed over some (possibly infinite) index set I. It is then easy to see that

$$
\mathbf{A} := (\coprod_{k \in I} A_0^k \xrightarrow{\coprod_{k \in I} a_j^k} \coprod_{k \in I} A_j^k)_{1 \leq j \leq n}
$$

is the coproduct of them, where $\coprod_{k \in I} A_i^k$ denotes $\mathbb{G}$-coproducts (hence the $\mathbb{CS}$-coproduct is constructed componentwise).

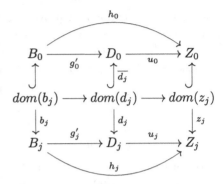

Fig. 8. Admissibility of u

For each j the partial morphism $\coprod_{k \in I} a_j^k$ is defined to be equal to a_j^r on each A_0^r. The unique mediator for a family $(\mathbf{f}^k : \mathbf{A}^k \to \mathbf{B})$ can be shown to be a $\mathbb{CS}$-morphism by similar arguments as above for $\mathbf{u}$. It is well-known that all colimits can be constructed from binary pushouts and coproducts [1], hence $\mathbb{CS}$ is indeed cocomplete. $\qquad\square$

The equivalent characterization of hereditaryness in Proposition 3 uses the predicates pushout, pullback, monomorphism, and commutativity, of which we have shown that validity in $\mathbb{CS}$ is equivalent to componentwise validity. By jumping back and forth from a comprehensive system to its components, this yields

Corollary 1. $\mathbb{CS}$ is a hereditary pushout category. $\qquad\square$

Although it is not the focus of this paper, we mention another important consequence for the application of graph transformations in $\mathbb{CS}$:

Corollary 2. $\mathbb{CS}$ is a weak adhesive HLR category [4] w.r.t. the class of all monomorphisms.

Proof. Heindel proves in [8], Prop. 8.1 that this conclusion can be drawn from Corollary 1, if pushouts are always stable under pullbacks, i.e. the implication

---

4 A more general proof has been given in [11], if $\mathbb{G}$ is a (variant of an) adhesive category, such that the result carries over to these base structures, as well.

"top face pushout, all side faces pullbacks $\Rightarrow$ bottom face pushout" holds for all choices of vertical morphisms in Fig. 3. But this implication is true in $\mathbb{CS}$ by Proposition 6 and because this holds in $\mathbb{G}$ [6]. □

This corollary guarantees validity of the classical theorems for DPO rewriting such as Local Church Rosser, Parallelism, or Local Confluence Theorem to hold in $\mathbb{CS}$, as well.[5]

4   The Partial Map Category of Comprehensive Systems Admits All Pushouts

The goal of this section is to prove that SPO rewriting is well possible for comprehensive systems $\mathbb{CS}$ by showing that the category $\mathbf{Par}(\mathbb{CS})$ possesses all pushouts. This will follow mainly from a result of Hayman and Heindel:

Proposition 7 (Existence of Pushouts of Partial Maps, [7]). *Let $\mathbb{C}$ be a category with pullbacks in which for each span $C \xleftarrow{g} A \xrightarrow{f} B$ of morphisms there is a cospan $C \xrightarrow{f'} D \xleftarrow{g'} B$ making the resulting square commutative. $\mathbf{Par}(\mathbb{C})$ has all pushouts if and only if $\mathbb{C}$ is a hereditary pushout category and inverse image functions have upper adjoints.* □

Section 4.1 is devoted to define inverse image functions and upper adjoints and Sect. 4.2 will show that they exist in $\mathbb{CS}$.

4.1   Upper Adjoints in General ...

Let in a category $\mathbb{C}$ with pullbacks an object A be given. There is the semilattice $Sub(A)$ of subobjects of A, which consists of all monomorphisms $m : M \rightarrowtail A$ modulo the equivalence relation $m \equiv (m' : M' \rightarrowtail A)$, where $m \equiv m'$, if and only if there is an isomorphism $i : M \to M'$ such that $m' \circ i = m$. In the sequel, subobjects will be denoted by small letters $m, n, \ldots$ and we implicitly assume their domain to be the corresponding upper case letter $M, N, \ldots$.

The following definitions are well-known from the literature, e.g. [6]. We write $m \sqsubseteq m'$, if there is a (necessarily unique) morphism $f : M \to M'$ with $m' \circ f = m$. In such a way $\sqsubseteq$ becomes a partial order and in fact a semilattice with meets, because $\mathbb{C}$ has pullbacks and the meet of m and m' is the diagonal in the pullback of m and m'. Furthermore one says that $\mathbb{C}$ *has images*, if for each $\mathbb{C}$-morphism $f : X \to A$ there is a least (w.r.t. $\sqsubseteq$) $m \in Sub(A)$ such that $f = m \circ e$ for some $e : X \to M$ and we write $Im(f) := M$. Finally, if $\mathbb{C}$ has coproducts, then, for a family $(m_k)_{k \in I}$ of subobjects of A, $\bigcup_{k \in I} m_k$ denotes the image of $[m_k]_{k \in I} : \coprod_{k \in I} M_k \to A$, the latter being the unique coproduct mediator of the family $(m_k : M_k \to A)_{k \in I}$. Thus, in particular, for all $p \in I$

$$m_p \sqsubseteq \bigcup_{k \in I} m_k. \tag{3}$$

---

[5] Whereas we obtain this result as a corollary from hereditariness, it is proved directly for underlying adhesive categories in [11].

Definition 5 (Inverse Images and Upper Adjoints). *[7] Let $f : A \to B$ be given in a category $\mathbb{C}$ with pullbacks. We denote by $f^{-1} : Sub(B) \to Sub(A)$ the inverse image function which assigns to $m \in Sub(B)$ its pullback along f.[6] A monotone[7] function $\mathcal{U} : Sub(A) \to Sub(B)$ is called an upper adjoint of f^{-1}, if for all $n \in Sub(A)$ and $m \in Sub(B)$:*

$$f^{-1}(m) \sqsubseteq n \iff m \sqsubseteq \mathcal{U}(n) \tag{4}$$

Note that $\mathcal{U}$ is unique, if it exists [7], and that f^{-1} is monotone, since pulling back (between comma categories) is functorial and preserves monomorphisms.

4.2   ...and in $\mathbb{CS}$

Proposition 8. *$\mathbb{CS}$ has images and the pullback functors preserve them.*

Proof. Let $\mathbf{f} : \mathbf{A} \to \mathbf{A}' = (A_i \xrightarrow{f_i} A'_i)_{0 \leq i \leq n}$ be a $\mathbb{CS}$-arrow. We use $\mathbb{G}$'s epi-mono-factorizations [6] to decompose f_0 and $(f_j)_{1 \leq j \leq n}$ in Fig. 5 accordingly. In particular $f_0 =: m_0 \circ e_0$. Then the pullback of m_0 and $\sqsubseteq'_j$ and its unique mediator u w.r.t. df and $e_0 \circ \sqsubseteq_j$ yields the situation in Fig. 9a, where the left upper square is a pullback by the pullback decomposition lemma.

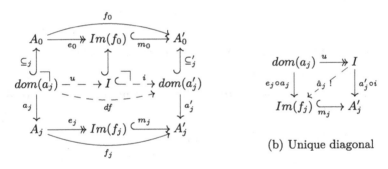

(a) Images componentwise

(b) Unique diagonal

Fig. 9. Images in $\mathbb{CS}$

In $\mathbb{G}$ pullbacks preserve epimorphisms, i.e. u is an epimorphism and the square in Fig. 9b has a unique diagonal [6] $\hat{a}_j : I \to Im(f_j)$, such that everything commutes. Adding this diagonal in Fig. 9a yields $\mathbf{Im}(\mathbf{f}) := (Im(f_0) \xrightarrow{\hat{a}_j} Im(f_j))_{1 \leq j \leq n}$ and the inclusion $\mathbf{Im}(\mathbf{f}) \hookrightarrow \mathbf{A}'$. Moreover, $\mathbf{Im}(\mathbf{f})$ can be shown to

---

[6] Since we are working with equivalence classes, is easy to see that f^{-1} is independent of the choice of representative of m and independent of the choice of pullback.

[7] A function $U : (X, \leq_X) \to (Y, \leq_Y)$ between two partially ordered sets is called monotone, if it preserves the order, i.e. $\forall x, x' \in X : x \leq_X x' \Rightarrow U(x) \leq_Y U(x')$.

be the image of $\mathbf{f} : \mathbf{A} \to \mathbf{A}'$ in $\mathbb{CS}$, because it was set up by componentwise epi-mono-factorization (in $\mathbb{G}$), in which the mono-part is componentwise the least subobject of the respective codomains of f_0 and f_j.

Pullback functors preserve images in $\mathbb{CS}$ because of the essential uniqueness of epi-mono-factorisations, of preservation of monomorphisms and epimorphisms [6] under pullbacks in $\mathbb{G}$, and componentwise pullback construction (cf. Proposition 6). □

Proposition 9 (Upper Adjoints in $\mathbb{CS}$). *Let $\mathbf{f} : \mathbf{A} \to \mathbf{B}$ and $\mathbf{n} \in Sub(\mathbf{A})$, then $\mathcal{U}(\mathbf{n}) := \bigcup\{\mathbf{m} \in Sub(\mathbf{B}) \mid f^{-1}(\mathbf{m}) \sqsubseteq \mathbf{n}\}$ is the upper adjoint of f^{-1}.*

Proof. To prove that $\mathcal{U}$ is monotone, assume $\mathbf{n}, \mathbf{n}' \in Sub(\mathbf{A})$ with $\mathbf{n} \sqsubseteq \mathbf{n}'$. Hence $X := \{\mathbf{m} \in Sub(\mathbf{B}) \mid f^{-1}(\mathbf{m}) \sqsubseteq \mathbf{n}\} \subseteq \{\mathbf{m} \in Sub(\mathbf{B}) \mid f^{-1}(\mathbf{m}) \sqsubseteq \mathbf{n}'\} =: X'$ and thus there is the mediator $\mathbf{u} : \coprod_{\mathbf{m} \in X} dom(\mathbf{m}) \to \coprod_{\mathbf{m} \in X'} dom(\mathbf{m})$, such that $\mathcal{U}(\mathbf{n}')$ becomes a factor in a decomposition of $\coprod_{\mathbf{m} \in X} dom(\mathbf{m}) \to A$. Since $\mathcal{U}(\mathbf{n})$ is the least of these factors, we obtain $\mathcal{U}(\mathbf{n}) \sqsubseteq \mathcal{U}(\mathbf{n}')$.

In equivalence (4) "$\Rightarrow$" follows immediately from (3), such that it remains to prove "$\Leftarrow$". For this it is sufficient to show $f^{-1}(\mathcal{U}(\mathbf{n})) \sqsubseteq n$ for all $\mathbf{n} \in Sub(\mathbf{A})$, because f^{-1} is monotone. Let for this $\mathbf{n} \in Sub(\mathbf{A})$ be arbitrary, $\mathcal{U}(\mathbf{n}) := \bigcup_{k \in I} \mathbf{m_k} : J \hookrightarrow \mathbf{B}$, and fix some $p \in I$. Then there is the coproduct injection $\mathbf{i_p} : \mathbf{M_p} \to \coprod_{k \in I} \mathbf{M_k}$ and by the definition of $\bigcup_{k \in I} \mathbf{m_k}$ in Sect. 4.1, we obtain the diagram

$$\mathbf{M_p} \xrightarrow[\mathbf{i_p}]{} \coprod_{k \in I} \mathbf{M_k} \xrightarrow{e} J \xrightarrow[\bigcup_{k \in I} \mathbf{m_k}]{} \mathbf{B} \qquad (5)$$

with the arc m_p from $\mathbf{M_p}$ to $\mathbf{B}$.

which is mapped by f^{-1} (interpreted as pullback functor) to the upper part of the following diagram:

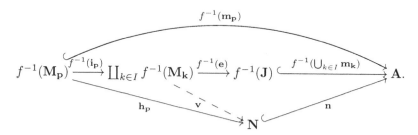

In this diagram, $\mathbf{h_p}$ exists with $\mathbf{n} \circ \mathbf{h_p} = f^{-1}(\mathbf{m_p})$, because $p \in I$ and thus $f^{-1}(\mathbf{m_p}) \sqsubseteq \mathbf{n}$ by the definition of $\mathcal{U}$. Note that the coproduct in (5) is mapped to $\coprod_{k \in I} f^{-1}(\mathbf{M_k})$ and $f^{-1}(\mathbf{i_p})$ are the respective coproduct injections, because pullbacks preserve coproducts in $\mathbb{G}$ and both are constructed componentwise, cf. Proposition 6. We obtain $\mathbf{v}$ as the unique mediator out of this coproduct w.r.t. all arrows $\{\mathbf{h_p} \mid p \in I\}$, i.e. $\mathbf{v} \circ f^{-1}(\mathbf{i_p}) = \mathbf{h_p}$ and hence for all $p \in I$: $\mathbf{n} \circ \mathbf{v} \circ f^{-1}(\mathbf{i_p}) = f^{-1}(\mathbf{m_p}) = f^{-1}(\bigcup_{k \in I} \mathbf{m_k}) \circ f^{-1}(\mathbf{e}) \circ f^{-1}(\mathbf{i_p})$, the last equality by functoriality of f^{-1}. By universality of coproducts this yields $\mathbf{n} \circ \mathbf{v} = f^{-1}(\bigcup_{k \in I} \mathbf{m_k}) \circ f^{-1}(\mathbf{e})$.

By Proposition 8, the latter term in this equation is the image factorisation of $\mathbf{n} \circ \mathbf{v}$ and hence $f^{-1}(\mathcal{U}(n)) = f^{-1}(\bigcup_{k \in I} \mathbf{m_k}) \sqsubseteq \mathbf{n}$, the former being the least, the latter being some subobject of $\mathbf{A}$ factoring through $\mathbf{n} \circ \mathbf{v}$.   □

4.3   The Main Theorem

Theorem 1. *Par(*$\mathbb{CS}$*) has all pushouts.*

Proof. Because $\mathbb{CS}$ has all pushouts by Proposition 6 and thus span-completions, this follows from Proposition 7, Corollary 1 and Proposition 9.   □

Due to space limitations we can not provide an example in which the full power of SPO rewriting compared to the DPO approach can be demonstrated. Instead we provide a simple example, which reveals one additional helpful aspect of our definition of $\mathbb{CS}$-morphisms. The proper construction of arbitrary pushouts in Par($\mathbb{CS}$) (with non-injective rules and/or partial matches) is elaborated in [7].

Assume that a comprehensive system is in state $\mathbf{G}$ (the bottom left system in Fig. 10) which is apparently inconsistent w.r.t $\mathbf{IMC}$ in Sect. 1, because the person in $G_{2/3}$ possesses no phone. If it turns out that this person is the same as the one in G_1, we can restore global consistency by applying rule $\rho : \mathbf{L} \rightharpoonup \mathbf{R} \in Arr(\text{Par}(\mathbb{CS}))$, which deletes a binary commonality (dashed) among two person records and adds a new ternary commonality which comprises these two records. The application of the rule yields a comprehensive system $\mathbf{H}$ (bottom right), which satisfies $\mathbf{IMC}$. Despite being relatively simple, this example demonstrates an advantage of our approach: We do not need *Negative Application Conditions* to prevent repeated application of the rule, because there is no longer a corresponding match of $\mathbf{L}$ into $\mathbf{H}$. This is the case, because l_1

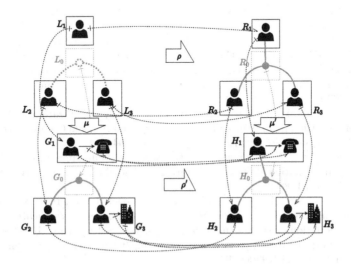

Fig. 10. SPO rule application

(the projection of L_0 into L_1, cf. Definition 3) is undefined on the only element in L_0, but the respective projection h_1 in **H** is defined on the hypothetically matched element (in H_0), i.e. the required reflection of definedness is violated, cf. Definition 4.

5 Related and Future Work

The best reference for *Single Pushout Rewriting* is [12], see also [2]. *Comprehensive systems* are basically a functor category invented in [20] and generalized in [11], its basic ideas originating from the theory of triple graphs [18]. Pushouts in partial map categories and especially hereditariness of colimits have been thoroughly investigated in [7,8].

Our approach still lacks the proof that it is practically applicable, but we hope that SPO rules can serve as a basis for repair rules [15] in order to maintain consistency of informationally overlapping multimodels. We must also find an appropriate way of SPO rule typing, which can not rely on pure slice categories, because a typing morphism should not be forced to reflect definedness. And there should be a thorough characterisation of conflict-freeness in $\mathbb{CS}$, which is difficult, because upper adjoints in $\mathbb{CS}$ are not constructed componentwise. The situation is as in the following quotation: "The contents of this [paper] should rather be considered a starting point ... than the final document of this research issue"[8].

Acknowledgment. We want to thank Tobias Heindel for pointing to the basic theoretical results needed to produce the results in this paper. Moreover, we would like to thank the referees for their constructive criticism.

References

1. Adámek, J., Herrlich, H., Strecker, G.E.: Abstract and Concrete Categories. KAT-MAT: University Bremen (2004). http://katmat.math.uni-bremen.de/acc/acc.pdf
2. Burmeister, P., Monserrat, M., Rosselló, F., Valiente, G.: Algebraic transformation of unary partial algebras II: single-pushout approach. Theor. Comput. Sci. **216**(1–2), 311–362 (1999)
3. Diskin, Z., König, H., Lawford, M.: Multiple model synchronization with multiary delta lenses with amendment and K-Putput. Formal Aspects Comput. **31**(5), 611–640 (2019). https://doi.org/10.1007/s00165-019-00493-0
4. Ehrig, H., Ehrig, K., Prange, U., Taentzer, G.: Fundamentals of Algebraic Graph Transformation. MTCSAES. Springer, Heidelberg (2006). https://doi.org/10.1007/3-540-31188-2
5. Feldmann, S., Kernschmidt, K., Wimmer, M., Vogel-Heuser, B.: Managing inter-model inconsistencies in model-based systems engineering: application in automated production systems engineering. J. Syst. Softw. **153**, 105–134 (2019). https://doi.org/10.1016/j.jss.2019.03.060. http://www.sciencedirect.com/science/article/pii/S0164121219300639

---

[8] This is an almost identical citation of the last statement in Michael's PhD Thesis!

6. Goldblatt, R.: Topoi: The Categorial Analysis of Logic. Dover Publications, New York (1984)
7. Hayman, J., Heindel, T.: On pushouts of partial maps. In: Giese, H., König, B. (eds.) ICGT 2014. LNCS, vol. 8571, pp. 177–191. Springer, Cham (2014). https://doi.org/10.1007/978-3-319-09108-2_12
8. Heindel, T.: A category theoretical approach to the concurrent semantics of rewriting: adhesive categories and related concepts. Ph.D. thesis, University of Duisburg-Essen (2009). http://duepublico.uni-duisburg-essen.de/servlets/DerivateServlet/Derivate-24329/diss.pdf
9. Heindel, T.: Hereditary pushouts reconsidered. In: Ehrig, H., Rensink, A., Rozenberg, G., Schürr, A. (eds.) ICGT 2010. LNCS, vol. 6372, pp. 250–265. Springer, Heidelberg (2010). https://doi.org/10.1007/978-3-642-15928-2_17
10. Kennaway, R.: Graph rewriting in some categories of partial morphisms. In: Ehrig, H., Kreowski, H.-J., Rozenberg, G. (eds.) Graph Grammars 1990. LNCS, vol. 532, pp. 490–504. Springer, Heidelberg (1991). https://doi.org/10.1007/BFb0017408
11. Kosiol, J., Fritsche, L., Schürr, A., Taentzer, G.: Adhesive subcategories of functor categories with instantiation to partial triple graphs. In: Guerra, E., Orejas, F. (eds.) ICGT 2019. LNCS, vol. 11629, pp. 38–54. Springer, Cham (2019). https://doi.org/10.1007/978-3-030-23611-3_3
12. Löwe, M.: Extended algebraic graph transformation. Ph.D. thesis, Technical University of Berlin, Germany (1991). http://d-nb.info/910935696
13. Löwe, M., Tempelmeier, M.: Single-pushout rewriting of partial algebras. In: Plump, D. (ed.) Proceedings of GCM Co-located with ICGT/STAF, L'Aquila, Italy. CEUR Workshop Proceedings, vol. 1403, pp. 82–96 (2015). http://ceur-ws.org/Vol-1403/paper7.pdf
14. Löwe, M.: Algebraic approach to single-pushout graph transformation. Theoret. Comput. Sci. **109**(1), 181–224 (1993)
15. Rabbi, F., Lamo, Y., Yu, I.C., Kristensen, L.M.: A diagrammatic approach to model completion. In: Dingel, J., Kokaly, S., Lucio, L., Salay, R., Vangheluwe, H. (eds.) Proceedings of the 4th Workshop on the Analysis of Model Transformations Co-located with MODELS 2015, Ottawa, Canada. CEUR Workshop Proceedings, vol. 1500, pp. 56–65. CEUR-WS.org (2015). http://ceur-ws.org/Vol-1500/paper7.pdf
16. Raoult, J.: On graph rewritings. Theor. Comput. Sci. **32**, 1–24 (1984). https://doi.org/10.1016/0304-3975(84)90021-5
17. Robinson, E., Rosolini, G.: Categories of partial maps. Inf. Comput. **79**(2), 95–130 (1988)
18. Schürr, A.: Specification of graph translators with triple graph grammars. In: Mayr, E.W., Schmidt, G., Tinhofer, G. (eds.) WG 1994. LNCS, vol. 903, pp. 151–163. Springer, Heidelberg (1995). https://doi.org/10.1007/3-540-59071-4_45
19. Stevens, P.: Bidirectional transformations in the large. In: MODELS 2017, pp. 1–11 (2017)
20. Stünkel, P., König, H., Lamo, Y., Rutle, A.: Multimodel correspondence through inter-model constraints. In: Conference Companion of the 2nd International Conference on Art, Science, and Engineering of Programming, Nice, France, 09–12 April 2018, pp. 9–17 (2018). http://doi.acm.org/10.1145/3191697.3191715
21. Stünkel, P., König, H., Lamo, Y., Rutle, A.: Towards multiple model synchronization with comprehensive systems. In: Wehrheim, H., Cabot, J. (eds.) FASE 2020. LNCS, vol. 12076, pp. 335–356. Springer, Cham (2020). https://doi.org/10.1007/978-3-030-45234-6_17

22. Stünkel, P., König, H., Lamo, Y., Rutle, A.: Towards multiple model synchronization with comprehensive systems: extended version. Technical report, University of Applied Sciences, FHDW Hannover (2020). https://fhdwdev.ha.bib.de/public/papers/02020-01.pdf
23. Trollmann, F., Albayrak, S.: Extending model synchronization results from triple graph grammars to multiple models. In: Van Van Gorp, P., Engels, G. (eds.) ICMT 2016. LNCS, vol. 9765, pp. 91–106. Springer, Cham (2016). https://doi.org/10.1007/978-3-319-42064-6_7

Initial Conflicts for Transformation Rules with Nested Application Conditions

Leen Lambers[1]($\boxtimes$)(iD) and Fernando Orejas[2](iD)

[1] Hasso Plattner Institute, University of Potsdam, Potsdam, Germany
leen.lambers@hpi.de
[2] Universitat Politècnica de Catalunya, Barcelona, Spain
orejas@lsi.upc.edu

Abstract. We extend the theory of initial conflicts in the framework of $\mathcal{M}$-adhesive categories to transformation rules with ACs. We first show that for rules with ACs, conflicts are in general neither inherited from a bigger context any more, nor is it possible to find a finite and complete subset of finite conflicts as illustrated for the category of graphs. We define initial conflicts to be special so-called symbolic transformation pairs, and show that they are minimally complete (and in the case of graphs also finite) in this symbolic way. We show that initial conflicts represent a proper subset of critical pairs again. We moreover demonstrate that (analogous to the case of rules without ACs) for each conflict a unique initial conflict exists representing it. We conclude with presenting a sufficient condition illustrating important special cases for rules with ACs, where we do not only have initial conflicts being complete in a symbolic way, but also find complete (and in the case of graphs also finite) subsets of conflicts in the classical sense.

Keywords: Graph transformation · Critical pairs · Conflicts

1 Introduction

Detecting and analyzing conflicts is an important issue in software analysis and design, which has been addressed successfully using powerful techniques from graph transformation (see, e.g., [12,15,17,24]), most of them based on critical pair analysis. The *power of critical pairs* is a consequence of the fact that: a) they are complete, in the sense that they represent all conflicts; b) there is a finite number of them; and c) they can be computed statically. The main problem is that their computation has exponential complexity in the size of the preconditions of the rules. For this reason, some significantly smaller subsets of critical pairs that are still complete have been defined [1,19,21], clearing the way for a more efficient computation. In particular, recently, in [19], a new approach

F. Orejas has been supported by the Salvador de Madariaga grant PRX18/00308 and by funds from the Spanish Research Agency (AEI) and the European Union (FEDER funds) under grant GRAMM (ref. TIN2017-86727-C2-1-R).

F. Gadducci and T. Kehrer (Eds.): ICGT 2020, LNCS 12150, pp. 109–127, 2020.
https://doi.org/10.1007/978-3-030-51372-6_7

Table 1. Critical pairs versus initial conflicts

	Plain rules	Rules with NACs	Rules with ACs
Critical pairs (CPs)	Subset of conflicts, complete [27]	Subset of conflicts, complete [17,20]	Symbolic, complete [6,9]
Initial conflicts	Subset of conflicts, min. complete, proper subset of CPs [1,19]	Symbolic (Definition 10), min. complete, regular (Theorem 6) & conservative extension of CPs (Theorem 7)	Symbolic (Definition 10), min. complete, proper subset of CPs (Theorem 3)

for conflict detection was introduced based on a different intuition. Instead of considering conflicts in a minimal context, as for critical pairs, we used the notion of initiality to characterize a complete set of minimal conflicts, showing that *initial conflicts* form a strict subset of critical pairs. In particular, we have that every conflict is represented by a unique initial conflict, as opposed to the fact that each conflict may be represented by many critical pairs.

Most of the work on critical pairs only applies to *plain* graph transformation systems, i.e. transformation systems with unconditional rules. Nevertheless, in practice, we often need to limit the application of rules, defining some kind of *application conditions* (ACs). In this sense, in [17,20] we defined critical pairs for rules with negative application conditions (NACs), and in [6,9] for the general case of ACs, where conditions are as expressive as arbitrary first-order formulas on graphs. However, to our knowledge, no work has addressed up to now the problem of finding significantly smaller subsets of critical pairs for this kind of rules. In this paper we generalize the theory of initial conflicts to rules with ACs in the framework of $\mathcal{M}$-adhesive transformation systems. In particular, the main contributions of this paper (as summarized in Table 1) are:

- The definition of the *notion of initial conflict for rules with ACs*, based on a notion of *symbolic transformation pair*, showing that the set of initial conflicts is a *proper subset* of the set of critical pairs and that it is *minimally complete*[1], in the sense that, no smaller set of symbolic transformation pairs exists that is also complete. In particular, the *cardinality* of the set of initial conflicts is, at most, the cardinality of the set of initial conflicts for the plain case, when disregarding the ACs, plus one. Moreover, as in the plain case, every conflict is an instance of a *unique* initial conflict.
- The identification of a class of *regular* initial conflicts that demonstrate a certain kind of regularity in their application conditions. This allows us to unfold them into a *complete (and in the case of graphs also finite) subset of conflicts*. In particular, we show that, in the case of rules with NACs, initial conflicts are regular, implying that our initial conflicts represent a *conservative extension* of the critical pair theory for rules with NACs.

The paper is organized as follows. We describe *related work* in Sect. 2 and, in Sect. 3, we present some *preliminary material*, where we also include some new

---

[1] Provided that the considered category has initial conflicts for the plain case.

results. More precisely, in Subsect. 3.1 and Subsect. 3.2 we briefly reintroduce the framework of $\mathcal{M}$-adhesive categories and of rules with ACs; in Subsect. 3.3 we reintroduce critical pairs for rules with ACs following [6,9]; in Subsect. 3.4 we reintroduce initial conflicts for plain rules, and in Sect. 4 we introduce initial parallel independent transformation pairs. This result is used in Sect. 4, where we present the main results of the paper about *initial conflicts for rules with ACs*. Then, in Sect. 5 we show our results on *unfolding initial conflicts*. Finally, we conclude in Sect. 6 discussing some future work. Detailed proofs can be found in the full version of the paper [23].

2 Related Work

Most work on checking *confluence* for rule-based rewriting systems is based on the seminal paper from Knuth and Bendix [14], who reduced the problem of checking local confluence to checking the joinability of a finite set of *critical pairs*. This technique has been extensively studied in the area of term rewriting systems (see, e.g., [25]), and it was introduced in the area of *graph transformation* by Plump [27] in the context of term-graph and hypergraph rewriting. Moreover, he also proved that (local) confluence of graph transformation systems is undecidable, even for terminating systems, as opposed to what happens in the area of term rewriting systems. However, recently, in [2] it is shown that confluence of terminating DPO transformation of graphs with interfaces is decidable. The authors explain that the reason is that interfaces play the same role as variables in term rewriting systems, where confluence is undecidable for terminating ground (i.e., without variables) systems, but decidable for non-ground ones.

Computing critical pairs in graph transformation, as introduced by Plump [27], is exponential in the size of the preconditions of the rules. For this reason, different *proper subsets of critical pairs* with a considerably reduced size were studied that are still complete [1,19,21], clearing the way for a more efficient computation. The notion of *essential critical pair* [21] for graph transformation systems already allowed for a significant reduction, and, the notion of *initial conflict* [19], introduced for the more general $\mathcal{M}$-adhesive systems, allowed for an even larger reduction. However, not all $\mathcal{M}$-adhesive categories have initial conflicts. In [19] it is shown that typed graphs do have them and [1] extended that result proving that arbitrary $\mathcal{M}$-adhesive categories satisfying some given conditions also have initial conflicts.

A recent line of work concentrates on the development of *multi-granular conflict detection techniques* [3,18,24]. An extensive literature survey shows [24] that conflict detection is used at different levels of granularity depending on its application field. The overview shows that conflict detection can be used for the analysis and design phase of software systems (e.g. for finding inconsistencies in requirement specifications), for model-driven engineering (e.g. supporting model version management), for testing (e.g. generation of interesting test cases), or for optimizing rule-based computations (e.g. avoiding backtracking). These multi-granular techniques are presented for rules without application conditions (ACs).

Our work builds further foundations for providing multi-granular techniques also in the case of rules with ACs in the future.

The use of (negative) *application conditions*, to limit the application of graph transformation rules, was introduced in [8,10], while the more general approach, using *nested conditions*, was introduced by Habel and Penneman [11]. Checking confluence for graph transformation systems with application conditions (ACs) has been studied in [17,20] for the case of negative application conditions (NACs), and in [6,9] for the more general case of ACs. However, it is an open issue to find proper subsets of critical pairs of considerably reduced size in the general case.

3   Preliminaries

We start with a brief introduction of $\mathcal{M}$-adhesive categories, *rules with nested application conditions (ACs)* (cf. Subsect. 3.2), and the main parts of *critical pair theory* for this type of rules [6,9] (cf. Subsect. 3.3). Thereafter, we reintroduce the notion of *initial conflicts* [19] for *plain* rules, i.e. rules without ACs (cf. Subsect. 3.4). We also introduce the notion of *initial parallel independent transformation pairs* as a counterpart (cf. Subsect. 3.5), since it will play a particular role when defining initial conflicts for rules with ACs in Subsect. 3.4. We assume that the reader is acquainted with the basic theory of DPO graph transformation and, in particular, the standard definitions of typed graphs and typed graph morphisms (see, e.g., [5]) and its associated category, **Graphs**$_{TG}$.

3.1   Graphs and High-Level Structures

Our results do not only apply to a specific class of graph transformation systems, like standard (typed) graph transformation systems, but to systems over any $\mathcal{M}$-adhesive category [5]. The idea behind considering $\mathcal{M}$-adhesive categories is to avoid similar investigations for different instantiations like e.g. Petri nets or hypergraphs. An $\mathcal{M}$-adhesive category is a category $\mathcal{C}$ with a distinguished morphism class $\mathcal{M}$ of monomorphisms satisfying certain properties. The most important one is the van Kampen (VK) property stating a certain kind of compatibility of pushouts and pullbacks along $\mathcal{M}$-morphisms. Moreover, additional properties are needed in our context: initial pushouts, describing the existence of a special "smallest" pushout over a morphism, $\mathcal{E}'$-$\mathcal{M}$ pair factorizations, extending the classical epi-mono factorization to a pair of morphisms with the same codomain. The definitions of these properties can be found in [6,7].

Assumption 1. *We assume that $\langle \mathcal{C}, \mathcal{M} \rangle$ is an $\mathcal{M}$-adhesive category with a unique $\mathcal{E}'$-$\mathcal{M}$ pair factorization (needed for Lemma 1, Definition 5, Theorem 3, Corollary 1) and binary coproducts (needed for Lemma 3, Definition 8, Theorem 1). For the Local Confluence Theorem for initial conflicts of rules with ACs we in addition need initial pushouts (cf. Subsect. 4.4).*

Remark 1. Most categories of structures used for specification are $\mathcal{M}$-adhesive and satisfy these additional properties [5], including the category $\langle \text{Graphs}_{TG}, \mathcal{M} \rangle$ with $\mathcal{M}$ being the class of all injective typed graph morphisms.

3.2 Rules with Application Conditions and Parallel Independence

Nested application conditions [11] (in short, application conditions, or just ACs) generalize the corresponding notions in [4,10,15], where a negative (positive) application condition, short NAC (PAC), over a graph P, denoted $\neg\exists a$ ($\exists a$) is defined in terms of a morphism $a : P \to C$. Informally, a morphism $m : P \to G$ satisfies $\neg\exists a$ ($\exists a$) if there does not exist a morphism $q : C \to G$ extending a to m (if there exists q extending a to m). Then, an AC is either the special condition true or a pair of the form $\exists(a, \mathrm{ac}_C)$ or $\neg\exists(a, \mathrm{ac}_C)$, where ac_C is an additional AC on C. Intuitively, a morphism $m : P \to G$ satisfies $\exists(a, \mathrm{ac}_C)$ if m satisfies a and the corresponding extension q satisfies ac_C. Moreover, ACs may be combined with the usual logical connectors. For a concrete definition of ACs we address the reader to [11] or [6].

ACs are used to restrict the application of rules to a given object. The idea is to equip the precondition (or left hand side) of rules with an application condition. Then we can only apply a given rule to an object G if the corresponding match morphism satisfies the AC of the rule. However, for technical reasons, we also introduce the application of rules *disregarding* the associated ACs.

Definition 1 (rules and transformations). *A rule $\rho = \langle p, \mathrm{ac}_L \rangle$ consists of a plain* rule $p = \langle L \hookleftarrow I \to R \rangle$ *with $I \hookrightarrow L$ and $I \hookrightarrow R$ morphisms in $\mathcal{M}$ and an application condition ac_L over L.*

$$
\begin{array}{ccccc}
\mathrm{ac}_L \blacktriangleright & L & \hookleftarrow I \longrightarrow & R \\
& m \downarrow \ (1) & \downarrow \ (2) & \downarrow m^* \\
& G & \hookleftarrow D \hookrightarrow & H
\end{array}
$$

A direct transformation $t : G \Rightarrow_{\rho,m,m^*} H$ consists of two pushouts (1) and (2), called DPO, with match m and comatch m^* such that $m \models \mathrm{ac}_L$. $G \hookleftarrow D \hookrightarrow H$ is called the derived span *of t. An AC-disregarding direct transformation $G \Rightarrow_{\rho,m,m^*} H$ consists of DPO (1) and (2), where m does not necessarily need to satisfy ac_L. Given a set of rules $\mathcal{R}$ for $\langle \mathcal{C}, \mathcal{M} \rangle$, the triple $\langle \mathcal{C}, \mathcal{M}, \mathcal{R} \rangle$ is an $\mathcal{M}$-adhesive system.*

Remark 2. In the rest of the paper we assume that each rule (resp. transformation or $\mathcal{M}$-adhesive system) comes with ACs. Otherwise, we state that we have a *plain* rule (resp. transformation or $\mathcal{M}$-adhesive system), which can be seen as a special case, in the sense that the ACs are (equivalent to) *true*.

ACs can be shifted over morphisms and rules as shown in the following lemma (for constructions see [7][2] and [7,11], respectively).

---

[2] Since this construction entails the enumeration of jointly epimorphic morphism pairs, its computation has exponential complexity in the size of the precondition of the rule and the size of the AC.

Lemma 1 (shift ACs over morphisms [7]). *There is a transformation* Shift *from morphisms and ACs to ACs such that for each AC,* ac_P, *and each morphism* $b: P \to P'$, Shift *transforms* ac_P *via b into an AC* $\text{Shift}(b, ac_P)$ *over* P' *such that for each morphism* $n: P' \to H$ *it holds that* $n \circ b \models ac_P \Leftrightarrow n \models \text{Shift}(b, ac_P)$.

Lemma 2 (shift ACs over rules [7,11]). *There is a transformation* L *from rules and ACs to ACs such that for every rule* $\rho: L \hookleftarrow I \hookrightarrow R$ *and every AC on* R, ac_R, L *transforms* ac_R *via* ρ *into the AC* $\text{L}(\rho, ac_R)$ *on* L, *such that for every direct transformation* $G \Rightarrow_{\rho, m, m^*} H$, $m \models \text{L}(\rho, ac_R) \Leftrightarrow m^* \models ac_R$.

For *parallel independence*, when working with rules with ACs, we need not only that each rule does not delete any element which is part of the match of the other rule, but also that the resulting transformation defined by each rule application still satisfies the ACs of the other rule application.

Definition 2 (transformation pairs and parallel independence). *A transformation pair* $H_1 \Leftarrow_{\rho_1, o_1} G \Rightarrow_{\rho_2, o_2} H_2$ *is parallel independent if there exists a morphism* $d_{12}: L_1 \to D_2$ *such that* $k_2 \circ d_{12} = o_1$ *and* $c_2 \circ d_{12} \models ac_{L_1}$ *and there exists a morphism* $d_{21}: L_2 \to D_1$ *such that* $k_1 \circ d_{21} = o_2$ *and* $c_1 \circ d_{21} \models ac_{L_2}$.

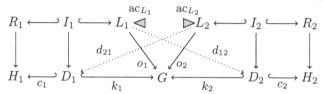

A transformation pair is *in conflict* or *conflicting* if it is parallel dependent. We distinguish different conflict types, generalizing straightforwardly the conflict characterization introduced for rules with NACs [20]. The transformation pair $H_1 \Leftarrow_{\rho_1, o_1} G \Rightarrow_{\rho_2, o_2} H_2$ is a *use-delete* (resp. *delete-use*) conflict if in Definition 2 the commuting morphism d_{12} (resp. d_{21}) does not exist, i.e. the second (resp. first) rule deletes something used by the first (resp. second) one. Moreover, it is an *AC-produce* (resp. *produce-AC*) conflict if in Definition 2 the commuting morphism d_{12} (resp. d_{21}) exists, but an extended match is produced by the second (resp. first) rule that does not satisfy the rule AC of the first (resp. second) rule. If a transformation pair is an *AC-produce* or *produce-AC* conflict, then we also say that it is an *AC conflict* or *AC conflicting*.

Remark 3 (use-delete XOR AC-produce). A use-delete (resp. delete-use) conflict cannot occur simultaneously to an AC-produce (resp. produce-AC) conflict, since the AC of the first (resp. second) rule can only be violated iff there exists an extended match for the first (resp. second) rule. However, a use-delete (resp. delete-use) conflict may occur simultaneously to a produce-AC (resp. AC-produce) conflict.

For grasping the notion of completeness of transformation pairs w.r.t. a property like parallel (in-)dependence, it is first important to understand how a given transformation can be extended to another transformation. In particular, an *extension diagram* describes how a transformation $t\colon G_0 \Rightarrow^* G_n$ can be extended to a transformation $t'\colon G'_0 \Rightarrow^* G'_n$ via the same rules and an *extension morphism* $k_0\colon G_0 \to G'_0$ that maps G_0 to G'_0 as shown in the following diagram on the left. For each rule application and transformation step, we have two double pushout diagrams as shown on the right, where the rule ρ_{i+1} is applied to G_i and G'_i.

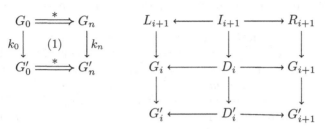

We introduce two notions of completeness, distinguishing $\mathcal{M}$-completeness from regular completeness, depending on the membership of the extension morphism in $\mathcal{M}$. It is known that critical pairs (resp. initial conflicts) for *plain rules* are $\mathcal{M}$-complete (resp. complete) w.r.t. parallel dependence [5,19]. In Subsect. 3.3, we reintroduce the fact that critical pairs for rules with ACs are $\mathcal{M}$-complete w.r.t. parallel dependence, but as symbolic transformation pairs. We learn in Sect. 4 that initial conflicts for rules with ACs are also complete in this symbolic way.

Definition 3 (($\mathcal{M}$-)completeness of transformation pairs). *A set of transformation pairs $\mathcal{S}$ for a pair of rules $\langle \rho_1, \rho_2 \rangle$ is complete (resp. $\mathcal{M}$-complete) w.r.t. parallel (in-)dependence if there is a pair $P_1 \Leftarrow_{\rho_1,o_1} K \Rightarrow_{\rho_2,o_2} P_2$ from $\mathcal{S}$ and an extension diagram via extension morphism m (resp. $m \in \mathcal{M}$) for each parallel (in-)dependent direct transformation pair $H_1 \Leftarrow_{\rho_1,m_1} G \Rightarrow_{\rho_2,m_2} H_2$.*

$$P_1 \xLeftarrow{\rho_1,o_1} K \xRightarrow{\rho_2,o_2} P_2$$
$$\big\downarrow \qquad \big\downarrow m \qquad \big\downarrow$$
$$H_1 \xleftarrow[\rho_1,m_1]{} G \xrightarrow[\rho_2,m_2]{} H_2$$

Fig. 1. ($\mathcal{M}$-)completeness of transformation pairs

3.3 Critical Pairs

Critical pairs for plain rules are just transformation pairs, where morphisms o_1 and o_2 are in $\mathcal{E}'$ (i.e., roughly, K is an overlapping of L_1 and L_2). In the category of **Graphs** they lead to finite and complete subsets of finite conflicts [4] (assumed that the rule graphs are also finite). However, when rules include ACs, we cannot use the same notion of critical pair since, as we show in Theorem 2, in general, for any two rules with ACs, there is no complete set of transformation pairs that is finite. To avoid this problem, our critical pairs for rules with ACs also include ACs, as in [6,9], where they are proved to be $\mathcal{M}$-*complete*, and they are also finite in the category of **Graphs** (assumed again that the rules are finite).

In particular, critical pairs are based on the notion of *symbolic transformation pairs*, which are pairs of *AC-disregarding transformations* on some object K with two special ACs on K. These two ACs, ac_K (*extension AC*) and ac_K^* (*conflict-inducing AC*), are used to characterize which embeddings of this pair, via some morphism $m : K \to G$, give rise to a transformation pair that is parallel dependent. If $m \models \mathrm{ac}_K$, then $m \circ o_1 : L_1 \to G$ and $m \circ o_2 : L_2 \to G$ are two morphisms, satisfying the associated ACs of ρ_1 and ρ_2, respectively. Moreover, if $m \models \mathrm{ac}_K^*$, then the two transformations $H_1 \Leftarrow_{\rho_1, m \circ o_1} G \Rightarrow_{\rho_2, m \circ o_2} H_2$ are parallel dependent. Symbolic transformation pairs allow us to present critical pairs as well as initial conflicts (cf. Subsect. 3.4) in a compact and unified way, since they both are instances of symbolic transformation pairs. Finally, note that each symbolic transformation pair $stp_K : \langle tp_K, \mathrm{ac}_K, \mathrm{ac}_K^* \rangle$ is by definition uniquely determined (up to isomorphism and equivalence of the extension AC and conflict-inducing AC) by its underlying AC-disregarding transformation pair.

Definition 4 (symbolic transformation pair). *Given rules* $\rho_1 = \langle p_1, \mathrm{ac}_{L_1} \rangle$ *and* $\rho_2 = \langle p_2, \mathrm{ac}_{L_2} \rangle$, *a symbolic transformation pair* $stp_K : \langle tp_K, \mathrm{ac}_K, \mathrm{ac}_K^* \rangle$ *for* $\langle \rho_1, \rho_2 \rangle$ *consists of a pair* $tp_K : P_1 \Leftarrow_{\rho_1, o_1} K \Rightarrow_{\rho_2, o_2} P_2$ *of AC-disregarding transformations together with ACs* ac_K *and* ac_K^* *on* K *given by:*

$$\mathrm{ac}_K = \mathrm{Shift}(o_1, \mathrm{ac}_{L_1}) \wedge \mathrm{Shift}(o_2, \mathrm{ac}_{L_2}), \text{ called extension AC, and}$$
$$\mathrm{ac}_K^* = \neg(\mathrm{ac}_{K,d_{12}}^* \wedge \mathrm{ac}_{K,d_{21}}^*), \text{ called conflict-inducing AC}$$

with $\mathrm{ac}_{K,d_{12}}^*$ *and* $\mathrm{ac}_{K,d_{21}}^*$ *given as follows:*

$$\text{if } (\exists \, d_{12} \text{ with } k_2 \circ d_{12} = o_1) \text{ then } \mathrm{ac}_{K,d_{12}}^* = \mathrm{L}(p_2^*, \mathrm{Shift}(c_2 \circ d_{12}, \mathrm{ac}_{L_1}))$$
$$\text{else } \mathrm{ac}_{K,d_{12}}^* = \text{false}$$
$$\text{if } (\exists \, d_{21} \text{ with } k_1 \circ d_{21} = o_2) \text{ then } \mathrm{ac}_{K,d_{21}}^* = \mathrm{L}(p_1^*, \mathrm{Shift}(c_1 \circ d_{21}, \mathrm{ac}_{L_2}))$$
$$\text{else } \mathrm{ac}_{K,d_{21}}^* = \text{false}$$

where $p_1^* = \langle K \xleftarrow{k_1} D_1 \xrightarrow{c_1} P_1 \rangle$ *and* $p_2^* = \langle K \xleftarrow{k_2} D_2 \xrightarrow{c_2} P_2 \rangle$ *are defined by the corresponding double pushouts.*

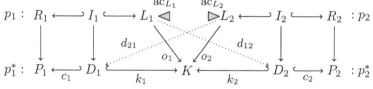

A *critical pair* is now a symbolic transformation pair in a minimal context such that there exists at least one extension to a pair of transformations being parallel dependent.

Definition 5 (critical pair). *Given rules* $\rho_1 = \langle p_1, ac_{L_1} \rangle$ *and* $\rho_2 = \langle p_2, ac_{L_2} \rangle$, *a critical pair for* $\langle \rho_1, \rho_2 \rangle$ *is a symbolic transformation pair* $stp_K : \langle tp_K, ac_K, ac_K^* \rangle$, *where the match pair* (o_1, o_2) *of* tp_K *is in* $\mathcal{E}'$, *and there exists a morphism* $m : K \rightarrow G \in \mathcal{M}$ *such that* $m \models ac_K \wedge ac_K^*$ *and* $m_i = m \circ o_i$, *for* $i = 1, 2$, *satisfy the gluing conditions, i.e.* m_i *has a pushout complement w.r.t.* p_i.

Definition 6 (($\mathcal{M}$-)completeness of symbolic transformation pairs). *A set of symbolic transformation pairs* $\mathcal{S}$ *for a pair of rules* $\langle \rho_1, \rho_2 \rangle$ *is complete (resp. $\mathcal{M}$-complete) w.r.t. parallel dependence if there is a symbolic transformation pair* $stp_K : \langle tp_K : P_1 \Leftarrow_{\rho_1, o_1} K \Rightarrow_{\rho_2, o_2} P_2, ac_K, ac_K^* \rangle$ *from* $\mathcal{S}$ *and an extension diagram as depicted in Fig. 1 with* $m : K \rightarrow G$ *(resp. $m : K \rightarrow G \in \mathcal{M}$) and* $m \models ac_K \wedge ac_K^*$ *for each parallel dependent direct transformation* $H_1 \Leftarrow_{\rho_1, m_1} G \Rightarrow_{\rho_2, m_2} H_2$.

In [6,9] it is shown that the set of *critical pairs* for a pair of rules is $\mathcal{M}$-complete w.r.t. parallel dependence.

3.4   Initial Conflicts for Plain Rules

Initial conflicts for plain rules follow an alternative approach to the original idea of critical pairs. Instead of considering all conflicting transformations in a minimal context (materialized by a pair of jointly epimorphic matches), initial conflicts use the notion of *initiality of transformation pairs* to obtain a more declarative view on the minimal context of critical pairs. Each initial conflict is a critical pair but not the other way round. Moreover, all initial conflicts for plain rules are complete w.r.t. parallel dependence and they still satisfy the Local Confluence Theorem for plain rules. Consequently, initial conflicts for plain rules represent an important, proper subset of critical pairs for performing static conflict detection as well as local confluence analysis.

Definition 7 (initial transformation pair and initial conflict). *Given a pair of plain direct transformations* $tp : H_1 \Leftarrow_{p_1, m_1} G \Rightarrow_{p_2, m_2} H_2$, *then* $tp^I :$ $H_1^I \Leftarrow_{p_1, m_1^I} G^I \Rightarrow_{p_2, m_2^I} H_2^I$ *is an initial transformation pair for* tp *if it can be embedded into* tp *via extension diagrams (1) and (2) and extension morphism* f^I, *as in the left diagram below, such that for each transformation pair* $tp' :$ $H_1' \Leftarrow_{p_1, m_1'} G' \Rightarrow_{p_2, m_2'} H_2'$ *that can be embedded into* tp *via extension diagrams (3) and (4) and extension morphism* f, *as in the left diagram below, it holds that* tp^I *can be embedded into* tp' *via unique extension diagrams (5) and (6) and unique vertical morphism* f'^I *s.t.* $f \circ f'^I = f^I$.

  Given a plain $\mathcal{M}$-adhesive system with initial transformation pairs for conflicts, an initial conflict *is a conflict* tp *isomorphic to* tp^I.

$$H_1^I \xLeftarrow{p_1,m_1^I} G^I \xRightarrow{p_2,m_2^I} H_2^I \qquad H_1^I \xLeftarrow{p_1,m_1^I} G^I \xRightarrow{p_2,m_2^I} H_2^I$$

$$g_1^I \downarrow \quad (1) \; f^I \quad (2) \quad \downarrow g_2^I \qquad g_1'^I \downarrow \quad (5) \; f'^I \quad (6) \quad \downarrow g_2'^I$$

$$H_1 \xLeftarrow{p_1,m_1} G \xRightarrow{p_2,m_1} H_2 \qquad H_1' \xLeftarrow{p_1,m'_1} G' \xRightarrow{p_2,m'_2} H_2'$$

$$g_1 \downarrow \quad (3) \; f \quad (4) \quad \downarrow g_2$$

$$H_1 \xLeftarrow{p_1,m_1} G \xRightarrow{p_2,m_2} H_2$$

The idea of representing all conflicts by a (finite) set of initial conflicts is based on the requirement of the *existence of initial transformation pairs* for parallel dependent or *conflicting* plain transformation pairs. This requirement holds for the category of typed graphs [19] and for any arbitrary $\mathcal{M}$-adhesive category fulfilling some extra conditions [1].

For plain $\mathcal{M}$-adhesive systems, initial conflicts are critical pairs [19]. Moreover, they are complete and minimal as transformation pairs w.r.t. parallel dependence, whereas critical pairs for plain rules are $\mathcal{M}$-complete [4].

3.5   Initial Parallel Independent Transformation Pairs for Plain Rules

In this section, we show the existence of initial transformation pairs for *parallel independent transformation pairs* (Fig. 2), allowing us to define a *complete* subset also w.r.t. parallel independence. The proof requires that binary coproducts exist.

Lemma 3 (existence of initial transformation pair for parallel independent transformation pair). *Given a pair of parallel independent plain direct transformations $tp : H_1 \Leftarrow_{p_1,m_1} G \Rightarrow_{p_2,m_2} H_2$, then $tp_{L_1+L_2} : R_1 + L_2 \Leftarrow_{p_1,i_1} L_1+L_2 \Rightarrow_{p_2,i_2} L_1+R_2$, where $i_1 : L_1 \to L_1+L_2$ and $i_2 : L_2 \to L_1+L_2$ are the coproduct morphisms, is initial for tp.*

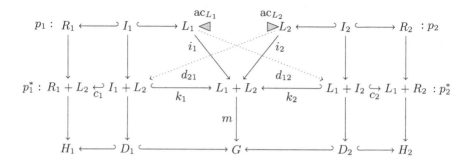

Fig. 2. Initial parallel independent transformation pair $tp_{L_1+L_2}$

Uniqueness of initial transformation pairs up to isomorphism implies that for each pair of *plain rules* $\langle p_1, p_2 \rangle$ there is a *unique initial parallel independent transformation pair* $tp_{L_1+L_2} : R_1 + L_2 \Leftarrow_{p_1,i_1} L_1 + L_2 \Rightarrow_{p_2,i_2} L_1 + R_2$.

Definition 8 (initial parallel independent transformation pair). *A pair of parallel independent plain transformations* $tp : H_1 \Leftarrow_{p_1,m_1} G \Rightarrow_{p_2,m_2} H_2$ *is an* initial parallel independent transformation pair *if it is isomorphic to the transformation pair* $tp_{L_1+L_2} : R_1 + L_2 \Leftarrow_{p_1,i_1} L_1 + L_2 \Rightarrow_{p_2,i_2} L_1 + R_2$.

The one-element set consisting of the initial parallel independent transformation pair for a given pair of rules is *complete w.r.t. parallel independence.*

Theorem 1 (completeness of initial parallel independent transformation pairs). *The set consisting of the initial parallel independent transformation pair* $tp_{L_1+L_2} : R_1 + L_2 \Leftarrow_{p_1,i_1} L_1 + L_2 \Rightarrow_{p_2,i_2} L_1 + R_2$ *for a pair of plain rules* $\langle p_1, p_2 \rangle$ *is* complete w.r.t. parallel independence.

4   Initial Conflicts

We start with showing why it is not possible to straightforwardly generalize the idea of initial conflicts from plain rules to rules with ACs. On the one hand, *conflict inheritance*, which was the basis for showing completeness of initial conflicts for plain rules, does not hold any more. Moreover, it is *impossible* in general to find a *finite and complete subset of finite conflicts* for rules with ACs (cf. Subsect. 4.2). This motivates again the need for having *symbolic transformation pairs* as introduced in Definition 4, allowing us to define *initial conflicts* (cf. Subsect. 4.3) as a set of specific symbolic transformation pairs, being complete w.r.t. parallel dependence indeed (as shown in Subsect. 4.4).

4.1   Conflict Inheritance

Conflicts are in general not inherited (as opposed to the case of plain rules [19]), i.e., not each (initial) transformation pair that can be embedded into a conflicting one will be conflicting again. This may happen in particular for AC conflicts. Use-delete (resp. delete-use) conflicts for rules with ACs are still inherited.

Lemma 4 (Use-delete (delete-use) conflict inheritance). *Given a pair of direct transformations tp in use-delete (resp. delete-use) conflict and another pair of direct transformations tp' that can be embedded into tp via extension morphism f and corresponding extension diagrams, then tp' is also in use-delete (resp. delete-use) conflict.*

Example 1 (No inheritance for AC conflicts). Consider rules $p_1 : \bigcirc \leftarrow \bigcirc \rightarrow \bigcirc\!\!\rightarrow\!\!\bigcirc$ (with AC *true*), producing an outgoing edge with a node, and $p_2 : \bigcirc \leftarrow \bigcirc \rightarrow \bigcirc\bigcirc$ with NAC $\neg \exists n : \bigcirc \rightarrow \bigcirc\bigcirc\bigcirc$, producing a node only if two other nodes do not exist already. Consider graph $G = \bigcirc\bigcirc$, holding two nodes. Applying both rules to G (with the matches sharing one node in G) we obtain a produce-AC conflict since the first rule creates a third node, forbidden by the second rule. Now both rules can be applied similarly to the shared node in the subgraph $G' = \bigcirc$ of G obtaining parallel independent transformations, illustrating that AC-conflicts are not inherited.

4.2   Complete Subset of Conflicts

We show that in $\mathcal{M}$-adhesive categories, in particular in the category of graphs, it is in general impossible to find a finite and complete subset of conflicts for finite rules with ACs. If it would always exist, we could derive that each first-order formula is equivalent to a finite disjunction of atomic formulas.

Theorem 2. *Given finite rules $\rho_1 = \langle p_1, \mathrm{ac}_{L_1} \rangle$ and $\rho_2 = \langle p_2, \mathrm{ac}_{L_2} \rangle$ for the $\mathcal{M}$-adhesive category* **Graphs**, *in general, there is no finite set of finite transformation pairs $\mathcal{S}$ for ρ_1 and ρ_2 that is complete w.r.t. parallel dependence.*

4.3   Initial Conflicts

We generalize the notion of *initial conflicts* for plain rules to rules with ACs. In particular, we introduce them as special symbolic transformation pairs. They are *conflict-inducing* meaning that there needs to exist an unfolding of the symbolic transformation pair into a concrete conflicting transformation pair. Moreover, their AC-disregarding transformation pair needs to be an initial conflict or initial parallel independent transformation pair. We also show the *relationship between initial conflicts and critical pairs* as reintroduced in Subsect. 3.3, demonstrating that initial conflicts represent a proper subset of critical pairs.

Definition 9 (unfolding of symbolic transformation pair). *Given a symbolic transformation pair $stp_K : \langle tp_K, \mathrm{ac}_K, \mathrm{ac}_K^* \rangle$ for rule pair $\langle \rho_1, \rho_2 \rangle$, then its unfolding $\mathcal{U}(stp_K)$ consists of all transformation pairs $H_1 \Leftarrow_{\rho_1, m_1} G \Rightarrow_{\rho_2, m_2} H_2$ representing the lower row of the extension diagrams via some extension morphism $m : K \to G$ as shown in Fig. 1 (with AC-disregarding transformation pair tp_K in the upper row). Moreover, we say that stp_K is* conflict-inducing *if its unfolding includes a conflicting transformation pair.*

Definition 10 (initial conflict). *Consider an $\mathcal{M}$-adhesive system with initial transformation pairs for conflicts along plain rules. An* initial conflict *for rules $\rho_1 = \langle p_1, \mathrm{ac}_{L_1} \rangle$ and $\rho_2 = \langle p_2, \mathrm{ac}_{L_2} \rangle$ is a conflict-inducing symbolic transformation pair $stp_K : \langle tp_K, \mathrm{ac}_K, \mathrm{ac}_K^* \rangle$ with the AC-disregarding transformation pair tp_K being initial, i.e. either tp_K is an initial conflict for rules p_1 and p_2 (in this case stp_K is called a* use-delete/delete-use initial conflict*) or it is the initial parallel independent transformation pair $tp_{L_1+L_2}$ for rules p_1 and p_2 (in this case $stp_K = stp_{L_1+L_2} = \langle tp_{L_1+L_2}, \mathrm{ac}_{L_1+L_2}, \mathrm{ac}_{L_1+L_2}^* \rangle$ is called the* AC initial conflict*).*

Note that the unfolding of a conflict-inducing symbolic transformation pair (and in particular of an AC initial conflict) may also include parallel independent transformation pairs. All conflicts in the unfolding of an AC initial conflict are AC conflicts, and never use-delete/delete-use conflicts (because otherwise we would get a contradiction using Lemma 4).

Example 2 (initial conflict). Consider again the rules from Example 1. Applying both rules to $L_1 + L_2 = \bigcirc\,\bigcirc$ (with disjoint matches) we obtain the AC initial conflict $stp_K = stp_{L_1+L_2} = \langle tp_{L_1+L_2}, \mathrm{ac}_{L_1+L_2}, \mathrm{ac}^*_{L_1+L_2}\rangle$. Thereby $\mathrm{ac}_{L_1+L_2}$ is equivalent to $\neg\exists(\bigcirc\,\bigcirc \to \bigcirc\,\bigcirc\,\bigcirc) \wedge \neg\exists(\bigcirc\,\bigcirc \to \bigcirc_{,2}\ \bigcirc\,\bigcirc)$, expressing that when during extension both nodes are merged, no two additional nodes, otherwise not one additional node should be given. Moreover, $\mathrm{ac}^*_{L_1+L_2}$ is equivalent to $\exists(\bigcirc\,\bigcirc \to \bigcirc_{,2}\ \bigcirc) \vee \exists(\bigcirc\,\bigcirc \to \bigcirc\,\bigcirc)$, expressing that either both nodes are not merged during extension, otherwise one additional node should be present for a conflict to arise. Both transformation pairs (the conflicting one from $G = \bigcirc\,\bigcirc$ as well as the parallel independent one from its subgraph $G' = \bigcirc$, sharing the merged node in their matches) described in Example 1 belong to its unfolding.

Each initial conflict is in particular also a critical pair.

Theorem 3 (initial conflict is critical pair). *Consider an $\mathcal{M}$-adhesive system with initial transformation pairs for conflicts along plain rules. Each initial conflict $stp_K : \langle tp_K, \mathrm{ac}_K, \mathrm{ac}^*_K\rangle$ is a critical pair.*

The reverse direction of Theorem 3 does not hold. In general, critical pairs $stp_K : \langle tp_K, \mathrm{ac}_K, \mathrm{ac}^*_K\rangle$ where tp_K represents a use-delete/delete-use conflict (but is not initial yet) are represented by the initial conflict $stp_I : \langle tp_I, \mathrm{ac}_I, \mathrm{ac}^*_I\rangle$ with tp_I the unique initial conflict for tp_K as plain transformation pair. Moreover, critical pairs $stp_K : \langle tp_K, \mathrm{ac}_K, \mathrm{ac}^*_K\rangle$ where tp_K is parallel independent as plain transformation pair are represented by one initial conflict $stp_{L_1+L_2} : \langle tp_{L_1+L_2}, \mathrm{ac}_{L_1+L_2}, \mathrm{ac}^*_{L_1+L_2}\rangle$ with $tp_{L_1+L_2}$ the initial parallel independent transformation pair.

Example 3 (initial conflicts: proper subset of critical pairs). Consider again the rules from Example 1 and their application to $G' = \bigcirc$. The symbolic transformation pair $stp_{G'} : \langle tp_{G'}, \mathrm{ac}_{G'}, \mathrm{ac}^*_{G'}\rangle$ is a critical pair, but not an initial conflict.

4.4 Completeness

We show that initial conflicts are complete (not $\mathcal{M}$-complete as in the case of critical pairs) w.r.t. parallel dependence as symbolic transformation pairs.

Theorem 4 (completeness of initial conflicts). *Consider an $\mathcal{M}$-adhesive system with initial transformation pairs for conflicts along plain rules. The set of initial conflicts for a pair of rules $\langle \rho_1, \rho_2\rangle$ is complete w.r.t. parallel dependence.*

Remark 4 (uniqueness of initial conflicts). For each conflict a *unique* (up-to-isomorphism) initial conflict exists representing it, since this property is inherited from the one for plain rules [19] and the fact that the initial parallel independent pair of transformations is unique w.r.t. a given rule pair.

Initial conflicts are also minimally complete, i.e. we are able to generalize the corresponding result for plain rules to rules with ACs.

5   Unfoldings of Initial Conflicts

We show a *sufficient condition* for being able to unfold initial conflicts into a *complete set of conflicts* that is *finite* if the set of initial conflicts is finite (cf. Subsect. 5.1). We demonstrate moreover that this sufficient condition is fulfilled for the special case of having merely *NACs* as rule application conditions (cf. Subsect. 5.2). Finally, we show that in this case we obtain in particular specific critical pairs for rules with negative application conditions (NACs) as introduced in [20] again. In this sense we show explicitly that initial conflicts as introduced in this paper represent a conservative extension of the critical pair theory for rules with NACs.

5.1   Finite and Complete Unfolding

We introduce *regular initial conflicts* leading to $\mathcal{M}$-complete subsets of conflicts by unfolding them in some particular way (cf. *disjunctive unfolding* in Definition 11). The idea is that the extension and conflict-inducing AC (ac_K and ac_K^*, respectively) of such a conflict $stp_K : \langle tp_K, \mathrm{ac}_K, \mathrm{ac}_K^* \rangle$ have a specific form that is amenable to finding $\mathcal{M}$-complete unfoldings. We expect the condition $\mathrm{ac}_K \wedge \mathrm{ac}_K^*$ to consist of a *disjunction of positive literals* (conditions of the form $\exists(a_i : K \to C_i, c_i)$) with a *negative remainder* (i.e. a condition $c_i = \wedge_{j \in J} \neg \exists(b_j : C_i \to C_j, d_j)$). Intuitively, this means that there is a finite number of possibilities to unfold the symbolic conflict into a concrete conflict by adding some specific positive context (expressed by the morphism a_i). The negative remainder c_i ensures that by adding this positive context to the context K of the symbolic transformation pair within the initial conflict, we indeed find a concrete conflict when not extending further at all. Moreover, it expresses under which condition the corresponding concrete representative conflict leads to further conflicts by extension. Finally, the subsets of $\mathcal{M}$-complete conflicts built using the disjunctive unfolding are *finite* if the set of initial conflicts it is derived from is finite.

Definition 11 (regular initial conflict, disjunctive unfolding). *Consider an $\mathcal{M}$-adhesive system with initial transformation pairs for conflicts along plain rules. Given an initial conflict $stp_K : \langle tp_K, \mathrm{ac}_K, \mathrm{ac}_K^* \rangle$ for rules $\langle \rho_1, \rho_2 \rangle$, then we say that it is* regular *if $\mathrm{ac}_K \wedge \mathrm{ac}_K^*$ is equivalent to a condition $\vee_{i \in I} \exists(a_i : K \to C_i, c_i)$ with $c_i = \wedge_{j \in J} \neg \exists(b_j : C_i \to C_j, d_j)$ a condition on C_i, b_j*

non-isomorphic and I some non-empty index set. Given a regular initial conflict stp_K : $\langle tp_K, ac_K, ac_K^* \rangle$, *then* $\mathcal{U}^{\mathcal{D}}(stp_K) = \cup_{i \in I} \{ tp_{C_i} : D_{1,i} \Leftarrow_{\rho_1, a_i \circ o_1} C_i \Rightarrow_{\rho_2, a_i \circ o_2} D_{2,i} \}$ *is the* disjunctive unfolding *of* stp_K.

Remark 5 (disjunctive unfolding). The disjunctive unfolding of a regular conflict is non-empty, but might consist of less elements than literals in the disjunction $\vee_{i \in I} \exists (a_i : K \to C_i, c_i)$: if a morphism a_i does not satisfy the gluing condition of the derived spans, then also every extension morphism starting from there will not satisfy the gluing condition, and we can safely ignore that case from the disjunctive unfolding.

Theorem 5 (finite and complete unfolding). *Consider an $\mathcal{M}$-adhesive system with initial transformation pairs for conflicts along plain rules. Given a rule pair $\langle \rho_1, \rho_2 \rangle$ with set S of initial conflicts such that each initial conflict stp in S is regular, then $\cup_{stp \in S} \mathcal{U}^{\mathcal{D}}(stp)$ is $\mathcal{M}$-complete w.r.t. parallel dependence. Moreover, $\cup_{stp \in S} \mathcal{U}^{\mathcal{D}}(stp)$ is finite if S is finite.*

It is possible to automatically check if some initial conflict is regular by using dedicated automated reasoning [22] as well as symbolic model generation for ACs [28] as follows. The reasoning mechanism [22] is shown to be refutationally complete ensuring that if the condition $ac_K \wedge ac_K^*$ of some initial conflict is unsatisfiable, this will be detected eventually. Moreover, the related symbolic model generation mechanism [28] is able to automatically transform each condition $ac_K \wedge ac_K^*$ into some disjunction $\vee_{i \in I} \exists (a_i : K \to C_i, c_i)$ with c_i a negative remainder if such an equivalence holds.

5.2 Unfolding for Rules with NACs

We show that for rules with NACs initial conflicts are regular. This means that in this special case there exists a complete subset of conflicts that is finite (in the case of graphs and assuming finite rules). This conforms to the findings in [17,20], where an $\mathcal{M}$-complete set of critical pairs – as specific subset of conflicts – for graph transformation rules with NACs was introduced [20] (and generalized to $\mathcal{M}$-adhesive transformation systems [17]).

Theorem 6 (regular initial conflicts for rules with NACs). *Consider an $\mathcal{M}$-adhesive system with initial transformation pairs for conflicts along plain rules. Given some initial conflict stp_K : $\langle tp_K, ac_K, ac_K^* \rangle$ for a pair of rules $\langle \rho_1, \rho_2 \rangle$ with $ac_{L_i} = \wedge_{j \in J} \neg \exists n_j : L_i \to N_j$ for $i = 1, 2$ and J some finite index set, then it is regular. In particular, $ac_K \wedge ac_K^*$ is equivalent to a condition $\vee_{i \in I} \exists (a_i : K \to C_i, c_i)$ with $c_i = \wedge_{q \in Q} \neg \exists n_q$ a condition on C_i and I some non-empty index set.*

The negative remainder c_i of each literal in $\vee_{i \in I} \exists (a_i : K \to C_i, c_i)$ of a regular initial conflict for rules with NACs thus consists of a set of NACs. Intuitively this means that we obtain for each initial conflict an $\mathcal{M}$-complete subset of concrete conflicts by adding the context described by a_i. As long as no NAC from c_i is violated we can extend such a concrete conflict to further ones.

Corollary 1 (complete unfolding: rules with NACs). *Consider an* $\mathcal{M}$-*adhesive system, with initial transformation pairs for conflicts along plain rules. Given a rule pair* $\langle \rho_1, \rho_2 \rangle$ *with* $\mathrm{ac}_{L_i} = \wedge_{j \in J} \neg \exists n_j : L_i \to N_j$ *for* $i = 1, 2$, *then* $\cup_{stp \in \mathcal{S}} \mathcal{U}^{\mathcal{D}}(stp)$ *is* $\mathcal{M}$-*complete w.r.t. parallel dependence.*

We show moreover that the initial conflict definition is a *conservative extension* of the critical pair definition for rules with NACs as given in [17,20], i.e., we show that each conflict in the disjunctive unfolding of an initial conflict as chosen in the proof of Theorem 6 is a critical pair for rules with NACs.

Theorem 7 (conservative unfolding). *In an* $\mathcal{M}$-*adhesive system with initial transformation pairs for conflicts along plain rules, if* $stp_K : \langle tp_K, \mathrm{ac}_K, \mathrm{ac}_K^* \rangle$ *is an initial conflict for rules* $\langle \rho_1, \rho_2 \rangle$ *with* $\mathrm{ac}_{L_i} = \wedge_{j \in J} \neg \exists n_j : L_i \to N_j$ *for* $i = 1, 2$ *and* J *some finite index set, then each conflict as chosen in the proof of Theorem 6 in* $\mathcal{U}^{\mathcal{D}}(stp)$ *is in particular a critical pair for* $\langle \rho_1, \rho_2 \rangle$ *as given in [17, 20].*

Example 4 (conservative unfolding). Consider again the rules from Example 1 (having only NACs as ACs) and their application to the graph $G = \bigcirc \bigcirc$. The corresponding transformation pair tp_G is a critical pair for rules with NACs as given in [17,20]. This is because it is in particular a conflicting pair of transformations, and the morphism violating the NAC (since finding the three nodes) and therefore causing the conflict after applying the first rule to $G = \bigcirc \bigcirc$ obtaining some graph $H_1 = \bigcirc\!\!\rightarrow\!\!\bigcirc \bigcirc$ is jointly surjective together with the corresponding co-match. As argued already in Example 2 this critical pair for rules with NACs belongs to the unfolding (and in particular to the disjunctive unfolding) of the unique AC initial conflict $stp_{L_1+L_2} : \langle tp_{L_1+L_2}, \mathrm{ac}_{L_1+L_2}, \mathrm{ac}_{L_1+L_2}^* \rangle$.

6   Conclusion and Outlook

In this paper we have *generalized the theory of initial conflicts* (from plain rules, i.e. rules without application conditions) to *rules with application conditions* (ACs) in the framework of $\mathcal{M}$-adhesive transformation systems. We build on the notion of symbolic transformation pairs, since it turns out that it is not possible to find a complete subset of concrete conflicting transformation pairs in the case of rules with ACs. We have shown that initial conflicts are (minimally) complete w.r.t. parallel dependence as symbolic transformation pairs. Moreover, initial conflicts represent (analogous to the case of plain rules) proper subsets of critical pairs in the sense that for each critical pair (or also for each conflict), there exists a unique initial conflict representing it. We concluded the paper by showing sufficient conditions for finding unfoldings of initial conflicts that lead to (finite and) complete subsets of conflicts (as in the case of rules with NACs). Thereby we have shown that initial conflicts for rules with ACs represent a conservative extension of the critical pair theory for rules with NACs.

As future work we aim at finding *further interesting classes* allowing finite and (minimally) complete unfoldings into *subsets of conflicts*. This will serve as a guideline to be able to *develop and implement efficient conflict detection* techniques for rules with (specific) ACs, which has been an open challenge until today.

We are moreover planning to develop (semi-)automated detection of unfoldings of initial conflicts of rules with arbitrary ACs using dedicated automated reasoning and model finding for graph conditions [22,26,28]. It would be interesting to investigate in which *use cases* initial conflicts (or critical pairs) are useful already as symbolic transformation pairs, and in which use cases we rather need to consider unfoldings indeed. This is in line with the research on multi-granular conflict detection [3,18,24] investigating different levels of granularity that can be interesting from the point of view of applying conflict detection to different use cases. Finally, we plan to investigate conflict detection in the light of initial conflict theory for *attributed graph transformation* [5,13,16], and in particular the case of rules with so-called attribute conditions more specifically. It would also be interesting to further investigate initial conflicts for transformation rules (with ACs) not following the DPO approach.

Acknowledgement. We thank Jens Kosiol for pointing out that the set of initial conflicts for plain rules is not only complete, but also minimally complete. We were able to transfer this result to rules with ACs in this paper. Many thanks also to the reviewers for their detailed and constructive comments helping to finalize the paper.

References

1. Azzi, G.G., Corradini, A., Ribeiro, L.: On the essence and initiality of conflicts in m-adhesive transformation systems. J. Log. Algebr. Methods Program. **109**, 100482 (2019)
2. Bonchi, F., Gadducci, F., Kissinger, A., Sobociński, P., Zanasi, F.: Confluence of graph rewriting with interfaces. In: Yang, H. (ed.) ESOP 2017. LNCS, vol. 10201, pp. 141–169. Springer, Heidelberg (2017). https://doi.org/10.1007/978-3-662-54434-1_6
3. Born, K., Lambers, L., Strüber, D., Taentzer, G.: Granularity of conflicts and dependencies in graph transformation systems. In: de Lara, J., Plump, D. (eds.) ICGT 2017. LNCS, vol. 10373, pp. 125–141. Springer, Cham (2017). https://doi.org/10.1007/978-3-319-61470-0_8
4. Ehrig, H., Ehrig, K., Habel, A., Pennemann, K.: Theory of constraints and application conditions: from graphs to high-level structures. Fundam. Inform. **74**(1), 135–166 (2006)
5. Ehrig, H., Ehrig, K., Prange, U., Taentzer, G.: Fundamentals of Algebraic Graph Transformation. Monographs in Theoretical Computer Science. An EATCS Series. Springer, Heidelberg (2006). https://doi.org/10.1007/3-540-31188-2
6. Ehrig, H., Golas, U., Habel, A., Lambers, L., Orejas, F.: $\mathcal{M}$-adhesive transformation systems with nested application conditions. Part 2: embedding, critical pairs and local confluence. Fundam. Inform. **118**(1–2), 35–63 (2012). https://doi.org/10.3233/FI-2012-705
7. Ehrig, H., Golas, U., Habel, A., Lambers, L., Orejas, F.: $\mathcal{M}$-adhesive transformation systems with nested application conditions. Part 1: parallelism, concurrency and amalgamation. Math. Struct. Comput. Sci. **24**(4) (2014). https://doi.org/10.1017/S0960129512000357
8. Ehrig, H., Habel, A.: Graph grammars with application conditions. In: Rozenberg, G., Salomaa, A. (eds.) The Book of L, pp. 87–100. Springer, Heidelberg (1986). https://doi.org/10.1007/978-3-642-95486-3_7

9. Ehrig, H., Habel, A., Lambers, L., Orejas, F., Golas, U.: Local confluence for rules with nested application conditions. In: Ehrig, H., Rensink, A., Rozenberg, G., Schürr, A. (eds.) ICGT 2010. LNCS, vol. 6372, pp. 330–345. Springer, Heidelberg (2010). https://doi.org/10.1007/978-3-642-15928-2_22

10. Habel, A., Heckel, R., Taentzer, G.: Graph grammars with negative application conditions. Fundam. Inform. **26**(3/4), 287–313 (1996). https://doi.org/10.3233/FI-1996-263404

11. Habel, A., Pennemann, K.: Correctness of high-level transformation systems relative to nested conditions. Math. Struct. Comput. Sci. **19**(2), 245–296 (2009). https://doi.org/10.1017/S0960129508007202

12. Hausmann, J.H., Heckel, R., Taentzer, G.: Detection of conflicting functional requirements in a use case-driven approach: a static analysis technique based on graph transformation. In: Tracz, W., Young, M., Magee, J. (eds.) Proceedings of the 24th International Conference on Software Engineering, ICSE 2002, pp. 105–115. ACM (2002). https://doi.org/10.1145/581339.581355

13. Hristakiev, I., Plump, D.: Attributed graph transformation via rule schemata: Church-Rosser theorem. In: Milazzo, P., Varró, D., Wimmer, M. (eds.) STAF 2016. LNCS, vol. 9946, pp. 145–160. Springer, Cham (2016). https://doi.org/10.1007/978-3-319-50230-4_11

14. Knuth, D., Bendix, P.: Simple word problems in universal algebras. In: Leech, J. (ed.) Computational Problems in Abstract Algebra, pp. 263–297. Pergamon Press, Oxford (1970)

15. Koch, M., Mancini, L.V., Parisi-Presicce, F.: Graph-based specification of access control policies. J. Comput. Syst. Sci. **71**(1), 1–33 (2005). https://doi.org/10.1016/j.jcss.2004.11.002

16. Kulcsár, G., Deckwerth, F., Lochau, M., Varró, G., Schürr, A.: Improved conflict detection for graph transformation with attributes. In: Rensink, A., Zambon, E. (eds.) Proceedings GaM@ETAPS 2015. EPTCS, vol. 181, pp. 97–112 (2015). https://doi.org/10.4204/EPTCS.181.7

17. Lambers, L.: Certifying rule-based models using graph transformation. Ph.D. thesis, Berlin Institute of Technology (2009). http://opus.kobv.de/tuberlin/volltexte/2010/2522/

18. Lambers, L., Born, K., Kosiol, J., Strüber, D., Taentzer, G.: Granularity of conflicts and dependencies in graph transformation systems: a two-dimensional approach. J. Log. Algebr. Methods Program. **103**, 105–129 (2019). https://doi.org/10.1016/j.jlamp.2018.11.004

19. Lambers, L., Born, K., Orejas, F., Strüber, D., Taentzer, G.: Initial conflicts and dependencies: critical pairs revisited. In: Heckel, R., Taentzer, G. (eds.) Graph Transformation, Specifications, and Nets. LNCS, vol. 10800, pp. 105–123. Springer, Cham (2018). https://doi.org/10.1007/978-3-319-75396-6_6

20. Lambers, L., Ehrig, H., Orejas, F.: Conflict detection for graph transformation with negative application conditions. In: Corradini, A., Ehrig, H., Montanari, U., Ribeiro, L., Rozenberg, G. (eds.) ICGT 2006. LNCS, vol. 4178, pp. 61–76. Springer, Heidelberg (2006). https://doi.org/10.1007/11841883_6

21. Lambers, L., Ehrig, H., Orejas, F.: Efficient conflict detection in graph transformation systems by essential critical pairs. Electr. Notes Theor. Comput. Sci. **211**, 17–26 (2008)

22. Lambers, L., Orejas, F.: Tableau-based reasoning for graph properties. In: Giese, H., König, B. (eds.) ICGT 2014. LNCS, vol. 8571, pp. 17–32. Springer, Cham (2014). https://doi.org/10.1007/978-3-319-09108-2_2

23. Lambers, L., Orejas, F.: Initial conflicts for transformation rules with nested application conditions (2020). arXiv:2005.05901 [cs.LO]

24. Lambers, L., Strüber, D., Taentzer, G., Born, K., Huebert, J.: Multi-granular conflict and dependency analysis in software engineering based on graph transformation. In: Chaudron, M., Crnkovic, I., Chechik, M., Harman, M. (eds.) Proceedings of the 40th International Conference on Software Engineering, ICSE 2018, pp. 716–727. ACM (2018). https://doi.org/10.1145/3180155.3180258

25. Ohlebusch, E.: Advanced Topics in Term Rewriting. Springer, Heidelberg (2002). https://doi.org/10.1007/978-1-4757-3661-8

26. Pennemann, K.: Development of correct graph transformation systems. Ph.D. thesis, University of Oldenburg, Germany (2009). http://oops.uni-oldenburg.de/volltexte/2009/948/

27. Plump, D.: Hypergraph rewriting: Critical pairs and undecidability of confluence. In: Sleep, R., Plasmeijer, R., van Eekelen, M. (eds.) Term Graph Rewriting: Theory and Practice, pp. 201–213. Wiley, Chichester (1993)

28. Schneider, S., Lambers, L., Orejas, F.: Symbolic model generation for graph properties. In: Huisman, M., Rubin, J. (eds.) FASE 2017. LNCS, vol. 10202, pp. 226–243. Springer, Heidelberg (2017). https://doi.org/10.1007/978-3-662-54494-5_13

Patch Graph Rewriting

Roy Overbeek[(✉)] and Jörg Endrullis

Vrije Universiteit Amsterdam, Amsterdam, The Netherlands
{r.overbeek,j.endrullis}@vu.nl

Abstract. The basic principle of graph rewriting is the stepwise replacement of subgraphs inside a host graph. A challenge in such replacement steps is the treatment of the *patch graph*, consisting of those edges of the host graph that touch the subgraph, but are not part of it.

We introduce *patch graph rewriting*, a visual graph rewriting language with precise formal semantics. The language has rich expressive power in two ways. First, rewrite rules can flexibly constrain the permitted shapes of patches touching matching subgraphs. Second, rules can freely transform patches. We highlight the framework's distinguishing features by comparing it against existing approaches.

Keywords: Graph rewriting · Embedding · Visual language

1 Introduction

When matching a graph pattern P inside a host graph G, G can be partitioned into (i) a *match* M, a subgraph of G isomorphic to the pattern P; (ii) a *context* C, the largest subgraph of G disjoint from M; and (iii) a *patch* J, the graph consisting of the edges that are neither in M nor in C. So the patch consists of edges that are either (a) between M and C, in either direction, or (b) between vertices of M not captured by the pattern P. For example, if P and G are respectively

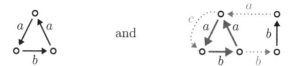

and

then the thick green subgraph is the (only) match M of P in G. The black subgraph of G is the context C, and the dotted red subgraph is the patch J. Metaphorically, patch J patches match M and context C together.

In graph rewriting, subgraphs of some host graph are stepwise replaced by other subgraphs. A requirement for such replacements is that they are properly re-embedded in the host graph. We contend that the patch is the most distinctive and interesting aspect of graph rewriting. This is because its shape is generally unpredictable, making it challenging to specify what constitutes a

© Springer Nature Switzerland AG 2020
F. Gadducci and T. Kehrer (Eds.): ICGT 2020, LNCS 12150, pp. 128–145, 2020.
https://doi.org/10.1007/978-3-030-51372-6_8

proper re-embedding of a subgraph replacement. This contrasts strongly with the situation for string and term rewriting, in which the embeddings of substrings and subterms are highly regular.

Most existing approaches to graph rewriting are rather uniform and coarse-grained in their treatment of the patch. For instance, suppose that we wish to delete the match M from G. What should happen to the edges of patch J, which would be left "dangling" by such a removal? The popular double-pushout (DPO) [9] approach to graph rewriting conservatively dictates that the application is not allowed in the first place: nodes connected to the patch *must* be preserved by the rewrite step, and the patch shall remain connected as before. The single-pushout (SPO) [18] variant, by contrast, permissively answers that such a deletion is always possible. As a side-effect, however, any resulting dangling patch edges are discarded.

In this paper, we introduce the *patch graph rewriting* (*PGR*) language. It has the following features:

- *Pluriform, fine-grained control over patches.* Rules themselves encode which kinds of patches are allowed around matches, as well as how they should be transformed for the re-embedding, using a *unified* notation. Thus, these policies are distinctly not decided on the level of the framework.
- *An intuitive visual language.* Despite their expressive power and formal semantics, patch rewrite rules admit a visual representation that we believe to be highly intuitive.
- *Lightweight formal semantics.* The formal details of PGR are based on elementary set and graph theory, and therefore accessible to a wide audience. In particular, an understanding of category theory is not required to understand these details, unlike for many dominant approaches in graph rewriting.

The remainder of our paper is structured as follows. To fix ideas and emphasize the visual language of PGR, we first provide an intuitive exposition in Sect. 2, and then follow with a formal introduction in Sect. 3. We show the usefulness of PGR by modeling wait-for graphs and deadlock detection in Sect. 4. We compare PGR to other approaches in Sect. 5. In Sect. 6, finally, we mention some future research directions for PGR.

2   Intuitive Semantics

We start with an intuitive introduction of PGR, to be made formally precise in Sect. 3. The graph G in Fig. 1 will serve as our leading example.

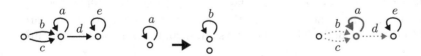

Fig. 1. Graph G.   **Fig. 2.** A simple rule.   **Fig. 3.** Match, context and patch.

We begin by considering the rewrite rule in Fig. 2. Figure 3 contains a depiction of G in which the match, context and patch are highlighted: the thick green subgraph is the match for the left-hand side of the rule, the solid black subgraph is the context for this rule application, and the red dotted edges form the patch. In PGR, the rewrite rule in Fig. 2 cannot yet be applied in G. This is because without further annotations, the rule may only be applied if the patch is empty, that is, if the node with the a-loop has no additional edges. In effect, this rule only allows replacing an isolated node with an a-loop by two isolated nodes, one of which has a b-loop.

Fig. 4. An annotated rule.   Fig. 5. Applying the rule on the left.

The rule in Fig. 2 can be generalized to allow for patch edges from and to the context by annotating the left-hand side of the rule as shown in Fig. 4. We call such annotations *patch type edges*. They can be thought of as placeholders for sets of patch edges:

(i) The dotted arrow with source ⟨1⟩ is a placeholder for an arbitrary number of edges from the context to the node with the a-loop.

(ii) Likewise, the outgoing dotted arrow with target ⟨2⟩ is a placeholder for an arbitrary number of edges going into the context.

The rule is now applicable to all nodes that have an a-loop and no other loop, allowing the node to be connected to the context through an arbitrary number of edges. In particular, then, the rule is applicable to the match highlighted in Fig. 3, and it gives rise to the step shown in Fig. 5.

Although we see how patch type edge annotations on the left-hand side can be used to *constrain* the set of permitted patches around a match, it does not tell us what to do with patch edges if a match is found. To indicate such transformations, the solution is simply to reuse the patch type edges in the right-hand side of the rule. The rule shown in Fig. 4 does not reuse any of the patch type edges, explaining why the corresponding patch edges are deleted in Fig. 5.

One way to preserve the incoming edges bound to ⟨1⟩ and the outgoing edges bound to ⟨2⟩ is shown in Fig. 6. As the visual representation suggests, the incoming edges bound to ⟨1⟩ get redirected to target the upper node of the right-hand side, and the sources of the outgoing edges bound to ⟨2⟩ are redirected to the lower node. The respective sources and targets of the edges are

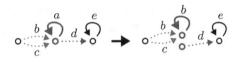

Fig. 6. Redirecting patch edges.   **Fig. 7.** Applying the rule on the left.

defined to remain unchanged. Applying the rule in G results in the rewrite step depicted in Fig. 7.

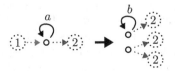

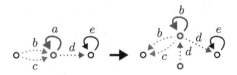

Fig. 8. Duplicating patch edges.   **Fig. 9.** Applying the rule on the left.

Using this visual language, it is easy to duplicate, remove, and redirect edges in the patch. The rule displayed in Fig. 8 removes the incoming patch edges bound to (1), and duplicates the patch edges bound to (2): one copy for the upper node of the right-hand side, and two copies for the lower node. The resulting rewrite step is shown in Fig. 9.

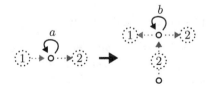

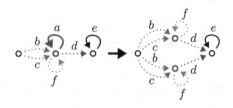

Fig. 10. Complex transformation.   **Fig. 11.** Applying the rule on the left.

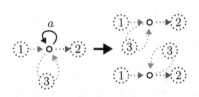

Fig. 12. Node duplication.   **Fig. 13.** Applying the rule on the left.

Patch graph rewriting also allows for some more exotic transformations, such as inverting patch edges or pulling edges from the context into the pattern. The rule in Fig. 10 reverses the direction of ⦂1⦂ and pulls ⦂2⦂ into the pattern, giving rise to the step in Fig. 11.

All of the above rules are only applicable to nodes that have an a-loop and no other loop. If we want the rule to be applicable to nodes that have additional loops, this can be expressed as in Fig. 12. This rule is now applicable to any node with an a-loop. It makes a copy of the node, as well as all edges incident to it (except for the a-loop, which is removed). If we slightly modify G to include an f-loop on the middle node, the rule gives rise to the rewrite step in Fig. 13.

In this brief visual introduction, we have focused on the transformation of the patch. The left-hand sides of the rules has each time been a single node with an a-loop. Its generalization to other patterns is largely obvious, but some edge cases arise. For instance, what could be the semantics of the rule ○⦂1⦂→○ ➤ ○→⦂1⦂ which redirects patch edges between nodes of the pattern into the context? We now turn to the formal semantics of path rewriting, which makes all preceding transformations precise and excludes such edge cases.

3   Formal Semantics

Notation 1 (Preliminaries). For functions $f : A_f \to B_f$ and $g : A_g \to B_g$ with disjoint domains (but possibly overlapping codomains), we write $f \cup g$ for the function $(f \cup g) : (A_f \cup A_g) \to (B_f \cup B_g)$ given by the union of f and g's underlying graphs. If typing permits, we generalize functions f to tuples (x, y) and sets S in the obvious way, i.e., $f((x,y)) = (f(x), f(y))$ and $f(S) = \{ f(x) \mid x \in S \}$.

We define directed, edge-labeled multigraphs in the standard way.

Definition 2 (Graph). *A graph $G = (V, E, s, t, \ell)$ with edge labels from L consists of a finite set of* vertices *(or nodes) V, a finite set of edges E, a* source *map $s : E \to V$, a* target *map $t : E \to V$, and a* labeling *$\ell : E \to L$. For $e \in E$, we say that $s(e)$, $t(e)$ and $\ell(e)$ are the* source, target *and* label *of e, respectively.*

For convenience, we will write $x \xrightarrow{\alpha} y \in E$ to denote an edge $e \in E$ such that $s(e) = x$, $t(e) = y$ and $\ell(e) = \alpha$.

Definition 3 (Basic Graph Notions). *We define the following basic graph notions:*

(i) An unlabeled graph *$G = (V, E, s, t)$ is a graph (V, E, s, t, ℓ) over a singleton label set. In this case we suppress the edge labels.*

(ii) A graph is simple *if for all $e, e' \in E$, $s(e) = s(e')$, $t(e) = t(e')$ and $\ell(e) = \ell(e')$ together imply $e = e'$.*

(iii) We say that graphs G and H are disjoint *if $V_G \cap V_H = \varnothing = E_G \cap E_H$.*

(iv) For disjoint edge sets $E \cap E' = \varnothing$, we define the graph union *as follows:*

$$(V, E, s, t, \ell) \cup (V', E', s', t', \ell') = (V \cup V', E \cup E', s \cup s', t \cup t', \ell \cup \ell').$$

To rename vertices and edges of a graph, we introduce "graph renamings". A renaming is a graph isomorphism, where the domain of the renaming is allowed to be a superset of the set of vertices/edges of the graph. In this way, the same renaming can be applied to various graphs with different vertex and edge sets.

Definition 4 (Graph Renaming). *A graph renaming ϕ for a graph G consists of two bijective functions $\phi_V : V_1 \to V_2$ and $\phi_E : E_1 \to E_2$ such that $V_G \subseteq V_1$ and $E_G \subseteq E_1$.*

The ϕ-renaming of G, denoted $\phi(G)$, is the graph (V, E, s, t, ℓ) defined by

$$V = \phi_V(V_G) \qquad s(\phi_E(e)) = \phi_V(s_G(e)) \qquad \ell(\phi_E(e)) = \ell_G(e)$$
$$E = \phi_E(E_G) \qquad t(\phi_E(e)) = \phi_V(t_G(e))$$

for every $e \in E_G$.

Definition 5 (Graph Isomorphism). *Graphs G and H are isomorphic, denoted $G \approx H$, if there is a graph renaming ϕ such that $H = \phi(G)$.*

Let L be a finite nonempty set of labels. In the sequel, we tacitly assume that all graphs have labels from L.

As motivated by the preceding sections, we allow to compose a context graph C and a match graph M by a "patch" J that may add edges between the nodes of C and M, as well as between the nodes of M.

Definition 6 (Patch). *Let C and M be disjoint graphs. A patch for C and M is a graph J such that $E_J \cap (E_C \cup E_M) = \varnothing$ and $V_J = s(E_J) \cup t(E_J)$, and*

$$(s(e), t(e)) \in (V_C \times V_M) \cup (V_M \times V_C) \cup (V_M \times V_M)$$

for every edge $e \in E_J$. In this relation mediated by J, we call C the context graph *and M the* match graph.

Definition 7 (Patch Composition). *Let J be a patch for a context graph C and a match graph M. The* patch composition *of C and M through patch J, denoted by $C \cdot_J M$, is the graph union $C \cup J \cup M$.*

Example 8. Consider the following graphs C, M and G, respectively:

The composition of C and M through patch $J = \{2 \xrightarrow{a} 3,\ 6 \xrightarrow{b} 2,\ 4 \xrightarrow{b} 5,\ 4 \xrightarrow{b} 6\}$ is G, in which C functions as the context graph and M functions as the match graph (w.r.t. J).

Before we consider the formal definition of rewriting, let us discuss the basic principle and motivate some of the design choices. As a first approximation, a graph rewrite rule $L \to R$ is a pair of graphs that behave like patterns. Since the edge and vertex identities in such rules are arbitrary (not to be confused with the edge labeling), we close the rule under isomorphism. The rule should also be applicable in contexts in which a patch connects a context and the pattern of the rule. The rule $L \to R$ thus give rise to rewrite steps of the form $C \cdot_J L' \to C \cdot_{J'} R'$ for graphs C, valid patches J, J' and graphs $L' \approx L$ and $R' \approx R$.

Additionally, we would like to exert control over the shape of patches in two ways. A graph rewriting rule should enable one to (a) constrain the choices for the patch J, and (b) define the patch J' in terms of J. For these purposes, we introduce the concepts of a patch type and a scheme.

Definition 9 (Patch Type). *A patch type T for a graph G is an unlabeled patch for G and the trivial graph with node set $\{\Box\}$. Here, the trivial graph functions as the context graph.*

Let J be a patch for a context graph C and match graph M, and T a patch type for M. A patch edge $(j_s \xrightarrow{\alpha} j_t) \in E_J$ (α any label) adheres to a patch type edge $(t_s \to t_t) \in E_T$ if the following conditions hold:

$$j_s \in V_C \Rightarrow t_s = \Box \qquad\qquad j_s \in V_M \Rightarrow j_s = t_s$$
$$j_t \in V_C \Rightarrow t_t = \Box \qquad\qquad j_t \in V_M \Rightarrow j_t = t_t$$

The patch J adheres to patch type T if there exists an adherence map from J to T, i.e., a function $f : E_J \to E_T$ such that e adheres to $f(e)$ for every $e \in E_J$.

The restriction to unlabeled patch type edges is motivated purely by simplicity. We intend to relax the definition in future work.

Proposition 10 (Unique Adherences). *Let the patch type T be a simple graph. If a patch J adheres to T, then the witnessing adherence map is unique.*

Intuitively, we use patch types to annotate the patterns of a rewrite rule. The result we call a scheme.

Definition 11 (Scheme). *A scheme is a pair (P, T) consisting of a graph P, called a pattern, and a patch type T for P.*

Example 12 (Depicting Schemes). We extend the representation for graphs to schemes (P, T) as shown on the right. The pattern P consists of the solid labeled arrows, and the patch type T consists of the dotted arrows. For dotted arrows without a source (or target), the source (or target) is implicitly the context graph node $\Box$. So T consists of the edges $\{\Box \to 1, \ 3 \to \Box, \ \Box \to 4, \ 4 \to \Box, \ 1 \to 3\}$.

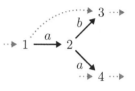

We are now ready to define a graph rewrite rule as a relation between schemes (P_L, T_L) and (P_R, T_R). We equip the rewrite rule with a "trace function" τ that relates edges in T_R back to edges in T_L, allowing us to interpret T_R as a transformation of T_L, in which patch edges may be freely moved, deleted, duplicated and inverted. For this we require the following constraint: if a patch type edge $e \in E_{T_R}$ connects to the context, the corresponding edge $\tau(e) \in E_{T_L}$ must also connect to the context. Without this constraint, it would not be clear how to interpret e's connection to the context.

Definition 13 (Quasi Patch Graph Rewrite Rule). *A* quasi patch graph rewrite rule $L \xrightarrow{\tau} R$ *is a pair of schemes* $L = (P_L, T_L)$ *and* $R = (P_R, T_R)$, *equipped with a* trace function $\tau : E_{T_R} \to E_{T_L}$ *that satisfies* $\square \in \{s(e), t(e)\} \implies \square \in \{s(\tau(e)), t(\tau(e))\}$ *for all* $e \in E_{T_R}$.

We normally require the left patch type T_L to be simple, so that the edges of T_L-adherent patches J adhere to a single edge in T_L (Proposition 10). As we shall see, this allows us to define a graph rewrite relation in which matches of a rule produce a unique result (modulo $\approx$).

Definition 14 (Patch Graph Rewrite Rule). *A patch graph rewrite rule is a quasi patch graph rewrite rule* $(P_L, T_L) \xrightarrow{\tau} (P_R, T_R)$ *in which* T_L *is simple.*

Since we restrict attention to unlabeled patch type graphs in this paper, we will use the opportunity to visualize the trace function τ by means of labels on patch type edges.

Example 15 (Depicting Rules). A depiction of a valid rewrite rule is:

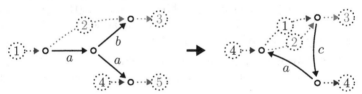

The trace function τ is visualized by means of labels on the type edges: τ maps type edges with label n on the right-hand side to the type edge with label n on the left-hand side. *Throughout the paper, colors are merely used as a supplementary visual aid.* (An application example will be given in Example 19.)

Definition 16 (Rule Isomorphism). *(Quasi) rewrite rules* $L_1 \xrightarrow{\tau_1} R_1$ *and* $L_2 \xrightarrow{\tau_2} R_2$ *are isomorphic, denoted* $L_1 \xrightarrow{\tau_1} R_1 \approx L_2 \xrightarrow{\tau_2} R_2$, *if there exists a graph renaming* ϕ *such that* $\phi_V(\square) = \square$, $\phi((L_1, R_1)) = (L_2, R_2)$, *and* $\phi_E \circ \tau_1 = \tau_2 \circ \phi_E$.

Definition 17 (Patch Graph Rewrite System). *A (quasi) patch graph rewrite system (PGR)* $\mathcal{R}$ *is a set of (quasi) rewrite rules. For* $\mathcal{R}$ *we define the isomorphism closure class* $\mathcal{R}^{\approx} = \{y \mid x \in \mathcal{R}, \ y \approx x\}$.

For a patch J, patch type T and adherence map $h : E_J \to E_T$, we define

$$ctx(e, h) = \begin{cases} \{\, s(e) \,\} & \text{if } s(h(e)) = \square \\ \{\, t(e) \,\} & \text{if } t(h(e)) = \square \\ \varnothing & \text{otherwise} \end{cases}$$

for every $e \in E_J$. So $ctx(e, h)$ contains the context node involved in the edge e, or is $\varnothing$ if the edge does not involve the context.

Definition 18 (Patch Graph Rewriting). *A (quasi) patch graph rewrite system $\mathcal{R}$ induces a rewrite relation $\to_{\mathcal{R}}$ on the set of graphs as follows:*
$C \cdot_J P_L \to_{\mathcal{R}} C \cdot_{J'} P_R$ *if*

(i) $(P_L, T_L) \xrightarrow{\tau} (P_R, T_R) \in \mathcal{R}^{\approx}$,
(ii) $h_L : E_J \to E_{T_L}$ *is an adherence map from patch J to patch type T_L,*
(iii) $h_R : E_{J'} \to E_{T_R}$ *is an adherence map from patch J' to patch type T_R, and*
(iv) for every $t \in E_{T_R}$ there exists a bijection $\sigma : h_R^{-1}(t) \to h_L^{-1}(\tau(t))$ such that
   $\ell_R(e) = \ell_L(\sigma(e))$ *and* $ctx(e, h_R) \subseteq ctx(\sigma(e), h_L)$ *for every* $e \in h_R^{-1}(t)$.

For such a rewrite step, we say that the graph $C \cdot_J P_L$ contains the redex *P_L.*

Example 19 (Application Example). The rule given in Example 15 gives rise to the following rewrite step:

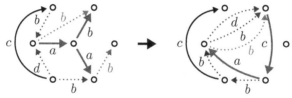

In the graph on the left we have highlighted the match (thick green) and the patch (dotted). We have additionally indicated the adherence map of the patch edges by reusing the colors of the rule definition.

We refer to Sect. 2 for many examples of simple rewrite rules and rewrite steps. A graph rewrite system modeling wait-for-graphs will be given in Sect. 4.

Remark 20 (Finding Redexes). Checking for the presence of a redex is simple. A graph G contains a redex with respect to rule $(P_L, T_L) \xrightarrow{\tau} (P_R, T_R) \in \mathcal{R}$ if and only if

1. there exists a subgraph M of G isomorphic to P_L, and
2. every edge $e \notin E_M$ incident to a $v \in V_M$ adheres to an edge in T_L.

Definition 18 can be understood in more operational terms as follows.

Lemma 21 (Constructing J'). *If conditions (i) and (ii) of Definition 18 are satisfied (fixing some adherence map h_L), the patch J' and adherence map h_R that satisfy conditions (iii) and (iv) are uniquely determined up to isomorphism. The patch J' can be constructed using the following procedure.*

   For every type edge $t = (t_s \to t_t) \in E_{T_R}$, consider every patch edge $j = (j_s \xrightarrow{\alpha} j_t) \in E_J$ for which $h_L(j) = \tau(t) = (t_s^{\tau} \to t_t^{\tau}) \in E_{T_L}$. There are five exclusive cases:

1. If $\square \notin \{t_s, t_t\}$, add a new edge $t_s \xrightarrow{\alpha} t_t$ to J'.
2. If $t_s = t_s^\tau = \square$, add a new edge $j_s \xrightarrow{\alpha} t_t$ to J'.
3. If $t_t = t_t^\tau = \square$, add a new edge $t_s \xrightarrow{\alpha} j_t$ to J'.
4. If $t_s = t_t^\tau = \square$, add a new edge $j_t \xrightarrow{\alpha} t_t$ to J'.
5. If $t_t = t_s^\tau = \square$, add a new edge $t_s \xrightarrow{\alpha} j_s$ to J'.

Here, the "new" edge j' is an edge not in C, P_R or the intermediate construction of J'. The adherence map h_R is defined such that $h_R(j') = t$ for each of the considered j' and t.

Non-quasi rules have the following desirable property.

Proposition 22 (Rule Determinism). *Let* $G = C \cdot_J P_L$. *If a rule* $(P_L, T_L) \xrightarrow{\tau} (P_R, T_R) \in \mathcal{R}^\approx$ *derives both* $G \to_\mathcal{R} C \cdot_{J'} P_R = G'$ *and* $G \to_\mathcal{R} C \cdot_{J''} P_R = G''$, *then* $G' \approx G''$.

Proof. This is a direct consequence of Proposition 10 and Lemma 21. □

In contrast to (non-quasi) graph rewrite rules, quasi rules are not generally deterministic. For instance, consider the quasi rewrite rule

which can match any graph G consisting of two nodes x and y and n edges from x to y. For each $e \in E_G$, the left adherence map h_L can either map e to the patch type edge labeled with 1, or to the type edge labeled with 2. Thus, 2^n choices for h_L are possible, and each choice causes a different subset of J to be deleted in a single rewrite step.

Notation 23 (Shorthand Notation). Given a pattern P, we often want to allow for any patch edges between the nodes of a subset $N \subseteq V_P$ as well as the context node $\square$. In the notation we have discussed so far, we would then need to draw the complete patch type graph induced by $N \cup \{\square\}$ (minus the loop on $\square$), which consists of $(|N| + 1)^2 - 1$ patch type edges.

To avoid spaghetti-like figures, we extend the visual presentation of schemes by allowing each node to be annotated with a set of names (written without set brackets). We say that a node *has name* x if x is in the set of names of this node. So a node can have 0 or more names. The name annotations are then shorthand for the following patch type edges:

(i) For every node n and name x of n, the node n has the two patch type edges $(\square, x) \cdots\blacktriangleright n \cdots\blacktriangleright (x, \square)$ from and to the context.

(ii) For every pair of nodes n, m and every name x of n and y of m, there is implicitly the patch type edge $n \cdots (x, y) \cdots\blacktriangleright m$ from n to m. Here n and m can be the same node, and x can be equal to y.

Observe that rules are non-quasi iff every node in the left-hand side has at most one name. We therefore require that distinct nodes do not share names.

As an example, a rule for merging two nodes can be written as

$$\underset{x}{\circ} \xrightarrow{a} \underset{y}{\circ} \quad \blacktriangleright \quad \underset{x,y}{\circ} \tag{1}$$

which is shorthand for

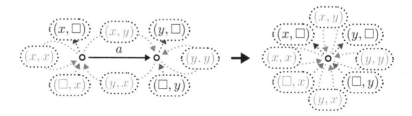

4   Modeling Wait-for Graphs and Deadlock Detection

We now give a more extensive and more realistic modeling example that showcases the expressive power of PGR.

A *wait-for graph* [14] is a hypergraph in which nodes represent processes, and hyperedges represent requests for resources. A hyperedge has a single source p, representing the process requesting the resources, and $M > 0$ targets distinct from p, representing the processes from which a resource is requested. The process p requires $0 < N \le M$ of these resources. Thus, for a fixed M, there are multiple types of hyperedges, each representing a particular N-out-of-M request. Processes can have at most one outgoing N-out-of-M request.

The following distributed system behavior is associated with wait-for graphs. A process without an outgoing request is said to be *unblocked*. An unblocked process can *grant* an incoming request, deleting the edge, or create a new N-out-of-M request. A process becomes unblocked when its pending N-out-of-M request is *resolved*, i.e., when N targeted processes have granted the request.

In order to better illustrate some of PGR's transformational power, we introduce one additional, noncanonical behavior. We consider a process p *overloaded* when it has $n > 2$ incoming requests. When p is overloaded, a clone of p, $c(p)$, may be created which takes over $n - 2$ of p's incoming requests. Because we assume that any outgoing request must be resolved before any incoming requests can be resolved, $c(p)$ replicates p's outgoing request, if p has one.

We first define a graph grammar that defines the class of valid wait-for graphs. Then, we will show how to augment the rule set in order to model the distributed system behavior. Finally, we explain how deadlocks can be detected. Throughout, we encode hypergraphs as multigraphs. Note that in this encoding, vertices representing processes are always free of loops, while vertices representing hyperedges always have loops. Hence, the given rules can appropriately discriminate between the two types of vertices.

4.1   Wait-for Graph Grammar

The starting graph of the grammar is the empty graph, denoted by $\varnothing$. Rule

$$\varnothing \quad \longrightarrow \quad \circ \qquad\qquad\qquad \text{(CREATE)}$$

models process creation, and rule

$$\text{(1-OF-1)}$$

allows constructing a valid 1-out-of-1 request between nodes. Labels z and s should be interpreted as 0 and the successor function, respectively, so that n s-loops encode that n requests are yet to be granted.

In the grammar, any N-out-of-M request can be extended to a valid N-out-of-$(M + 1)$ request using rule

$$\text{(EXT-0)}$$

and to a valid $(N + 1)$-out-of-$(M + 1)$ request using rule

$$\text{(EXT-1)}$$

These four rules suffice for generating any valid wait-for graph.

4.2   System Modeling

To model a distributed system, we need rule CREATE for process creation, as well as its inverse, DESTROY, for process destruction. Note that DESTROY constrains the process selected for destruction to be isolated in our framework (i.e., it is not associated with any pending requests), as desired.

Any N-out-of-M request is understood to be an atomic action. So for, e.g., modeling 2-out-of-2 requests, we need the rule

$$\text{(2-OF-2)}$$

Such rules can be easily simulated by a contiguous sequence of rewrite steps 1-OF-1 · EXT-0* · EXT-1*, in which the node making the request remains fixed. We omit the details.

A grant request may be modeled by

(GRANT)

and a request resolution by

(RESOLVE)

This leaves only the cloning behavior for an overloaded process p. This requires two rules: one for the case in which p is unblocked, and one for the case in which p is blocked. We use the shorthand notation introduced in Notation 23, so that named nodes r_i induce type edges among themselves and from and into the context.[1]

The case in which p is unblocked is modeled by rule

(CLONE-1)

and the case in which p is blocked is modeled by rule

(CLONE-2)

Cloning would be easier to express if PGR were to be extended with support for hyperedges and cardinality constrained type edges. We envision a rule like

(CLONE*)

to capture the same cases as rules CLONE-1 and CLONE-2 combined. We leave such an extension to future work. In particular, the precise semantics of hyperedge transformation would have to be determined.

---

[1] The type edges between distinctly named nodes $r_i \neq r_j$ are redundant in the considered scenario, since we know that these type edges will never have adherents.

4.3   Deadlock Detection

Deadlock detection on some valid wait-for graph G can be performed by restricting the rewrite system to rules GRANT, RESOLVE and DESTROY, yielding a terminating rewrite system. Then the network represented by G contains a deadlock if and only if the (unique) normal form of G is the empty graph $\varnothing$.

5   Comparison

We compare PGR to several other rewriting frameworks. We have selected these frameworks because of their popularity and/or because they bear certain similarities to our approach.

Double-Pushout (DPO). Ehrig et al.'s double-pushout approach (DPO) [9] is one of the most prominent approaches in graph rewriting.

A rewrite rule in the DPO approach is of the form $L \hookleftarrow K \rightarrow R$, where L is the subgraph to be replaced by subgraph R. The graph K is an "interface", used to identify a part of L with a part of R, and it can be thought of as describing which part of L is preserved by the rule. The identification is formally established through the inclusion $L \hookleftarrow K$ and the graph morphism $K \rightarrow R$. The morphism $K \rightarrow R$ may be non-injective, allowing it to merge nodes that are in the interface.

A DPO rewrite rule $L \xleftarrow{\varphi} K \xrightarrow{\psi} R$ is applied inside a graph G as follows [7,8]. Let $m : L \rightarrow G$ be a graph morphism, which we may assume to be injective [16]. The graph $m(L) \approx L$ is said to be a *match* for L. The arising rewrite step replaces $m(L)$ of G by a fresh copy $c(R)$ of R, redirecting edges left dangling by the removal of a $v \in m(L)$ to node $c(\psi(\varphi^{-1}(m^{-1}(v))))$. For the redirection of edges to work, nodes that leave dangling ends need to be part of the interface, that is, in $m(\varphi(K))$. This is known as the "gluing condition".[2] If the gluing condition is not met, the rewrite step is not permitted.

Using Notation 23, it is easy to see that PGR at least as expressive as DPO with respect to the generated rewrite relation. A DPO rule $L \xleftarrow{\varphi} K \xrightarrow{\psi} R$ can be directly simulated by a PGR rule $L \rightarrow R$ in which the nodes are annotated with their (set of) names in the interface: $v \in V_L$ is annotated with the names $\varphi^{-1}(v)$, and $v \in V_R$ is annotated with the names $\psi^{-1}(v)$.

However, DPO is stronger in one respect: a DPO rewrite step preserves the subgraph specified in K, whereas a PGR rewrite can be thought of as destroying and replacing the left-hand side of the rule. The consequences for metaproperties relating to parallelism and concurrency will have to be investigated.

Generalized DPO. In some variants of DPO, the inclusion $L \hookleftarrow K$ is generalized to a (possibly non-injective) morphism $\varphi : K \rightarrow L$. Intuitively, this allows a node v of L to be "split apart" in the interface K. Applying the DPO method to a host graph ensures that the patch graph edges incident to v will be incident to

---

[2] By the injectivity assumption for m, we need not consider what is known as the "identification condition".

one of v's split copies. It does not dictate how these edges should be distributed. Thus, such rewrite steps are non-deterministic.

Generalized DPO rules $L \xleftarrow{\varphi} K \xrightarrow{\psi} R$ can be translated to PGR rules $L \rightarrow R$ in the same way as discussed for DPO. Since φ is no longer required to be injective, nodes $v \in V_L$ can be annotated with multiple names $\varphi^{-1}(v)$, thereby leading to (non-deterministic) quasi rules (Definition 13).

Single-Pushout (SPO). The single-pushout (SPO) approach by Löwe [18] is the destructive sibling of DPO. It is operationally like DPO, but it drops the gluing condition. Any edges that would become dangling in the host graph are instead removed.

An SPO rule $L \xleftarrow{\varphi} K \xrightarrow{\psi} R$ can be simulated by a PGR rule $L \rightarrow R$ with annotations as discussed above for DPO, except that each node $v \in V_L$, for which $\varphi^{-1}(v)$ is empty, is now annotated by a fresh name. The rewrite step will then delete all patch edges connected to such a node.

DPO Rewriting in Context (DPO-C). The DPO Rewriting in Context (DPO-C) approach by Löwe [19,20] addresses the issue of non-determinism in generalized variants of DPO, using ingoing and outgoing arrow annotations to dictate how these edges should be distributed over split copies. The visual representation of DPO-C therefore bears some similarity to that of PGR. In addition, absence of arrow annotations also define negative application conditions like in PGR. However, the patch cannot be transformed as freely as in PGR. For instance, see rule (2) below.

AGREE. AGREE [3] and PBPO [4] by Corradini et al. extend DPO with the ability to erase and clone nodes, while also being able to specify how patch edges are distributed among the copies. For this purpose, a "filter" for the edges determines what kind of patch edges are to be dropped. This filter can distinguish different types of edges based on their source, target and label. Thereby AGREE and PBPO subsume mildly restricted versions of DPO, SPO, and other formalisms.

PGR has some features that are not present in AGREE and PBPO. First, in PGR rule applicability can be restricted by conditions on the permitted shape of the patch. Second, PGR allows (almost) arbitrary redirecting, moving and copying of patch edges outside the scope of AGREE and PBPO. For instance,

$$\boxed{1} \rightarrow \circ \rightarrow \boxed{2} \quad \boxed{6} \rightarrow \circ \rightarrow \boxed{7} \quad \Rightarrow \quad \boxed{6} \rightarrow \circ \rightarrow \boxed{2} \quad \boxed{1} \rightarrow \circ \rightarrow \boxed{7} \qquad (2)$$

cannot be expressed in the latter frameworks. Also inverting the direction of patch edges, or moving patch edges between nodes of the pattern is not possible in AGREE and PBPO.

On the other hand, AGREE and PBPO capture some transformations that cannot be expressed in PGR. First, AGREE and PBPO can express some global transformations, unlike PGR. Second, the "patch edge filter" in AGREE and PBPO can distinguish patch edges depending on their label and the "type" of the source/target in the context (here the type is given by some type graph).

Both features are not present in PGR as presented in this paper. However, PGR can be extended with constraints on the patch type edges. We leave the investigation of a suitable constraint language to future work.

Nagl's Approach. Nagl [21] has defined a very powerful graph transformation approach. Rather than identifying "gluing points" for the replacement of L by R in a host graph G (as in the previous approaches), rules are equipped with expressions that describe how R is to be embedded into the remainder graph $G^- = G - L$. An expression can, e.g., specify that an edge must be created from $u \in G^-$ to $v \in R$ if there existed a path (of certain length and containing certain labels) from u to a $w \in L$. Thus, the embedding context may no longer even be local.

While not all of these transformations are supported by PGR, the expressions are arguably much less intuitive than our representation, in which both application conditions and transformations are visualized in a unified manner.

Habel et al.'s Approach. Habel et al. [15] have introduced graph grammars with rule-dependent application conditions that also admit a very intuitive visual representation. These conditions are more powerful than PGR's application conditions, since they can extend arbitrarily far into the context graph. However, transformations are not included in the approach, unlike in PGR, in which the notations for application conditions and transformations are unified.

Drags. To generalize term rewriting techniques to the domain of graphs, Dershowitz and Jouannaud [6] have recently defined the *drag* data structure and framework. A drag is a multigraph in which nodes are labeled by symbols that have an arity equal to the node's outdegree. Nodes labeled with nullary variable symbols are called *sprouts*, and resemble output ports. In addition, the drag comes equipped with a finite number of *roots*, which resemble input ports.

A composition operation $\otimes_\xi$ for drags, parameterized by a two-way *switchboard* ξ identifying sprouts with roots, gives rise to a rewrite relation $W \otimes_\xi L \rightarrow W \otimes_\xi R$. For this rewrite relation to be well-defined, it is required, among others, that L and R have the same number of roots and the same multisets of variables.

Since drag rewriting imposes arity restrictions on nodes, it is more restrictive than patch rewriting concerning the shapes of the graphs that can be rewritten. As drag rewrite steps are local, we believe that PGR can simulate them, but we leave this investigation to future work.

6   Conclusion

We have introduced *patch graph rewriting*, a framework for graph rewriting that enriches the rewrite rules with a simple, yet powerful language for constraining and transforming the local embedding—or *patch*.

For future work, we plan to investigate various meta-properties central in graph rewriting [8], in particular confluence [12,17,22], termination [2,5], the concurrency theorem [10], decomposability and reversibility of rules. We intend to study these properties both globally, for all graphs, as well as locally [11,13], for a given language of graphs [23]. Furthermore, we are interested in extending the framework with constraint labels on patch type edges, and in allowing label transformations. We believe this could be useful for modeling a larger class of distributed algorithms [14]. Another interesting direction of research is an equational perspective on patch rewriting, as similarly investigated by Ariola and Klop for term graph rewriting [1].

Acknowledgments. This paper has benefited from discussions with Jan Willem Klop, Nachum Dershowitz, Femke van Raamsdonk, Roel de Vrijer, and Wan Fokkink. We thank Andrea Corradini and the anonymous reviewers for their useful suggestions. Both authors received funding from the Netherlands Organization for Scientific Research (NWO) under the Innovational Research Incentives Scheme Vidi (project. No. VI.Vidi.192.004).

References

1. Ariola, Z.M., Klop, J.W.: Equational term graph rewriting. Fundamenta Informaticae. **26**(3,4), 207–240 (1996). https://doi.org/10.3233/FI-1996-263401
2. Bruggink, H.J.S., König, B., Zantema, H.: Termination analysis for graph transformation systems. In: Diaz, J., Lanese, I., Sangiorgi, D. (eds.) TCS 2014. LNCS, vol. 8705, pp. 179–194. Springer, Heidelberg (2014). https://doi.org/10.1007/978-3-662-44602-7_15
3. Corradini, A., Duval, D., Echahed, R., Prost, F., Ribeiro, L.: AGREE – algebraic graph rewriting with controlled embedding. In: Parisi-Presicce, F., Westfechtel, B. (eds.) ICGT 2015. LNCS, vol. 9151, pp. 35–51. Springer, Cham (2015). https://doi.org/10.1007/978-3-319-21145-9_3
4. Corradini, A., Duval, D., Echahed, R., Prost, F., Ribeiro, L.: The PBPO graph transformation approach. J. Log. Algebr. Meth. Program. **103**, 213–231 (2019). https://doi.org/10.1016/j.jlamp.2018.12.003
5. Dershowitz, N., Jouannaud, J.-P.: Graph path orderings. In LPAR, volume 57 of EPiC Series in Computing, EasyChair, pp. 307–325 (2018). https://doi.org/10.29007/6hkk
6. Dershowitz, N., Jouannaud, J.-P.: Drags: a compositional algebraic framework for graph rewriting. Theor. Comput. Sci. **777**, 204–231 (2019). https://doi.org/10.1016/j.tcs.2019.01.029
7. Ehrig, H.: Tutorial introduction to the algebraic approach of graph grammars. In: Ehrig, H., Nagl, M., Rozenberg, G., Rosenfeld, A. (eds.) Graph Grammars 1986. LNCS, vol. 291, pp. 1–14. Springer, Heidelberg (1987). https://doi.org/10.1007/3-540-18771-5_40
8. Ehrig, H., Korff, M., Löwe, M.: Tutorial introduction to the algebraic approach of graph grammars based on double and single pushouts. In: Ehrig, H., Kreowski, H.-J., Rozenberg, G. (eds.) Graph Grammars 1990. LNCS, vol. 532, pp. 24–37. Springer, Heidelberg (1991). https://doi.org/10.1007/BFb0017375
9. Ehrig, H., Pfender, M., Schneider, H.J.: Graph-grammars: an algebraic approach. In: Proceedings of the 14th Annual Symposium on Switching and Automata Theory, Swat 1973, pp. 167–180. IEEE Computer Society (1973). https://doi.org/10.1109/SWAT.1973.11

10. Ehrig, H., Rosen, B.K.: Parallelism and concurrency of graph manipulations. Theoret. Comput. Sci. **11**(3), 247–275 (1980). https://doi.org/10.1016/0304-3975(80)90016-X

11. Endrullis, J., de Vrijer, R.C., Waldmann, J.: Local termination theory and practice. Logical Methods Comput. Sci. **6**(3) (2010)

12. Endrullis, J., Klop, J.W., Overbeek, R.: Decreasing diagrams for confluence and commutation. Logical Methods Comput. Sci. **16**(1) (2020). https://doi.org/10.23638/LMCS-16(1:23)2020

13. Endrullis, J., Zantema, H.: Proving non-termination by finite automata. In: Proceedings Conference on Rewriting Techniques and Applications, RTA 2015, LIPIcs, vol. 36, pp. 160–176. Schloss Dagstuhl - Leibniz-Zentrum fuer Informatik (2015). https://doi.org/10.4230/LIPIcs.RTA.2015.160

14. Fokkink, W.: Distributed Algorithms: An Intuitive Approach. MIT Press, Cambridge (2014)

15. Habel, A., Heckel, R., Taentzer, G.: Graph grammars with negative application conditions. Fundam. Inform. **26**(3/4), 287–313 (1996). https://doi.org/10.3233/FI-1996-263404

16. Habel, A., Müller, J., Plump, D.: Double-pushout graph transformation revisited. Math. Struct. Comput. Sci. **11**(5), 637–688 (2001). https://doi.org/10.1017/S0960129501003425

17. Heckel, R., Küster, J.M., Taentzer, G.: Confluence of typed attributed graph transformation systems. In: Corradini, A., Ehrig, H., Kreowski, H.-J., Rozenberg, G. (eds.) ICGT 2002. LNCS, vol. 2505, pp. 161–176. Springer, Heidelberg (2002). https://doi.org/10.1007/3-540-45832-8_14

18. Löwe, M.: Algebraic approach to single-pushout graph transformation. Theor. Comput. Sci. **109**(1&2), 181–224 (1993). https://doi.org/10.1016/0304-3975(93)90068-5

19. Löwe, M.: Double-pushout rewriting in context. In: Mazzara, M., Ober, I., Salaün, G. (eds.) STAF 2018. LNCS, vol. 11176, pp. 447–462. Springer, Cham (2018). https://doi.org/10.1007/978-3-030-04771-9_32

20. Löwe, M.: Double-pushout rewriting in context. In: Guerra, E., Orejas, F. (eds.) ICGT 2019. LNCS, vol. 11629, pp. 21–37. Springer, Cham (2019). https://doi.org/10.1007/978-3-030-23611-3_2

21. Nagl, M.: Set theoretic approaches to graph grammars. In: Ehrig, H., Nagl, M., Rozenberg, G., Rosenfeld, A. (eds.) Graph Grammars 1986. LNCS, vol. 291, pp. 41–54. Springer, Heidelberg (1987). https://doi.org/10.1007/3-540-18771-5_43

22. Plump, D.: Confluence of graph transformation revisited. In: Middeldorp, A., van Oostrom, V., van Raamsdonk, F., de Vrijer, R. (eds.) Processes, Terms and Cycles: Steps on the Road to Infinity. LNCS, vol. 3838, pp. 280–308. Springer, Heidelberg (2005). https://doi.org/10.1007/11601548_16

23. Rensink, A.: Canonical graph shapes. In: Schmidt, D. (ed.) ESOP 2004. LNCS, vol. 2986, pp. 401–415. Springer, Heidelberg (2004). https://doi.org/10.1007/978-3-540-24725-8_28

Hypergraph Basic Categorial Grammars

Tikhon Pshenitsyn[(✉)] [iD]

Department of Mathematical Logic and Theory of Algorithms,
Faculty of Mathematics and Mechanics, Lomonosov Moscow State University,
GSP-1, Leninskie Gory, Moscow 119991, Russian Federation
tpshenitsyn@lpcs.math.msu.su

Abstract. This work is an attempt to generalize categorial grammars, which deal with string languages, to hypergraphs. We consider a particular approach called basic categorial grammar (BCG) and introduce its natural extension to hypergraphs — *hypergraph basic categorial grammar* (HBCG). We show that BCGs can be naturally embedded in HBCGs. It turns out that, as BCGs are equivalent to context-free grammars, HBCGs are equivalent to hyperedge replacement grammars in generalized Greibach normal form. We also present several structural properties of HBCGs. From practical point of view, we show that HBCGs can be used to describe semantics of sentences of natural languages. We incorporate the lambda semantics into the new mechanism in the same way as it is done for BCGs and show that such an embedding allows one to describe semantics of sentences with cross-serial dependencies.

1 Introduction

Formal mechanisms serving to describe formal (string) languages include two large classes: generative grammars and categorial grammars. The former generate strings using rewriting rules (productions): a string is correct if it can be produced by a grammar. The most well-known example of such a formalism is context-free grammar (CFG). Categorial grammars, in opposite, take the whole string at first and then check whether it is correct as follows: there is a set of types and a uniform mechanism which defines what sequences of types are correct; a particular grammar contains a lexicon, i.e. a correspondence between symbols in an alphabet and types of the system. In order to check whether a string $a_1 \ldots a_n$ is correct one chooses types $T_1, \ldots, T_n$ such that a_i corresponds to T_i in the grammar and then checks if $T_1, \ldots, T_n$ is correct with respect to uniform rules of the formalism.

One of the most fundamental examples of categorial grammars is basic categorial grammar (BCG). It is introduced in works of Ajdukiewicz [2] and Bar-Hillel [3]. Types in BCGs are built of primitive types Pr using left and right divisions $\backslash, /$. There are two uniform rules of interaction between types: given $A, (A\backslash B)$ or $(B/A), A$ standing nearby each other within a sequence of types

---

The study was funded by RFBR, project number 20-01-00670.

F. Gadducci and T. Kehrer (Eds.): ICGT 2020, LNCS 12150, pp. 146–162, 2020.
https://doi.org/10.1007/978-3-030-51372-6_9

one can replace them by B. The sequence of types is said to be correct iff it can be reduced to some distinguished $s \in Pr$. It is proved by Gaifman [4] that this approach has the same generating power as context-free grammars.

BCGs can serve to describe natural languages. E.g. the sentence *Tim thinks Helen is smart* corresponds to the sequence of types NP, $(NP\backslash S)/S$, NP, $(NP\backslash S)/ADJ$, ADJ, which can be reduced to S; thus this sentence is grammatically correct[1]. Moreover, it is possible to combine BCGs with the λ-calculus and to model semantics of this sentence. Namely, if the λ-term $\lambda x.smart(x)$ is assigned to the adjective *smart*, the λ-term $\lambda s.\lambda x.think(s)(x)$ is assigned to the verb *thinks*, and the λ-term $\lambda f.\lambda x.f(x)$ is assigned to *is*, then the reductions that are done in order to obtain S from the sequence above can be treated as applications in the λ-calculus; hence, the sentence above is described by the meaning $think(smart(Helen))(Tim)$.

Let us return to generative grammars. The principles underlying them can be extended to graphs; a class of resulting formalisms is called graph grammars. In this paper we focus on a particular approach to generating graphs named hyperedge replacement grammar (HRG in short). An overview on HRGs can be found in [9]. We are interested in HRG because it is closely related to CFG: definitions of these formalisms are similar to each other; consequently, they share many crucial properties, e.g. the pumping lemma and the fixed-point theorem. Moreover, HRGs represent a natural extension of CFGs, since strings can be represented by string graphs and CFGs can be modeled using HRGs.

The question we are going to discuss in this paper is how to generalize basic categorial grammars to hypergraphs and to obtain a categorial mechanism related to HRGs. We present such a generalization — *hypergraph basic categorial grammars*. We extend notions of types, of reduction laws, and of semantics to hypergraphs. As expected, the resulting mechanism is closely related both to BCGs and HRGs, which is shown in Sects. 5 and 6. In Sect. 7 several structural properties of HBCGs are studied. In Sect. 8 we show how to enrich our mechanism with the lambda semantics. In Sect. 9 we show an application of our theory to linguistics.

2   Basic Categorial Grammars (for Strings)

The survey of categorial grammars including basic categorial grammars can be found in [8]. Here we introduce the main definitions to show connections with the new formalism.

Let us fix a countable set $Pr = \{p_i\}_{i=1}^{\infty}$ of primitive types.

Definition 2.1. *The set Tp of types is defined inductively as follows: it is the least set such that $Pr \subseteq Tp$ and for each $A, B \in Tp$ $B\backslash A, A/B$ are also in Tp.*

---

[1] In the types considered, NP stands for noun phrases, ADJ stands for adjectives, and S stands for sentences.

Throughout this paper small letters $p, q, \ldots$ and strings composed of them (e.g. np, cp) range over primitive types. Capital letters $A, B, \ldots$ usually range over types (however, graphs are often referred to as G and H).

There are two rules of BCGs:

1. $\Gamma, A, A \backslash B, \Delta \mapsto \Gamma, B, \Delta$
2. $\Gamma, B/A, A, \Delta \mapsto \Gamma, B, \Delta$

Here Γ, Δ are finite (possibly empty) sequences of types. Thus $\mapsto$ is a relation on $Tp^+ \times Tp^+$. We denote by $\overset{*}{\mapsto}$ its reflexive transitive closure. $\Gamma \overset{k}{\mapsto} \Delta$ denotes that Δ is obtained from Γ in k steps (the same notation is used for all the relations in this work).

Definition 2.2. A basic categorial grammar *is a tuple* $Gr = \langle \Sigma, s, \rhd \rangle$ *where* Σ *is a finite set (alphabet), s is a distinguished primitive type, and* $\rhd \subseteq \Sigma \times Tp$ *is a finite binary relation, i.e. it assigns a finite number of types to each symbol in the alphabet.*

The language $L(Gr)$ *generated by* Gr *is the set of all strings* $a_1 \ldots a_n$ *for which there are types* $T_1, \ldots, T_n$ *such that* $a_i \rhd T_i$, *and* $T_1, \ldots, T_n \overset{*}{\mapsto} s$.

3   Hypergraphs and Operations on Them

This section is concerned with definitions related to hypergraphs. All the notions except for compression are well known and widely accepted (see [9]). Note that we use a slightly different notation from that in [9].

3.1   Hypergraphs, Sub-hypergraphs

$\mathbb{N}$ includes 0. The set Σ^* is the set of all strings over the alphabet Σ including the empty string ε. The length $|w|$ of the word w is the number of symbols in w. Σ^+ denotes the set of all nonempty strings. The set $\Sigma^{\circledast}$ is the set of all strings consisting of distinct symbols. The set of all symbols contained in the word w is denoted by $[w]$. If $f : \Sigma \to \Delta$ is a function from one set to another, then it is naturally extended to a function $f : \Sigma^* \to \Delta^*$ ($f(\sigma_1 \ldots \sigma_k) = f(\sigma_1) \ldots f(\sigma_k)$).

Let C be some fixed set of labels for whom the function $type : C \to \mathbb{N}$ is considered.

Definition 3.1. A hypergraph G over C *is a tuple* $G = \langle V, E, att, lab, ext \rangle$ *where* V *is the set of* nodes, E *is the set of* hyperedges, $att : E \to V^{\circledast}$ *assigns a string (i.e. an ordered set) of* attachment *nodes to each edge,* $lab : E \to C$ *labels each edge by some element of C in such a way that* $type(lab(e)) = |att(e)|$ *whenever* $e \in E$, *and* $ext \in V^{\circledast}$ *is a string of* external *nodes.*

Components of a hypergraph G are denoted by $V_G, E_G, att_G, lab_G, ext_G$ *resp.*

In the remainder of the paper, hypergraphs are simply called graphs, and hyperedges are simply called edges. The set of all graphs with labels from C is denoted by $\mathcal{H}(C)$. In drawings of graphs black dots correspond to nodes, labeled squares correspond to edges, att is represented with numbered lines, and external nodes are depicted by numbers in brackets. If an edge has exactly two attachment nodes, it can be denoted by an arrow (which goes from the first attachment node to the second one).

Definition 3.2. *The function type (or $type_G$ to be exact) returns the number of nodes attached to some edge in a graph G: $type_G(e) := |att_G(e)|$. If G is a graph, then $type(G) := |ext_G|$.*

Example 3.1. There are two hypergraphs on Fig. 1, to the left and right of the symbol $\underset{\chi}{\longmapsto}$. E.g., the graph on the left has 5 edges; there are three edges within it for which *type* equals 2, one edge with *type* equal to 1, and an edge with *type* equal to 3.

Definition 3.3. *A sub-hypergraph (or just subgraph) H of a graph G is a hypergraph such that $V_H \subseteq V_G$, $E_H \subseteq E_G$, and for all $e \in E_H$ $att_H(e) = att_G(e)$, $lab_H(e) = lab_G(e)$.*

Definition 3.4. *If $H = \langle \{v_i\}_{i=1}^n, \{e_0\}, att, lab, v_1 \ldots v_n \rangle$, $att(e_0) = v_1 \ldots v_n$ and $lab(e_0) = a$, then H is called* a handle. *It is denoted by $\odot(a)$.*

Definition 3.5. *Let $H \in \mathcal{H}(C)$ be a graph, and let $f : E_H \rightarrow C$ be a relabeling function. Then $f(H) = \langle V_H, E_H, att_H, lab_{f(H)}, ext_H \rangle$ where $lab_{f(H)}(e) = f(e)$ for all e in E_H. It is required that $type(lab_H(e)) = type(f(e))$ for $e \in E_H$.*

If one wants to relabel only one edge e_0 within H with a label a, then the result can be denoted by $H[e_0 := a]$

Definition 3.6. *An isomorphism between graphs G and H is a pair of bijective functions $\mathcal{E} : E_G \rightarrow E_H$, $\mathcal{B} : V_G \rightarrow V_H$ such that $att_H \circ \mathcal{E} = \mathcal{B} \circ att_G$, $lab_G = lab_H \circ \mathcal{E}$, $\mathcal{B}(ext_G) = ext_H$. In this work, we do not distinguish between isomorphic graphs.*

3.2 Operations on Graphs

In graph formalisms certain graph transformation are in use. To generalize categorial grammars we present the following operation called compression.

Compression. Let G be a graph, and let H be a subgraph of G. Compression of H into an a-labeled edge within G is a procedure of transformation of G, which can be done under the following conditions:

(a) For each $v \in V_H$, if v is attached to some edge $e \in E_G \setminus E_H$ (i.e. $v \in [att(e)]$), then v has to be external in H ($v \in [ext_H]$).

(b) If $v \in V_H$ is external in G, then it is external in H ($[ext_G] \cap V_H \subseteq [ext_H]$).
(c) $type(H) = type(a)$.

Then the procedure is the following:

1. Remove all nodes of V_H except for those of ext_H from V_G;
2. Remove E_H from E_G;
3. Add a new edge $\widetilde{e}$;
4. Set $att(\widetilde{e}) = ext_H$, $lab(\widetilde{e}) = a$.

Let $G[\![a/H]\!]$ (or $G[\![a, \widetilde{e}/H]\!]$) denote the resulting graph. Formally, $G[\![a/H]\!] = \langle V', E', att', lab', ext_G \rangle$, where $V' = V_G \setminus (V_H \setminus ext_H)$, $E' = \{\widetilde{e}\} \cup (E_G \setminus E_H)$, $att'(e) = att_G(e)$, $lab'(e) = lab_G(e)$ for $e \neq \widetilde{e}$, and $att'(\widetilde{e}) = ext_H$, $lab'(\widetilde{e}) = a$.

Replacement. This procedure is defined in [9]. In short, the replacement of an edge e_0 in G with a graph H can be done if $type(e_0) = type(H)$ as follows:

1. Remove e_0;
2. Insert an isomorphic copy of H (namely, H and G have to consist of disjoint sets of nodes and edges);
3. For each i, fuse the i-th external node of H with the i-th attachment node of e_0.

To be more precise, the set of edges in the resulting graph is $(E_G \setminus \{e_0\}) \cup E_H$, and the set of nodes is $V_G \cup (V_H \setminus ext_H)$. The result is denoted by $G[H/e_0]$.

Proposition 3.1. *Compression and replacement are opposite:*

1. $G[\![a, e/H]\!][H/e] \equiv G$ *(for a subgraph H of G satisfying conditions (a) and (b); a is an arbitrary label);*
2. $G[H/e][\![a, e/H]\!] \equiv G$ *(provided $e : type(e) = type(H)$ and $a = lab_G(e)$).*

3.3   Hyperedge Replacement Grammars

Definition 3.7. *A hyperedge replacement grammar is a tuple $Gr = \langle N, \Sigma, P, S \rangle$, where N is a finite alphabet of nonterminal symbols, Σ is a finite alphabet of terminal symbols ($N \cap \Sigma = \emptyset$), P is a set of productions, and $S \in N$. Each production is of the form $A \to H$ where $A \in N$, $H \in \mathcal{H}(N \cup \Sigma)$ and $type(A) = type(H)$.*

Edges labeled by terminal (nonterminal) symbols are called *terminal (non-terminal) edges*.

If G is a graph, $e_0 \in E_G$, $lab(e_0) = A$ and $A \to H \in P$, then G directly derives $G[H/e_0]$ (denote $G \Rightarrow G[H/e_0]$). The transitive reflexive closure of $\Rightarrow$ is denoted by $\overset{*}{\Rightarrow}$. If $G \overset{*}{\Rightarrow} H$, then G is said to derive H. The corresponding sequence of production applications is called a derivation.

Definition 3.8. *The* language generated by an HRG $\langle N, \Sigma, P, S \rangle$ *is the set of graphs $H \in \mathcal{H}(\Sigma)$ such that $\odot(S) \overset{*}{\Rightarrow} H$. Two grammars are said to be* equivalent *iff they generate the same language.*

Further we simply write $A \overset{*}{\Rightarrow} G$ instead of $\odot(A) \overset{*}{\Rightarrow} G$.

4   Hypergraph Basic Categorial Grammars

In this section, we present definitions needed to extend BCGs to graphs. Firstly, we introduce the notion of a type; then we define a rewriting rule, which operates on graphs labeled by types; finally, we introduce the definitions of a hypergraph basic categorial grammar and of a language generated by it.

4.1   Types

We fix a countable set Pr of primitive types and a function $type : Pr \to \mathbb{N}$ such that for each $n \in \mathbb{N}$ there are infinitely many $p \in Pr$ for which $type(p) = n$. Types are constructed from primitive types using division. Simultaneously, we define the function $type$ on types.

Let us fix some symbol $ that is not included in all the sets considered. **NB!** This symbol is allowed to label edges with different number of attachment nodes. To be consistent with Definition 3.1 one can assume that there are countably many symbols $_n$ such that $type(\$_n) = n$.

Definition 4.1. *The set Tp_χ of types is the least set satisfying the following conditions:*

1. $Pr \subseteq Tp_\chi$;
2. *Let N ("numerator") be in Tp_χ. Let D ("denominator") be a graph such that exactly one of its edges (call it e_0) is labeled by \$, and other edges (possibly, there are none of them) are labeled by elements of Tp_χ; let also $type(N) = type(D)$. Then $T = \div(N/D)$ also belongs to Tp_χ, and $type(T) := type_D(e_0)$.*

In types, $ serves to "connect" a denominator and a numerator.

Example 4.1. The following structure is a type:

$$E_0 = \div \left(q \Big/ \begin{array}{c} {}^{(1)} \ {}^2 \ \boxed{q} \ {}^3 \ {}^{(3)} \\ \$ \ \ \ {}_1 \\ {}_p \ \ {}_{(2)} \end{array} \right).$$

Here p, q belong to Pr, $type(p) = 2$, $type(q) = 3$; $type(E_0) = 2$.

4.2   D-Isomorphism

In order to generalize the rules $A/B, B \mapsto A$ and $B, B\backslash A \mapsto A$, denominators of types are going to be "overlaid" on subgraphs of graphs. This idea is formalized by the notion of a d-isomorphism.

Definition 4.2. *A graph D that has exactly one edge labeled by $\$$ while other ones are labeled by elements of Tp_χ, is called* d-formed. *The only edge of D labeled by $\$$ is denoted head(D).*

Definition 4.3. *A graph-edge pair is a pair $(H; e_0)$ where H is a graph, and $e_0 \in E_H$.*

Definition 4.4. *A d-isomorphism between a d-formed graph D and an graph-edge pair $(H; e_0)$ is a pair of functions $(\epsilon; \beta)$ such that*

- $\epsilon : E_D \to E_H$, $\beta : V_D \to V_H$ *are bijections;*
- $att_H \circ \epsilon = \beta \circ att_D$;
- $\epsilon(head(D)) = e_0$;
- *For all $e \in E_D \setminus \{head(D)\}$ $lab_D(e) = lab_H(\epsilon(e))$;*
- $\beta(ext_D) = ext_H$.

4.3   Rule ($\div$)

The concept of hypergraph basic categorial grammars (HBCGs) is based on the mechanism of reduction of hypergraphs labeled by types. There is an inference rule, which is denoted by ($\div$), generalizing two rules for BCGs presented earlier.

The following dramatis personae participate in the rule ($\div$):

- $G \in \mathcal{H}(Tp_\chi)$;
- H — a subgraph of G, and $(H; e_0)$, which is a graph-edge pair;
- $lab(e_0) = \div(N/D)$;
- $(\epsilon; \beta)$ — a d-isomorphism between D and $(H; e_0)$.

The rule ($\div$) can be applied to $(H; e_0)$ if the following conditions are fulfilled:

1. If $v \in [att(e)]$ for some $e \in E_G \setminus E_H$ and $v \in V_H$, then $v \in [ext_H]$;
2. $[ext_G] \cap V_H \subseteq [ext_H]$.

Then the rule is of the form

$$G \underset{\chi}{\mapsto} G[\![N/H]\!]   (\div).$$

Requirements 1 and 2 guarantee that compression is possible.

If $G \overset{*}{\underset{\chi}{\mapsto}} H$, then G is said to be reducible to H (as usual, $\overset{*}{\underset{\chi}{\mapsto}}$ is the reflexive transitive closure of $\underset{\chi}{\mapsto}$).

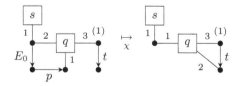

Fig. 1. Example of the application of ($\div$). Here E_0 is from Example 4.1. Note that the order of attachement node of the q-labeled edge is changed w.r.t. to E_0.

4.4   Hypergraph Basic Categorial Grammar

Definition 4.5. *A hypergraph basic categorial grammar (HBCG for short) Gr is a tuple $Gr = \langle \Sigma, s, \triangleright \rangle$ where Σ is a finite alphabet, s is a primitive type, and $\triangleright \subseteq \Sigma \times Tp_\chi$ is a binary relation (called a lexicon) which assigns a finite number of types to each symbol in the alphabet. Additionally, we require that the function type is defined on Σ such that $a \triangleright T$ implies $type(a) = type(T)$.*

The set $dict(Gr) = \{T \in Tp_\chi | \exists a \in \Sigma : a \triangleright T\} \cup \{s\}$ is called a *dictionary*.

Definition 4.6. *The language $L(Gr)$ generated by an HBCG $Gr = \langle \Sigma, s, \triangleright \rangle$ is the set of all hypergraphs $G \in \mathcal{H}(\Sigma)$ for which a function $f_G : E_G \to Tp_\chi$ exists such that:*

1. $lab_G(e) \triangleright f_G(e)$ whenever $e \in E_G$;
2. $f_G(G) \overset{*}{\underset{\chi}{\mapsto}} \odot(s)$.

All the definitions presented above are slightly more complicated and technical than that of BCGs; however, the concept of HBCGs is closely related both to HRGs and BCGs.

Example 4.2. Let us consider an example of an HRG from [11] (a little bit modified) generating abstract meaning representations, which contains four rules:

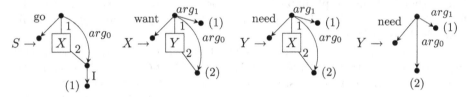

Here S is the initial symbol. This HRG can be converted into an equivalent HBCG as follows. Let s, x, y, i, a_0, a_1 be primitive types. Then the following lexicon defines an HBCG that generates the same language as the HRG above:

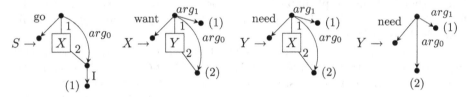

$I \triangleright i.$

To be more precise, the HBCG is of the form $\langle\{want, need, go, I, arg_0, arg_1\}, s, \triangleright\rangle$. All the primitive types have *type* being equal to 2.

The conversion can be done since there is a terminal edge in the right-hand side of each production (see more in Sect. 6). A more thorough example of an HBCG is given in Sect. 9.

5   Embedding of BCG

To justify that HBCGs appropriately extend BCGs, we present an embedding of the latter into the former in a natural and a simple way.

A function $tr : Tp \to Tp_\chi$ presented below embeds string types into graph types:

- $tr(p) := p, p \in Pr, type(p) = 2;$

- $tr(A/B) := \div \left(tr(A) \middle/ \begin{array}{c} (1) \bullet \xrightarrow{\quad \$ \quad} \bullet \xrightarrow{\quad tr(B) \quad} \bullet \ (2) \end{array} \right)$

- $tr(B\backslash A) := \div \left(tr(A) \middle/ \begin{array}{c} (1) \bullet \xrightarrow{\quad tr(B) \quad} \bullet \xrightarrow{\quad \$ \quad} \bullet \ (2) \end{array} \right)$

Recall that a string graph induced by a word $w = a_1 \ldots a_n$ is a graph of the form $\langle\{v_i\}_{i=0}^n, \{e_i\}_{i=1}^n, att, lab, v_0 v_n\rangle$ where $att(e_i) = v_{i-1}v_i$, $lab(e_i) = a_i$. This graph is denoted by $w^\bullet$.

$tr(Tp)$ denotes the set of translations of all types. If $\Gamma = T_1, \ldots, T_n$ is a sequence of types, then $tr(\Gamma) := (tr(T_1) \ldots tr(T_n))^\bullet$. If $Gr = \langle\Sigma, s, \triangleright\rangle$ is a BCG, then $tr(Gr)$ is the HBCG $\langle\Sigma, s, \triangleright_\chi\rangle$ where $a \triangleright_\chi T \Leftrightarrow T = tr(A)$ and $a \triangleright A$. Two propositions below establish connection between BCGs and HBCGs.

Proposition 5.1. *If $G \in \mathcal{H}(tr(Tp))$, $C \in Tp$ and $G \overset{*}{\underset{\chi}{\mapsto}} \odot(tr(C))$, then $G = tr(\Gamma)$ for some Γ such that $\Gamma \overset{*}{\mapsto} C$.*

Proof. Proof by induction on the number of steps in the derivation $G \overset{n}{\underset{\chi}{\mapsto}} \odot(tr(C))$.

Basis. If $n = 0$, then $G = \odot(tr(C))$ and $C \overset{0}{\mapsto} C$. **Step.** Let the first reduction be applied to $(H; e_0)$ within G. If $lab(e_0) = tr(A/B)$, then $H = \langle\{v_0, v_1, v_2\}, \{e_0, e_1\}, att, lab, v_0 v_2\rangle$ where $att(e_i) = v_i v_{i+1}, i = 0, 1$ and $lab(e_1) = tr(B)$. Since $G[\![\tilde{e}, tr(A)/H]\!] \overset{*}{\underset{\chi}{\mapsto}} \odot(tr(C))$, by the induction hypothesis we obtain $G[\![\tilde{e}, tr(A)/H]\!] = tr(\Delta)$ such that $\Delta = \Phi, A, \Psi \overset{*}{\mapsto} C$. Then $G = G[\![\tilde{e}, tr(A)/H]\!][H/\tilde{e}] = tr(\Phi, A/B, B, \Psi)$ and $\Phi, A/B, B, \Psi \overset{*}{\mapsto} C$. The case $lab(e_0) = tr(B\backslash A)$ is treated similarly. $\square$

Proposition 5.2. *If $\Gamma \overset{*}{\mapsto} C$, then $tr(\Gamma) \overset{*}{\underset{\chi}{\mapsto}} \odot(tr(C))$.*

It is proved by a straightforward conversion of the reduction process for strings into the reduction process for graphs. These propositions yield

Theorem 5.1. *If Gr is a BCG, then $L(tr(Gr)) = \{w^\bullet | w \in L(Gr)\}$.*

6 Equivalence of HRGs and HBCGs

It is well known that CFGs and BCGs are equivalent; one of the simplest proofs involves Greibach normal form for CFGs. In this section, we show that this proof can be generalized to a wide class of graph grammars in a natural way.

6.1 Greibach Normal Form for HRGs

Firstly, one has to extend the notion of the (weak) Greibach normal form. There are a few works in which variants of such extension are introduced, see [10,12]. However, normal forms presented in these works are more strict than it is needed for our purposes. In this paper, we use the following

Definition 6.1. *The HRG Gr is in the* weak Greibach normal form (WGNF) *iff there is exactly one terminal edge in the right-hand side of each production. Formally, $\forall (X \to H) \in P_{Gr} \; \exists! e_0 \in E_H : lab_H(e_0) \in \Sigma_{Gr}$ and for $e \neq e_0$ $lab_H(e) \in N_{Gr}$.*

Note that not each language generated by some HRG can be generated by an HRG in GNF. It follows from

Example 6.1. Consider an HRG $Gr = \langle \{S\}, \{a\}, P, S \rangle$ where P contains two productions (for $type(S) = type(a) = 0$):

– $S \to \langle \{v_0\}, \{e_0\}, att, lab, \varepsilon \rangle$, $lab(e_0) = S$, $att(e_0) = \varepsilon$;
– $S \to \circledcirc(a)$.

This grammar produces graphs that have exactly one edge labeled by a and arbitrarily many isolated nodes. If there is an equivalent $Gr' = \langle N, \{a\}, P', S' \rangle$ in GNF, then each right-hand side of each production in P' contains exactly one terminal edge. Note that if $S' \overset{k}{\Rightarrow} G, G \in \mathcal{H}(\{a\})$ in Gr', then G has k terminal edges; hence k has to equal 1 and therefore $S' \to G \in P'$. However, there are infinitely many graphs in $L(Gr)$ while $|P'| < \infty$. □

The characterization of languages generated by HRGs in the WGNF is a subject of the further study.

It turns out that HBCGs generate the same class of languages as HRGs in the normal form presented. This is proved below.

6.2 The Equivalence Theorems

Definition 6.2. *The set $st(T)$ of subtypes of a type T is defined inductively as follows:*

1. $st(p) = \{p\}, p \in Pr$;
2. $st(\div(N/D)) = \{\div(N/D)\} \cup st(N) \cup \left(\bigcup_{e \in E_D \setminus \{head(D)\}} st\,(lab_D(e)) \right)$.

Theorem 6.1. *For each HBCG an equivalent HRG in GNF exists.*

Proof. Let $Gr = \langle \Sigma, s, \rhd \rangle$ be an HBCG. We denote by N the set of subtypes of all types contained in the dictionary (formally, $N = \{R \in Tp_\chi | R \in st(T), T \in dict(Gr)\}$). The set P contains the following productions:

1. If $a \rhd T$, then $T \to \odot(a) \in P$.
2. If $a \rhd \div(N/D)$, then $N \to D[head(D) := a] \in P$.

It is argued that $Gr' = \langle N, \Sigma, P, s \rangle$ generates the same language as Gr. It suffices to show that $T \overset{*}{\Rightarrow} G$ for $G \in \mathcal{H}(\Sigma), T \in N$ if and only if $f(G) \overset{*}{\underset{\chi}{\mapsto}} \odot(T)$ for some $f : E_G \to Tp_\chi$ such that $lab(e) \rhd f(e)$. It is done in a straightforward way by induction on the number of steps in the derivation:

"Only if" part. Let $T \overset{k}{\Rightarrow} G$. Induction on k. **Basis**: if $k = 1$, then either $G = \odot(a)$ and $a \rhd T$ or $G = D[head(D) := a]$ is obtained in such a way that $a \rhd \div(T/D)$ and $|E_D| = 1$. Then $D[head(D) := \div(T/D)] \underset{\chi}{\mapsto} \odot(T)$. **Step** $(k > 1)$: let the first production applied be of the form $T \to D[head(D) := a]$ for $a \rhd \div(T/D)$ and let $D \setminus \{head(D)\} = \{e_1, \ldots, e_n\}, n \geq 1$. Let $G_i \in \mathcal{H}(\Sigma)$ be a graph that is obtained from $T_i = lab(e_i)$ in the derivation process $(i = 1, \ldots, n)$. Note that G_i is a subgraph of G. By induction hypothesis, $f_i(G_i) \overset{*}{\underset{\chi}{\mapsto}} \odot(T_i)$ for some appropriate $f_i : E_{G_i} \to Tp_\chi$. Then f_i can be combined into a single function f as follows: $f(e) := f_i(e)$ whenever $e \in G_i$ and $f(head(D)) := \div(T/D)$. Thus $f(G)$ can be reduced to $\widetilde{G} = D[head(D) := \div(T/D)]$ by induction hypothesis (its n subgraphs is compressed into n edges), and then $(\div)$ can be applied to $\div(N/D)$ in such a way that $\widetilde{G} \underset{\chi}{\mapsto} \odot(T)$.

The "if" part is proved similarly: one has to transform applications of $(\div)$ in Gr into productions in Gr'. □

Example 4.2 provides an example of application of the theorem above.

Theorem 6.2. *Each HRG Gr in GNF is equivalent to some HBCG.*

Proof. Let $Gr = \langle N, \Sigma, P, S \rangle$. Consider elements of N as elements of Pr with the same function *type* defined on them. Since Gr is in GNF, each production in P is of the form $\pi = X \to G$ where G contains exactly one terminal edge e_0 (say $lab_G(e_0) = a \in \Sigma$). We convert this production into the type $T_\pi := \div(X/G[e_0 := \$])$. Then we introduce the HBCG $Gr' = \langle \Sigma, S, \rhd \rangle$ where $\rhd$ is defined as follows: $a \rhd T_\pi$. Finally, note that, if one applies the transformation described in Theorem 6.1 to Gr', he obtains Gr, which implies that $L(Gr) = L(Gr')$. □

Corollary 6.1. *The problem of whether a given graph belongs to the language generated by a given HBCG is NP-complete.*

Proof. This problem is in NP since, if the answer is "YES", there is a certificate of polynomial size that justifies this; namely, this is a sequence of applications of ($\div$) (a derivation). Another explanation is that an HBCG can be converted into an equivalent HRG in polynomial time for whom the membership problem is in NP.

In [9], an NP-complete graph language generated by some HRG *ERG* is introduced. One notices that there is *at least one* terminal edge in each production in *ERG*; by adding nonterminal symbols corresponding to terminal ones one transforms *ERG* into an equivalent one in GNF, and then — to an HBCG using Theorem 6.2 (it all takes polynomial time). □

7   Structural Properties

In this section, we study some structural properties of HBCGs.

7.1   HBCGs with One Primitive Type

The set of primitive types Pr is countably infinite; i.e. we are allowed to use as many primitive types as we want. However, the following theorem shows that it suffices to have one primitive type only.

Theorem 7.1. *For each HBCG $Gr = \langle \Sigma, s, \rhd \rangle$ an equivalent one $\langle \Sigma, s, \rhd' \rangle$ exists such that types in its dictionary do not have primitive subtypes except for s.*

Proof. We consider the following *substitution* of primitive types. Let $s, p_1, \ldots, p_n$ be all the primitive types occurring in $dict(Gr)$ and let $\mathcal{T}$ be the set of all subtypes of all types in $dict(Gr)$. Denote $M := \max\{type(T) | T \in \mathcal{T}\}$. Then we define $F(p_k) = \div(s/D_k)$ ($k = 1, \ldots, n$):

- $V_{D_k} = \{v_1, v_2, \ldots, v_{t_k}, w_1, w_2, \ldots, w_{M+k}\}$ (for $t_k = type(p_k)$).
- $E_{D_k} = \{e_0, e_1\}$.
- $att_{D_k}(e_0) = v_1 v_2 \ldots v_{t_k}$, $att_{D_k}(e_1) = w_1 w_2 \ldots w_{M+k}$.
- $lab(e_0) = \$$, $lab(e_1) = T_k$ where
  - $T_k = \div(s/D'_k)$;
  - $D'_k = \langle \{u_1, \ldots, u_{M+k}\}, e', att', lab', u_1 \ldots u_t \rangle$ (for $t = type(s)$);
  - $att'(e') = u_1 \ldots u_{M+k}$, $lab'(e') = \$$.
- $ext_{D_k} = w_1 w_2 \ldots w_t$.

Behind this definition a simple idea stands: $F(p_k)$ has a huge edge in the denominator, which is larger than any edge existing in the lexicon.

Let $F(T)$ stand for a type obtained from $T \in \mathcal{T}$ by substituting each p_k with $F(p_k)$ (we do not change s). Now, if $a \rhd T$, then let $a \rhd' F(T)$. No more relations in $\rhd'$ exist. We argue that $Gr' = \langle \Sigma, s, \rhd' \rangle$ is a desired grammar. It follows from its definition that it contains only s as a primitive type. Clearly,

$L(Gr) \subseteq L(Gr')$: each derivation in Gr can be remade in Gr', if one considers $F(p_k)$ as an atomic, indivisible type corresponding to p_k.

To prove the reverse inclusion let us consider the set $T' = \{F(T)|T \in T\}$. Assume that for a graph $G \in \mathcal{H}(T)$ the graph $F(G)$ is reducible to $\odot(s)$ (here $F(G)$ is obtained from G by changing each label a by the label $F(a)$). At each step of the reduction process the rule $(\div)$ is applied either to a type of the form $F(\div(N/D))$, $\div(N/D) \in T$ or to $F(p_k)$. However, note that no edges in $F(G)$ have type exceeding M, whereas D_k requires to be overlaid on the edge of the type $M + k > M$. Consequently, $(\div)$ cannot be applied to $F(p_k)$, and $G \overset{*}{\underset{\chi}{\mapsto}} \odot(s)$.

If H belongs to $L(Gr')$, then there is a function $f : E_H \to T'$ such that $lab(e) \rhd' f(e)$ whenever $e \in E_H$ and $f(H) \overset{*}{\underset{\chi}{\mapsto}} \odot(s)$. Since $f(H) = F(G)$ for some $G \in \mathcal{H}(T)$, G is also reducible to $\odot(s)$. Finally, note that $g = F^{-1} \circ f$ satisfies the condition $lab(e) \rhd g(e)$ for $e \in E_H$ and $G = g(H) \overset{*}{\underset{\chi}{\mapsto}} \odot(s)$. Therefore, H belongs to $L(Gr)$. $\qquad\square$

7.2   Counters

One of the features HBCGs inherit from BCGs is so-called counters.

Definition 7.1. *Let $f : Pr \to \mathbb{Z}$ be some function. An f-counter $\#_f : Tp_\chi \to \mathbb{Z}$ is defined as follows:*

- $\#_f(p) = f(p)$;
- *If $T = \div(N/D)$ and $E_D = \{head(D), e_1, \ldots, e_n\}$, then*

$$\#_f(T) = \#_f(N) - \sum_{i=1}^{n} \#_f(lab(e_i)).$$

If G is labeled by types and $E_G = \{e_1, \ldots, e_n\}$, then $\#_f(G) := \sum_{i=1}^{n} \#_f(lab(e_i))$.

Proposition 7.1. *If $G \overset{k}{\underset{\chi}{\mapsto}} G'$, then $\#_f(G) = \#_f(G')$ for each f.*

Proof. Induction on k. **Basis:** if $k = 0$, then $G = G'$.

Step: let $G \underset{\chi}{\mapsto} G[\![N/H]\!] \overset{k-1}{\underset{\chi}{\mapsto}} G'$ where $\div(N/D)$ is the type involved in the first rule $(\div)$. Let us denote all the participants of the compression $G \underset{\chi}{\mapsto} G[\![N/H]\!]$ similarly to those in Sect. 4.3. Then, since D and $(H; e_0)$ are d-isomorphic, $\#_f(lab(e)) = \#_f(lab(\epsilon(e)))$ for $e \in E_D, e \neq head(D)$. Thus $\#_f(G) = \#_f(G[\![N/H]\!])$. The induction hypothesis completes the proof. $\qquad\square$

Counters can be used to check whether a graph G can be reduced to another one G': if $\#_f(G) \neq \#_f(G')$ for some f, then $G \overset{*}{\underset{\chi}{\not\mapsto}} G'$.

Example 7.1. We provide two specific examples of counters:

- $f = g_q, q \in Pr$: $g_q(p) = 1$ whenever $p = q$ and $g_q(p) = 0$ otherwise. E.g. for the graph E_0 from Example 4.1 $\#_{g_p}(E_0) = -1$, $\#_{g_q}(E_0) = 0$.
- $f = h_m, m \in \mathbb{N}$: $h_m(p) = 1$ whenever $type(p) = m$ and $h_m(p) = 0$ otherwise.

8   Lambda Semantics

The λ-calculus is a formal tool, which has a number of applications in functional programming and in formal semantics. In this paper, we do not provide the definitions of this mechanism and refer the reader to the paper [5], which is an overview of the λ-calculus.

In basic categorial grammars, one can assign λ-terms to types. $\alpha : A$ denotes a λ-term α assigned to a type A (i.e. this is the pair $(\alpha; A)$). The rules of reduction then have the following form:

1. $\Gamma, \alpha : A, \beta : A \backslash B, \Delta \mapsto \Gamma, \beta\alpha : B, \Delta$
2. $\Gamma, \beta : B/A, \alpha : A, \Delta \mapsto \Gamma, \beta\alpha : B, \Delta$

Here $\beta\alpha$ stands for the application of β to α. A linguistic example that shows how λ-terms describe semantics of a natural language was given in Sect. 1.

This approach can be generalized to hypergraphs and HBCGs. Let T be a type, i.e. belong to Tp_χ. By $\tau : T$ we denote a pair containing a λ-term τ.

Now we are going to incorporate the λ-calculus into the rule $(\div)$. Let objects involved in this rule be denoted as in Sect. 4.3. We additionally require that edges in E_D are numbered: $E_D = \{e_0, e_1, \ldots, e_k\}$ (and this numbering is fixed for a given type). If a λ-term τ is assigned to $\div(N/D)$ and for $i > 0$ $lab(\epsilon(e_i)) = \alpha_i : T_i$, then the rule $(\div)$ is of the form

$$G \underset{\chi}{\mapsto} G[\![\tau\alpha_1\alpha_2 \ldots \alpha_k : N/H]\!] \qquad (\div).$$

Here $\tau\alpha_1\alpha_2 \ldots \alpha_k = (((\tau\alpha_1)\alpha_2) \ldots)\alpha_k$. This means that λ-terms written on edges that are consumed by the denominator D are treated as arguments of the λ-term assigned to $\div(N/D)$.

An example of an application of HBCGs enriched with the λ-calculus to linguistics is presented in the next section.

9   Cross-serial Dependencies

It is well known that context-free languages in the usual sense fail to describe certain linguistic phenomena. One of them is so-called *cross-serial dependencies (CSD)* — a class of phenomena that that can be described by the language $\{ww|w \in \Sigma^*\}$ of reduplicated strings.

We focus here on the following example of CSD from the Russian language:

Оля,   Петя,   Вася финишировали первой,   вторым,   третьим.
Olya, Petya and Vasya   finished   the first, the second and the third.

The meaning of this sentence is: *Olya was the first who finished, Petya was the second who finished, and Vasya was the third who finished* (e.g. when speaking about a competition). In Russian, the ordinal numerals *the first, the second, the third* agree with nouns in gender which leads to CSD (note that *Olya* is a female name, and *Petya, Vasya* are male ones). Below we show how to generate such Russian sentences using an HBCG and how to model their semantics using the λ-calculus. In order to simplify the example, we ignore some features of the Russian language. We denote by pd_m (pd_f) a primitive type which stands for masculine (feminine resp.) predicate phrases in singular form in the instrumental case (such as ordinal numbers, e.g. **первым**); we denote by np_m (np_f) a primitive type corresponding to masculine (feminine) noun phrases in singular form in the nominative case (such as proper nouns, e.g. **Петя**); np_p (pd_p) denotes nouns (predicate phrases resp.) in a plural form; s denotes sentences (it is a distinguished type). Then the grammar generating sentences of the above form is the following:

$$\text{Вася, Петя (masculine)} \vartriangleright \div \left(Z \Big/ \underset{(1)\ (2)\ (3)\ (4)}{\bullet \overset{\$}{\to} \bullet \quad \bullet \overset{pd_m}{\to} \bullet} \right),\ \div \left(Z \Big/ \ \underset{(1)\quad(2)\ (3)\quad(4)}{\overset{1\ \boxed{Z}\ 4}{\bullet\overset{2\ \$\ 3}{\to}\bullet\ \bullet\overset{pd_m}{\to}\bullet}} \right)$$

$$\text{Оля, Маша (feminine)} \vartriangleright \div \left(Z \Big/ \underset{(1)\ (2)\ (3)\ (4)}{\bullet \overset{\$}{\to} \bullet \quad \bullet \overset{pd_f}{\to} \bullet} \right),\ \div \left(Z \Big/ \ \underset{(1)\quad(2)\ (3)\quad(4)}{\overset{1\ \boxed{Z}\ 4}{\bullet\overset{2\ \$\ 3}{\to}\bullet\ \bullet\overset{pd_f}{\to}\bullet}} \right)$$

первым, вторым, третьим, ... (masculine) $\vartriangleright pd_m$;
первой, второй, третьей, ... (feminine) $\vartriangleright pd_f$;
финишировали $\vartriangleright T_0 = tr((np_p \backslash s)/pd_p)$.

Here Z equals $\div \left(s \Big/ \underset{(1)\quad T_0\quad(2)}{\overset{1\ \boxed{\$}\ 4}{\bullet\overset{2\ \ \ 3}{\triangle}\bullet}} \right)$ Note that tr is defined in Sect. 5.

Let us denote the first type in the first row above as I_m, and the second one as T_m; by analogue, types in the second row are denoted by I_f and T_f resp.

In addition, we assign semantic types to the syntactic ones:

- **Вася** $\vartriangleright \lambda P.\lambda f.f(P)(Vasya) : I_m, \quad \lambda \zeta.\lambda P.\lambda f.\,(f(P)(Vasya) \wedge \zeta f) : T_m; \ldots$
- **Оля** $\vartriangleright \lambda P.\lambda f.f(P)(Olya) : I_f, \quad \lambda \zeta.\lambda P.\lambda f.\,(f(P)(Olya) \wedge \zeta f) : T_f; \ldots$
- **первым**:$first : pd_m$, **вторым**:$second : pd_m, \ldots$;
- **первой**:$first : pd_f$, **второй**:$second : pd_f, \ldots$;
- **финишировали**:$\lambda P.\lambda x.finish(P)(x) : T_0$.

Then the sentence (1) belongs to the language generated by this HBCG:

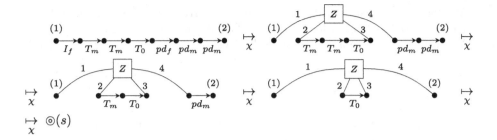

Let us track the λ-terms occuring in the derivation process step-by-step:

1. There is a term $\tau_1 = \lambda f. f(first)(Olya)$ assigned to Z after the first step;
2. $\zeta = \tau_1$ and $P = second$ are applied to $\lambda \zeta. \lambda P. \lambda g. (g(P)(Petya) \wedge \zeta g)$ in the second step; the result is $\lambda g. (g(second)(Petya) \wedge g(first)(Olya))$.
3. Similarly, after the third step the following λ-term is assigned to Z:

$$\lambda h. (h(third)(Vasya) \wedge (h(second)(Petya) \wedge h(first)(Olya))).$$

4. Finally, one obtains the following term:

$$finish(third)(Vasya) \wedge (finish(second)(Petya) \wedge finish(first)(Olya)).$$

Thus, this HBCG not only generates sentences of the form (1) (HRGs can deal with them as well) but also provides their semantical representation which is composed of λ-terms assigned to types.

10   Conclusions and Related Work

There are a few papers devoted to combining categorial grammars with graph tools. E.g. there is a recent work of Sebastian Beschke and Wolfgang Menzel [7] where it is shown how to enrich the Lambek calculus, which is another categorial approach (see [13]), with graph semantics. In the paper [14] (which is rather linguistic than mathematical) an extension of some concepts of the Lambek calculus to graphs is presented; namely, sentences are considered to be graph structures (functor-argumentor-structures), and then categorial graph grammars are introduced, which deal with these structures. Hypergraph basic categorial grammars introduced in our work, however, do not seem to be closely related to any of these approaches. Possibly, it is because our motivation for HBCGs is rather logical and mathematical: our main purpose was to directly combine concepts of BCGs and HRGs, so the resulting mechanism satisfies these requirements. Nevertheless, we hope that HBCGs also can be used (possibly, with some further modifications) in practical applications, e.g. in linguistics.

Note that there is a work [6] where HRGs are used to describe CSD of the Dutch language. Actually, examples in Sect. 9 have a similar structure with examples in [6]. Comparing [6] with this paper, we conclude that one of the crucial features distinguishing between HRGs and HBCGs from linguistic point of view is the λ-semantics, which can be naturally built into the latter.

There is a number of questions that remain open; we hope to study them in future works.

- We showed that the membership problem for HBCGs is NP-complete. How to restrict HBCGs in order to obtain efficient parsing algorithms?
- We introduced some applications of HBCGs to linguistics. We are interested in further developing a theory that would use HBCGs and the λ-calculus to model visual structures related to natural languages. Particularly, we desire to consider syntactic trees which linguists deal with from the point of view of our approach.
- How to generalize other string categorial approaches to hypergraphs?

Acknowledgments. I thank my scientific advisor prof. Mati Pentus for his careful attention to my study and anonymous reviewers for their valuable advice.

References

1. Aho, A.V., Ullman, J.D.: The Theory of Parsing, Translation, and Compiling. Prentice-Hall, Upper Saddle River (1972)
2. Ajdukiewicz, K.: Die syntaktische Konnexitat (Syntactic connexion). Studia Philosophica **1**, 1–27 (1935)
3. Bar-Hillel, Y.: A quasi-arithmetical notation for syntactic description. Language **29**, 47–58 (1953)
4. Bar-Hillel, Y., Gaifman, H., Shamir, E.: On categorial and phrase structure grammars. Bull. Res. Counc. Israel **9**, 1–6 (1960)
5. Barendregt, H., Barendsen, E.: Introduction to lambda calculus. Nieuw archief voor wisenkunde **4**, 337–372 (1984)
6. Bauer, D., Owen, R.: Hyperedge replacement and nonprojective dependency structures. In: TAG (2016)
7. Beschke, S., Menzel, W.: Graph algebraic combinatory categorial grammar, pp. 54–64 (2018)
8. Buszkowski, W.: Categorial grammars and their logics. In: Garrido, Á., Wybraniec-Skardowska, U. (eds.) The Lvov-Warsaw School. Past and Present. SUL, pp. 91–115. Springer, Cham (2018). https://doi.org/10.1007/978-3-319-65430-0_6
9. Drewes, F., Kreowski, H.-J., Habel, A.: Hyperedge replacement graph grammars (1997)
10. Engelfriet, J.: A Greibach normal form for context-free graph grammars. In: Kuich, W. (ed.) ICALP 1992. LNCS, vol. 623, pp. 138–149. Springer, Heidelberg (1992). https://doi.org/10.1007/3-540-55719-9_70
11. Gilroy, S., Lopez, A., Maneth, S.: Parsing graphs with regular graph grammars, pp. 199–208 (2017)
12. Jansen, C., Heinen, J., Katoen, J.-P., Noll, T.: A local Greibach normal form for hyperedge replacement grammars. In: Dediu, A.-H., Inenaga, S., Martín-Vide, C. (eds.) LATA 2011. LNCS, vol. 6638, pp. 323–335. Springer, Heidelberg (2011). https://doi.org/10.1007/978-3-642-21254-3_25
13. Lambek, J.: The mathematics of sentence structure. Am. Math. Monthly **65**(3), 154–170 (1958)
14. Robering, K.: Categorial graph grammar: a direct approach to functor-argumentor-structure. Theor. Linguist. **26**, 31–73 (2000)

Multilevel Typed Graph Transformations

Uwe Wolter[1]([✉])[ID], Fernando Macías[2][ID], and Adrian Rutle[3][ID]

[1] University of Bergen, Bergen, Norway
Uwe.Wolter@uib.no
[2] IMDEA Software Institute, Madrid, Spain
fernando.macias@imdea.org
[3] Western Norway University of Applied Sciences, Bergen, Norway
aru@hvl.no

Abstract. Multilevel modeling extends traditional modeling techniques with a potentially unlimited number of abstraction levels. Multilevel models can be formally represented by multilevel typed graphs whose manipulation and transformation are carried out by multilevel typed graph transformation rules. These rules are cospans of three graphs and two inclusion graph homomorphisms where the three graphs are multilevel typed over a common typing chain. In this paper, we show that typed graph transformations can be appropriately generalized to multilevel typed graph transformations improving preciseness, flexibility and reusability of transformation rules. We identify type compatibility conditions, for rules and their matches, formulated as equations and inequations, respectively, between composed partial typing morphisms. These conditions are crucial presuppositions for the application of a rule for a match—based on a pushout and a final pullback complement construction for the underlying graphs in the category **Graph**—to always provide a well-defined canonical result in the multilevel typed setting. Moreover, to formalize and analyze multilevel typing as well as to prove the necessary results, in a systematic way, we introduce the category **Chain** of typing chains and typing chain morphisms.

Keywords: Typing chain · Multilevel typed graph transformation · Pushout · Pullback complement

1 Introduction

Multilevel modeling (MLM) extends conventional techniques from the area of Model-Driven Engineering by providing model hierarchies with multiple levels of abstraction. The advantages of allowing multiple abstraction levels (e.g. reducing accidental complexity in software models and avoiding synthetic type-instance anti-patterns) and flexible typing (e.g. multiple typing, linguistic extension and deep instantiation), as well as the exact nature of the techniques used for MLM are well studied in the literature [1,4–6,8,10,17]. Our particular approach [19,20] to MLM facilitates the separation of concerns by allowing integration of different

© Springer Nature Switzerland AG 2020
F. Gadducci and T. Kehrer (Eds.): ICGT 2020, LNCS 12150, pp. 163–182, 2020.
https://doi.org/10.1007/978-3-030-51372-6_10

multilevel modeling hierarchies as separate aspects of the system to be modelled. In addition, we enhance reusability of concepts and their behaviour by allowing the definition of flexible transformation rules which are applicable to different hierarchies with a variable number of levels. In this paper, we present a revised and extended formalisation of these rules using graph theory and category theory.

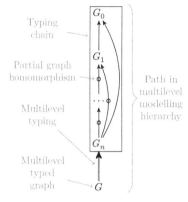

As models are usually represented abstractly as graphs, we outline in this paper the graph theoretic foundations of our approach to MLM using multilevel typed graphs, prior to introducing our formalisation of multilevel typed rule definition and application. Multilevel models are organized in hierarchies, where any graph G is *multilevel typed* over a *typing chain* of graphs (see Fig. 1). The typing relations of elements within each graph are represented via graph morphisms. Since we allow for deep instantiation [4–6,8], which refers to the ability to instantiate an element at any level below the level in which it is defined, these morphisms need to be *partial graph homomorphisms*. Moreover, more than one model can be typed by the same typing chain (or, conversely, models can be instantiated more than once), hence, all the *paths* that contain such typing relations constitute a full, tree-shaped *multilevel modelling hierarchy* (see Example 1). Finally, the topmost model G_0 in any hierarchy is fixed, and the typing relations of all models (and the elements inside them) must converge, directly or via a sequence of typing morphisms, into G_0. Therefore, the graph morphisms into G_0 are always total.

Fig. 1. MLM terminology

Multilevel typed graph transformation rules are cospans $L \xhookrightarrow{\lambda} I \xleftarrow{\rho} R$ of inclusion graph homomorphisms, with $I = L \cup R$, where the three graphs are multilevel typed over a common typing chain $\mathcal{MM}$. A match of the left-hand side L of the rule in a graph S, at the bottom of a certain hierarchy, multilevel typed over a typing chain $\mathcal{TG}$, is given by a graph homomorphism $\mu : L \to S$ and a flexible typing chain morphism from $\mathcal{MM}$ into $\mathcal{TG}$. The typing chain $\mathcal{MM}$ is local for the rules and is usually different from $\mathcal{TG}$ which is determined by the path from S to the top of the hierarchy (see Fig. 1).

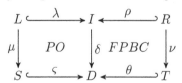

To apply these rules we rely on an adaptation of the Sesqui pushout (Sq-PO) approach [7] to cospans. We construct first the pushout and then the final pullback complement (FPBC) of the underlying graph homomorphisms in the category **Graph** as shown in Fig. 2. Based on these traditional constructions we want to build, in a canonical way, type compatible multilevel typings of the result graphs D and T over the typing chain $\mathcal{TG}$. For this to work, we need

Fig. 2. Rule structure and basic constructions for rule application

quite reasonable type compatibility conditions for rules and relatively flexible conditions for matches, formulated as equations and inequations, resp., between composed partial typing morphisms.

We introduce typing chain morphisms, and the corresponding category **Chain** of typing chains and typing chain morphisms, to formalize flexible matching and application of multilevel typed rules. The composition of partial graph homomorphisms is based on pullbacks in the category **Graph**, thus type compatibility conditions can be equivalently expressed by commutativity and pullback conditions in **Graph**. Therefore, we formalize and analyze multilevel typing as well as describe constructions and prove the intended results, in a systematic way, within the category **Chain**. Especially, we show that the first step in a rule application can be described by a pushout in **Chain**. Moreover, the second step is described as a canonical construction in **Chain**, however, it is an open question whether this is a final pullback construction in **Chain** or not.

A preliminary version of typing chains are an implicit constituent of the concept "deep metamodeling stack" introduced in [22] to formalize concepts like parallel linguistic and ontological typing, linguistic extensions, deep instantiation and potencies in deep metamodeling. We revised this earlier version and further developed it to a concept of its own which serves as a foundation of our approach to multilevel typed model transformations in [20,26]. Compared to [20], we present in this paper a radically revised and extended theory of multilevel typed graph transformations. In particular, the theory is now more powerful, since we drop the condition that typing chain morphisms have to be closed (see Definition 5). Moreover, we detail the FPBC step which is missing in [20]. Due to space limitations, we will not present the background results concerning the equivalence between the practice of individual direct typing – which are used in applications and implementations – and our categorical reformulation of this practice by means of typing chains. These equivalence results as well as examples and proofs can be found in [26].

2   Typing Chains and Multilevel Typing of Graphs

Graph denotes the category of (directed multi-) graphs $G = (G^N, G^A, sc^G, tg^G)$ and graph homomorphisms $\phi = (\phi^N, \phi^A) : G \to H$ [12]. We will use the term **element** to refer to both nodes and arrows.

Multilevel typed graphs are graphs typed over a typing chain, i.e., a sequence $[G_n, G_{n-1}, \ldots, G_1, G_0]$ of graphs where the elements in any of the graphs G_i, with $n \geq i \geq 1$, are, on their part, multilevel typed over the sequence $[G_{i-1}, \ldots, G_1, G_0]$. Paths in our MLM hierarchies give rise to typing chains. The indexes i refer to the abstraction levels in a modeling hierarchy where 0 denotes the most abstract top level.

Following well-established approaches in the Graph Transformations field [12], we define typing by means of graph homomorphisms. This enables us to establish and develop our approach by reusing, variating, and extending the wide range of constructions and results achieved in that field. Moreover, this paves the way to generalize the present "paradigmatic" approach, where models are just graphs, to more sophisticated kinds of diagrammatic models, especially those that take advantage of diagrammatic constraints [22,23].

We allow typing to jump over abstraction levels, i.e., an element in graph G_i may have no type in G_{i-1} but only in one (or more) of the graphs $G_{i-2}, \ldots, G_1, G_0$. Two different elements in the same graph may have their types located in different graphs along the typing chain. To formalize this kind of flexible typing, we use partial graph homomorphisms that we introduced already in [22].

Definition 1. *A **partial graph homomorphism** $\varphi : G \dashrightarrow H$ is given by a subgraph $D(\varphi) \sqsubseteq G$, called the **domain of definition** of φ, and a graph homomorphism $\varphi : D(\varphi) \to H$.*

Note that we use, in abuse of notation, the same name for both the partial and the corresponding total graph homomorphisms. To express transitivity of typing and later also compatibility of typing, we need as well the composition of partial graph homomorphisms as a partial order between partial graph homomorphisms.

Definition 2. *The **composition** $\varphi; \psi : G \dashrightarrow K$ of two partial graph homomorphisms $\varphi : G \dashrightarrow H$ and $\psi : H \dashrightarrow K$ is defined as follows:*

- $D(\varphi; \psi) := \varphi^{-1}(D(\psi))$,
- $(\varphi; \psi)^N(e) := \psi^N(\varphi^N(e))$ for all $e \in D(\varphi; \psi)^N$ and $(\varphi; \psi)^A(f) := \psi^A(\varphi^A(f))$ for all $f \in D(\varphi; \psi)^A$.

More abstractly, the composition of two partial graph homomorphisms is defined by the following commutative diagram of total graph homomorphisms.

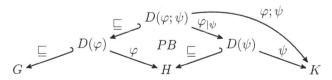

Note that $D(\varphi; \psi) = D(\varphi)$ if ψ is total, i.e., $H = D(\psi)$.

Definition 3. *For any two partial graph homomorphisms $\varphi, \phi : G \dashrightarrow H$ we have $\varphi \preceq \phi$ iff $D(\varphi) \sqsubseteq D(\phi)$ and φ, ϕ coincide on $D(\varphi)$.*

Now, we can define typing chains as a foundation for our investigation of multilevel typed graph transformations in the rest of the paper.

Definition 4. *A **typing chain** $\mathcal{G} = (\overline{G}, n, \tau^{\mathcal{G}})$ is given by a natural number n, a sequence $\overline{G} = [G_n, G_{n-1}, \ldots, G_1, G_0]$ of graphs of length $n + 1$ and a family $\tau^{\mathcal{G}} = (\tau^{\mathcal{G}}_{j,i} : G_j \dashrightarrow G_i \mid n \geq j > i \geq 0)$ of partial graph homomorphisms, called **typing morphisms**, satisfying the following properties:*

- **Total:** *All the morphisms $\tau^{\mathcal{G}}_{j,0} : G_j \to G_0$ with $n \geq j \geq 1$ are total.*

- **Transitive:** For all $n \geq k > j > i \geq 0$ we have $\tau_{k,j}^{\mathcal{G}}; \tau_{j,i}^{\mathcal{G}} \preceq \tau_{k,i}^{\mathcal{G}}$.
- **Connex:** For all $n \geq k > j > i \geq 0$ we have $D(\tau_{k,j}^{\mathcal{G}}) \cap D(\tau_{k,i}^{\mathcal{G}}) \sqsubseteq D(\tau_{k,j}^{\mathcal{G}}; \tau_{j,i}^{\mathcal{G}}) = (\tau_{k,j}^{\mathcal{G}})^{-1}(D(\tau_{j,i}^{\mathcal{G}}))$, moreover, $\tau_{k,j}^{\mathcal{G}}; \tau_{j,i}^{\mathcal{G}}$ and $\tau_{k,i}^{\mathcal{G}}$ coincide on $D(\tau_{k,j}^{\mathcal{G}}) \cap D(\tau_{k,i}^{\mathcal{G}})$.

Due to Definitions 2 and 3, transitivity and connexity together mean that $D(\tau_{k,j}^{\mathcal{G}}) \cap D(\tau_{k,i}^{\mathcal{G}}) = D(\tau_{k,j}^{\mathcal{G}}; \tau_{j,i}^{\mathcal{G}})$, i.e., we do have a (unique) total graph homomorphism $\tau_{k,j|i}^{\mathcal{G}} : D(\tau_{k,j}^{\mathcal{G}}) \cap D(\tau_{k,i}^{\mathcal{G}}) \rightarrow D(\tau_{j,i}^{\mathcal{G}})$ and the following commutative diagram of total graph homomorphisms

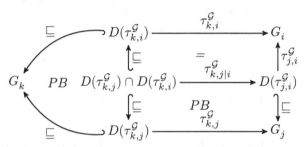

Remark 1. For any element **e** in any graph G_i in a typing chain, with $i > 0$, there exists a unique index $m_{\mathbf{e}}$, with $i > m_{\mathbf{e}} \geq 0$, such that **e** is in the domain of the typing morphism $\tau_{i,m_{\mathbf{e}}}^{\mathcal{G}}$ but not in the domain of any typing morphism $\tau_{i,j}^{\mathcal{G}}$ with $i > j > m_{\mathbf{e}}$. We call $\tau_{i,m_{\mathbf{e}}}^{\mathcal{G}}(\mathbf{e})$ the **direct type** of **e**. For any other index k, with $m_{\mathbf{e}} > k \geq 0$, we call $\tau_{i,k}^{\mathcal{G}}(\mathbf{e})$, if it is defined, a **transitive type** of **e**.

Example 1. Figure 3 depicts the typing morphisms between the graphs in a simplified sample hierarchy. The direct types for nodes and arrows are indicated with blue and cursive labels, respectively. All typing morphisms in the simple typing chain $\mathcal{TG}$, determined by the sequence [hammer_plant, generic_plant, Ecore] of graphs, are total except the one from hammer_plant to generic_plant, since the direct type of **has** is located in Ecore. We have chosen Ecore as the top-most graph since it provides implementation support through the Eclipse Modeling Framework [24]. This enables our approach to MLM to exploit the best from fixed-level and multi-level concepts [18].   □

To describe later the flexible matching of multilevel typed rules and the result of rule applications, we need a corresponding flexible notion of morphisms between typing chains.

Definition 5. A **typing chain morphism** $(\phi, f) : \mathcal{G} \rightarrow \mathcal{H}$ between two typing chains $\mathcal{G} = (\overline{G}, n, \tau^{\mathcal{G}})$ and $\mathcal{H} = (\overline{H}, m, \tau^{\mathcal{H}})$ with $n \leq m$ is given by

- a function $f : [n] \rightarrow [m]$, where $[n] = \{0, 1, 2, \ldots, n\}$, such that *(1)* $f(0) = 0$ and *(2)* $j > i$ implies $f(j) - f(i) \geq j - i$ for all $i, j \in [n]$, and
- a family of total graph homomorphisms $\phi = (\phi_i : G_i \rightarrow H_{f(i)} \mid i \in [n])$ such that

$$\tau_{j,i}^{\mathcal{G}}; \phi_i \preceq \phi_j; \tau_{f(j),f(i)}^{\mathcal{H}} \quad \text{for all } n \geq j > i \geq 0, \tag{1}$$

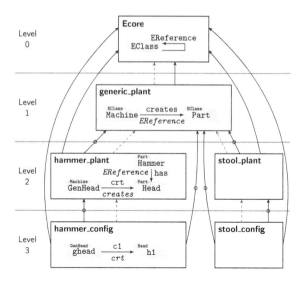

Fig. 3. Multilevel modeling hierarchy with typing morphisms

i.e., due to Definitions 2 and 3, we assume for any $n \geq j > i \geq 0$ the existence of a total graph homomorphism $\phi_{j|i}$ that makes the diagram of total graph homomorphisms displayed in Fig. 4 commutative.

A typing chain morphism $(\phi, f) : \mathcal{G} \to \mathcal{H}$ is **closed** iff $\tau_{j,i}^{\mathcal{G}}; \phi_i = \phi_j; \tau_{f(j),f(i)}^{\mathcal{H}}$ for all $n \geq j > i \geq 0$, i.e., the right lower square in Fig. 4 is a pullback.

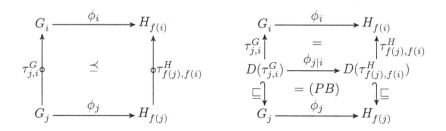

Fig. 4. Establishing a morphism between two typing chains, level-wise

Typing morphisms are composed by the composition of commutative squares.

Definition 6. *The **composition** $(\phi, f); (\psi, g) : \mathcal{G} \to \mathcal{K}$ of two typing chain morphisms $(\phi, f) : \mathcal{G} \to \mathcal{H}$, $(\psi, g) : \mathcal{H} \to \mathcal{K}$ between typing chains $\mathcal{G} = (\overline{G}, n, \tau^{\mathcal{G}})$, $\mathcal{H} = (\overline{H}, m, \tau^{\mathcal{H}})$, $\mathcal{K} = (\overline{K}, l, \tau^{\mathcal{K}})$ with $n \leq m \leq l$ is defined by $(\phi, f); (\psi, g) := (\phi; \psi_{\downarrow f}, f; g)$, where $\psi_{\downarrow f} := (\psi_{f(i)} : H_{f(i)} \to K_{g(f(i))} \mid i \in [n])$, and thus $\phi; \psi_{\downarrow f} := (\phi_i; \psi_{f(i)} : G_i \to K_{g(f(i))} \mid i \in [n])$.*

Chain denotes the category of typing chains and typing chain morphisms.

A natural way to define multilevel typing of a graph H over a typing chain $\mathcal{G}$ would be a family $\sigma = (\sigma_i : H \dashrightarrow G_i \mid n \geq i \geq 0)$ of partial graph homomorphisms satisfying certain properties. However, as shown in [26], those families are not appropriate to state adequate type compatibility requirements for rules and matches and to construct the results of rule applications. Therefore, we employ the sequence of the domains of definition of the σ_i's as a typing chain and describe multilevel typing by means of typing chain morphisms. The following Lemma describes how any sequence of subgraphs gives rise to a typing chain.

Lemma 1. *Any sequence $\overline{H} = [H_n, H_{n-1}, \dots, H_1, H_0]$ of subgraphs of a graph H, with $H_0 = H$, can be extended to a typing chain $\mathcal{H} = (\overline{H}, n, \tau^{\mathcal{H}})$ where for all $n \geq j > i \geq 0$ the corresponding **typing morphism** $\tau_{j,i}^{\mathcal{H}} : H_j \dashrightarrow H_i$ is given by $D(\tau_{j,i}^{\mathcal{H}}) := H_j \cap H_i$ and the span of total inclusion graph homomorphisms*

$$H_j \xleftarrow{\quad \sqsubseteq \quad} D(\tau_{j,i}^{\mathcal{H}}) = H_j \cap H_i \xhookrightarrow{\quad \tau_{j,i}^{\mathcal{H}} \quad} H_i \ .$$

*We call the typing chain $\mathcal{H} = (\overline{H}, n, \tau^{\mathcal{H}})$ an **inclusion chain on** H.*

A **multilevel typing of a graph** H over a typing chain $\mathcal{G} = (\overline{G}, n, \tau^{\mathcal{G}})$ is given by an inclusion chain $\mathcal{H} = (\overline{H}, n, \tau^{\mathcal{H}})$ on H (of the same length as $\mathcal{G}$) and a typing chain morphism $(\sigma^{\mathcal{H}}, id_{[n]}) : \mathcal{H} \to \mathcal{G}$.

3 Multilevel Typed Graph Transformations

Underlying Graph Transformation. To meet the characteristics of our application areas [19–21] we work with cospans $L \xrightarrow{\lambda} I \xleftarrow{\rho} R$ of inclusion graph homomorphisms, where $I = L \cup R$, as the **underlying graph transformation rule** of a multilevel typed rule. To apply such a rule [7,12,13], we have to find a match $\mu : L \to S$ of L in a graph S at the bottom-most level of an MLM hierarchy. To describe the effect of a rule application, we adapt the Sq-PO approach [7] to our cospan-rules: First, we construct a pushout and, second, a final pullback complement (FPBC) to create the graphs D and T, resp. (see Fig. 2). The details behind choosing cospan rules and Sq-PO, as opposed to span rules and double-pushout (DPO), are out of the scope of this paper. In short, however: (i) cospan rules are more suitable from an implementation point-of-view since they allow for first adding new elements then deleting (some of the) old elements [13], and (ii) having both old and new elements in I allows us to introduce constraints on new elements depending on old constraints involving elements to be deleted [23]. Moreover, we apply the rules using our variant of Sq-PO [7,13] since (i) the pushout complement in DPO, even if it exists, may not be unique, in contrast the FPBC, if it exists, is always unique (up to isomorphism), (ii) FPBC allows faithful deletion in unknown context, i.e., dangling edges may be deleted by applying the rules, however, the co-match ν is always total, i.e., if the match μ identifies elements to be removed with elements to be preserved, the FPBC will not exist and the application will not be allowed.

Multilevel Typed Rule. We augment the cospan rule to a **multilevel typed rule** by chosing a typing chain $\mathcal{MM} = (\overline{MM}, n, \tau^{\mathcal{MM}})$, the typing chain of the rule, together with **coherent** multilevel typings over $\mathcal{MM}$ of L and R, respectively. That is, we choose an inclusion chain $\mathcal{L} = (\overline{L}, n, \tau^{\mathcal{L}})$ on L, an inclusion chain $\mathcal{R} = (\overline{R}, n, \tau^{\mathcal{R}})$ on R and typing chain morphisms $(\sigma^{\mathcal{L}}, id_{[n]})$: $\mathcal{L} \to \mathcal{MM}$ with $\sigma^{\mathcal{L}} = (\sigma_i^{\mathcal{L}} : L_i \to MM_i \mid i \in [n])$, $(\sigma^{\mathcal{R}}, id_{[n]}) : \mathcal{R} \to \mathcal{MM}$ with $\sigma^{\mathcal{R}} = (\sigma_i^{\mathcal{R}} : R_i \to MM_i \mid i \in [n])$ (see Fig. 5), such that $L_i \cap R = L \cap R_i = L_i \cap R_i$ and, moreover, $\sigma_i^{\mathcal{L}}$ and $\sigma_i^{\mathcal{R}}$ coincide on the intersection $L_i \cap R_i$ for all $i \in [n]$.

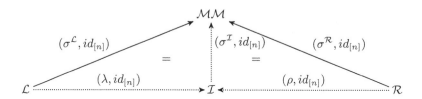

Fig. 5. Rule morphisms and their type compatibility

The inclusion chain $\mathcal{I} = (\overline{I}, n, \tau^{\mathcal{I}})$ on the union (pushout) $I = L \cup R$ is simply constructed by level-wise unions (pushouts): $I_i := L_i \cup R_i$ for all $i \in [n]$; thus, we have $I_0 = I$. Since **Graph** is an adhesive category [12], the construction of $\mathcal{I}$ by pushouts and the coherence condition ensure that we get for any $i \in [n]$ two pullbacks as shown in Fig. 6. The existence of these pullbacks implies, according to the following Lemma, that we can reconstruct the inclusion chains $\mathcal{L}$ and $\mathcal{R}$, respectively, as reducts of the inclusion chain $\mathcal{I}$.

Lemma 2. *Let be given two inclusion chains $\mathcal{G} = (\overline{G}, n, \tau^{\mathcal{G}})$ and $\mathcal{H} = (\overline{H}, m, \tau^{\mathcal{H}})$ with $n \leq m$ and a function $f : [n] \to [m]$ such that $f(0) = 0$ and $j > i$ implies $f(j) - f(i) \geq j - i$ for all $i, j \in [n]$. For any family $\phi = (\phi_i : G_i \to H_{f(i)} \mid i \in [n])$ of graph homomorphisms the following two requirements are equivalent:*

1. *For all $n \geq j > 0$ the left-hand square in Fig. 7 is a pullback.*
2. *The pair (ϕ, f) constitutes a closed typing chain morphism $(\phi, f) : \mathcal{G} \to \mathcal{H}$ where for all $n \geq j > i \geq 0$ the right-hand diagram in Fig. 7 consists of two pullbacks.*

Given a closed typing chain morphism $(\phi, f) : \mathcal{G} \to \mathcal{H}$ between inclusion chains, as described in Lemma 2, we call $\mathcal{G}$ the **reduct of $\mathcal{H}$ along $\phi_0 : G_0 \to H_0$** and $f : [n] \to [m]$ while $(\phi, f) : \mathcal{G} \to \mathcal{H}$ is called a **reduct morphism**. Note that the composition of two reduct morphisms is a reduct morphism as well.

Lemma 2 ensures that the families $(\lambda_i : L_i \to I_i \mid i \in [n])$ and $(\rho_i : R_i \to I_i \mid i \in [n])$ of inclusion graph homomorphisms establish reduct morphisms $(\lambda, id_{[n]}) : \mathcal{L} \to \mathcal{I}$ and $(\rho, id_{[n]}) : \mathcal{R} \to \mathcal{I}$, resp., as shown in Fig. 5.

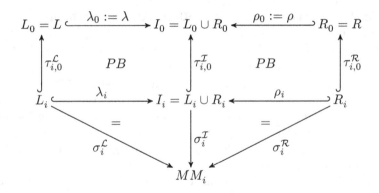

Fig. 6. Type compatibility of rule morphisms level-wise

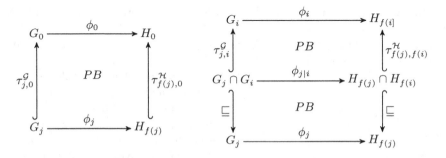

Fig. 7. Reduct of inclusion chains

Finally, we have to construct a typing chain morphism $(\sigma^{\mathcal{I}}, id_{[n]}) : \mathcal{I} \to \mathcal{MM}$ making the diagram in Fig. 5 commute: For all $i \in [n]$, we constructed the union (pushout) $I_i := L_i \cup R_i$. Moreover, $\sigma_i^{\mathcal{L}}$ and $\sigma_i^{\mathcal{R}}$ coincide on $L_i \cap R_i$, by coherence assumption, thus we get a unique $\sigma_i^{\mathcal{I}} : I_i \to MM_i$ such that (see Fig. 6)

$$\sigma_i^{\mathcal{L}} = \lambda_i; \sigma_i^{\mathcal{I}} \quad \text{and} \quad \sigma_i^{\mathcal{R}} = \rho_i; \sigma_i^{\mathcal{I}} \tag{2}$$

Since **Graph** is adhesive, Lemma 2 ensures that the family $\sigma^{\mathcal{I}} = (\sigma_i^{\mathcal{I}} : I_i \to MM_i \mid i \in [n])$ of graph homomorphisms establishes indeed a typing chain morphism $(\sigma^{\mathcal{I}}, id_{[n]}) : \mathcal{I} \to \mathcal{MM}$ while the Eq. 2 ensure that the diagram in Fig. 5 commutes indeed.

Example 2. Figure 8 shows a multilevel typed rule *CreatePart* from a case study [20]. This rule can be used to specify the behaviour of machines that create parts, by matching an existing type of machine that generates a certain type of parts, and in the instance at the bottom, generating such a part. META defines a typing chain $\mathcal{MM}$ of depth 3. It declares the graph ($\texttt{M1} \xrightarrow{\texttt{cr}} \texttt{P1}$) that becomes MM_2. The declaration of the direct types $\texttt{Machine}$, $\texttt{creates}$, $\texttt{Part}$ for the elements in MM_2 declares, implicitly, a graph $MM_1 := (\texttt{Machine} \xrightarrow{\texttt{creates}} \texttt{Part})$ that is in turn, implicitly, typed over $MM_0 :=$ ECore. All the morphisms in τ^{MM} are total and uniquely determined thus we have, especially, $\tau_{2,0}^{MM} = \tau_{2,1}^{MM}; \tau_{1,0}^{MM}$.

FROM and TO declare as well the left-hand side $L := ($ m1 $)$ and the right-hand-side $R := ($ m1 $\xrightarrow{\text{C}}$ p1 $)$, resp., of the rule and the direct types of the elements in L and R. These direct types are all located in MM_2 thus $L_2 = L$ and $R_2 = R$ where the direct types define nothing but the typing morphisms $\sigma_2^L : L_2 \to MM_2$ and $\sigma_2^R : R_2 \to MM_2$, resp. The other typing morphisms are obtained by "transitive closure", i.e., $\sigma_1^L := \sigma_2^L; \tau_{2,1}^{MM}$, $\sigma_0^L := \sigma_2^L; \tau_{2,0}^{MM}$ and $\sigma_1^R := \sigma_2^R; \tau_{2,1}^{MM}$, $\sigma_0^R := \sigma_2^R; \tau_{2,0}^{MM}$, thus we have $L = L_0 = L_1 = L_2$ and $R = R_0 = R_1 = R_2$.

For the "plain variant" of the rule *CreatePart* (in Fig. 15), MM consists only of the graphs $MM_1 = ($ M1 $\xrightarrow{\text{cr}}$ P1 $)$, $MM_0 =$ ECore and the trivial $\tau_{1,0}^{MM}$.

Multilevel Typed Match. In the multilevel typed setting all the graphs S, D, T are multilevel typed over a common typing chain $TG = (\overline{TG}, m, \tau^{TG})$, with $n \leq m$, that is determined by the path from S to the top of the current MLM hierarchy (see Fig. 1).

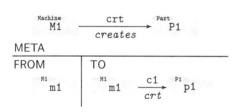

Fig. 8. *CreatePart*: a sample rule

A **match** of the multilevel typed rule into a graph S with a given multi-level typing over TG, i.e., an inclusion chain $S = (\overline{S}, m, \tau^S)$ with $S_0 = S$ and a typing chain morphism $(\sigma^S, id_{[m]}) : S \to TG$, is given by a graph homo-morphism $\mu : L \to S$ and a typing chain morphism $(\beta, f) : MM \to TG$ such that the following two conditions are satisfied:

- **Reduct:** L is the reduct of S along $\mu : L \to S$ and $f : [n] \to [m]$, i.e., $\mu_0 := \mu : L_0 = L \longrightarrow S_0 = S$ extends uniquely (by pullbacks) to a reduction morphism $(\mu, f) : L \to S$ with $\mu = (\mu_i : L_i \to S_{f(i)} \mid i \in [n])$ (see Fig. 9).
- **Type compatibility:** $(\sigma^L, id_{[n]}); (\beta, f) = (\mu, f); (\sigma^S, id_{[m]})$, i.e., we require

$$\sigma_i^L; \beta_i = \mu_i; \sigma_{f(i)}^S \text{ for all } n \geq i > 0. \tag{3}$$

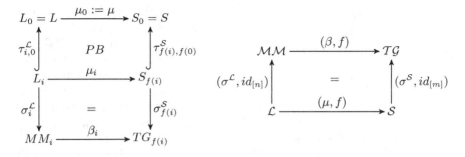

Fig. 9. Conditions for multilevel typEd Match

Application of a Multilevel Typed Rule – Objectives. The basic idea is
to construct for a given application of a graph transformation rule, as shown in
Fig. 2, a unique type compatible multilevel typing of the result graphs D and
T. The parameters of this construction are typing chains $\mathcal{MM}, \mathcal{TG}$; a coherent
multilevel typing of the graph transformation rule over $\mathcal{MM}$; a multilevel typing
of the graph S over $\mathcal{TG}$ and a typing chain morphism $(\beta, f) : \mathcal{MM} \to \mathcal{TG}$
extending the given match $\mu : L \to S$ of graphs to a multilevel typed match
satisfying the two respective conditions for multilevel typed matches.

Example 3 (Multilevel Typed Match). To achieve precision in rule application
the elements `Machine`, `creates`, `Part` in the original rule *CreatePart* are constants required to match syntactically with elements in the hierarchy. In such
a way, $MM_1 = ($ `Machine` $\xrightarrow{\text{creates}}$ `Part` $)$ has to match with `generic_plant`
while $MM_2 = ($ `M1` $\xrightarrow{\text{cr}}$ `P1` $)$ could match with `hammer_plant` or `stool_plant`.
We will observe later that for the plain version of the rule *CreatePart* in Fig. 15
we could match $MM_1 = ($ `M1` $\xrightarrow{\text{cr}}$ `P1` $)$ either with $TG_2 =$ `hammer_plant` or
$TG_1 =$ `generic_plant` in the hierarchy in Fig. 3, where the second match would
lead to undesired results (see Example 4).

Pushout step. As shown later, the pushout of the span $S \xleftarrow{\mu} L \xhookrightarrow{\lambda} I$
in **Graph** extends, in a canonical way, to a pushout of the span

$$S \xleftarrow{\quad (\mu, f) \quad} \mathcal{L} \xhookrightarrow{\quad (\lambda, id_{[n]}) \quad} \mathcal{I}$$

of reduct morphisms in **Chain** such that the result typing chain $\mathcal{D} = (\overline{D}, m, \tau^{\mathcal{D}})$
is an inclusion chain and the typing chain morphisms $(\varsigma, id_{[m]}) : S \hookrightarrow \mathcal{D}$ and
$(\delta, f) : \mathcal{I} \to \mathcal{D}$ become reduct morphisms (see the bottom in Fig. 10).

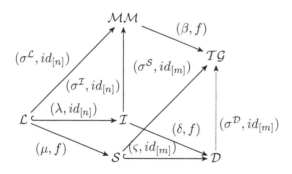

We get also a type compatible typing chain morphism from $\mathcal{D}$ into $\mathcal{TG}$: The back triangle in Fig. 10 commutes due to the type compatibility of the rule (see Fig. 5). The roof square commutes since the match is type compatible (see Fig. 9). This gives us (μ, f); $(\sigma^{\mathcal{S}}, id_{[m]}) = (\lambda, id_{[n]}); (\sigma^{\mathcal{I}}, id_{[n]}); (\beta, f)$, thus the universal property of the pushout bottom square pro-

Fig. 10. Pushout step

vides a unique chain morphism $(\sigma^{\mathcal{D}}, id_{[m]}) : \mathcal{D} \to \mathcal{TG}$ such that both type compatibility conditions $(\varsigma, id_{[m]}); (\sigma^{\mathcal{D}}, id_{[m]}) = (\sigma^{\mathcal{S}}, id_{[m]})$ and $(\delta, f); (\sigma^{\mathcal{D}}, id_{[m]}) = (\sigma^{\mathcal{I}}, id_{[n]}); (\beta, f)$ are satisfied.

Pullback Complement Step. As shown later, the final pullback complement $D \xleftarrow{\theta} T \xleftarrow{\nu} R$ in **Graph** extends, in a canonical way, to a sequence of reduct morphisms $\mathcal{D} \xleftarrow{(\theta, id_{[n]})} \mathcal{T} \xleftarrow{(\nu, f)} \mathcal{R}$ in **Chain** such that the bottom square in Fig. 11 commutes.

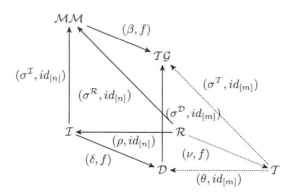

Fig. 11. Pullback complement step

Pushout of Reduct Morphisms – Two Steps. We discuss the intended pushout of the span

$$\mathcal{S} \xleftarrow{(\mu, f)} \mathcal{L} \xhookrightarrow{(\lambda, id_{[n]})} \mathcal{I}$$

of reduct morphisms in **Chain**. The reduct morphism $(\lambda, id_{[n]})$ is surjective w.r.t. levels, thus the pushout inclusion chain $\mathcal{D}$ should have the same length as $\mathcal{S}$. The rule provides, however, only information how to extend the subgraphs of $S_0 = S$ at the levels $f([n]) \subseteq [m]$. For the subgraphs in $\mathcal{S}$ at levels in $[m] \setminus f([n])$ the rule does not impose anything thus we let the subgraphs at those levels untouched. In terms of typing chain morphisms, this means that we factorize the reduct morphism (μ, f) into two reduct morphisms and that we will construct the resulting inclusion chain $\mathcal{D}$ in two pushout steps (see Fig. 12) where $\mathcal{S}_{\downarrow f} := (\overline{S}_{\downarrow f}, n, \tau^{\mathcal{S}}_{\downarrow f})$ with $\overline{S}_{\downarrow f} := [S_{f(n)}, S_{f(n-1)}, \ldots, S_{f(1)}, S_{f(0)=0}]$ and $\tau^{\mathcal{S}}_{\downarrow f} := (\tau^{\mathcal{S}}_{f(j),f(i)} : S_{f(j)} \multimap S_{f(i)} \mid n \geq j > i \geq 0)$ Note, that $\overline{S}_{\downarrow f} := [S_{f(n)}, \ldots, S_{f(0)}]$ is just a shorthand for the defining statement: $(\overline{S}_{\downarrow f})_i := S_{f(i)}$ for all $n \geq i \geq 0$.

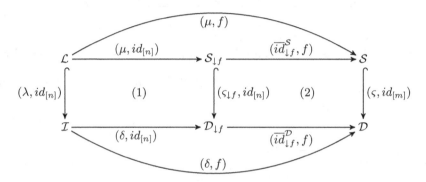

Fig. 12. Two pushout steps to construct the inclusion chain $\mathcal{D}$

The reduct morphism $(\overline{id}^{\mathcal{S}}_{\downarrow f}, f) : \mathcal{S}_{\downarrow f} \to \mathcal{S}$ is a level-wise identity and just embeds an inclusion chain of length $n + 1$ into an inclusion chain of length $m + 1$, i.e., $\overline{id}^{\mathcal{S}}_{\downarrow f} = (id_{f(i)} : S_{f(i)} \to S_{f(i)} \mid i \in [n])$. In the pushout step (1) we will construct a pushout of inclusion chains of equal length and obtain a chain $\mathcal{D}_{\downarrow f} := (\overline{D}_{\downarrow f}, n, \tau^{\mathcal{D}}_{\downarrow f})$ with $\overline{D}_{\downarrow f} = [D_{f(n)}, D_{f(n-1)}, \ldots, D_{f(1)}, D_{f(0)=0}]$ and $\tau^{\mathcal{D}}_{\downarrow f} = (\tau^{\mathcal{D}}_{f(j),f(i)} : D_{f(j)} \multimap D_{f(i)} \mid n \geq j > i \geq 0)$.

In the pushout step (2) we will fill the gaps in $\mathcal{D}_{\downarrow f}$ with the corresponding untouched graphs from the original inclusion chain $\mathcal{S}$.

Pushouts of Graphs for Inclusion Graph Homomorphisms. Our constructions and proofs rely on the standard construction of pushouts in **Graph** for a span of an inclusion graph homomorphism $\phi : G \hookrightarrow H$ and an arbitrary graph homomorphism $\psi : G \to K$ where we assume that H and K are disjoint. The pushout P is given by $P^N := K^N \cup H^N \setminus G^N$, $P^A := K^A \cup H^A \setminus G^A$ and $sc^P(e) := sc^K(e)$, if $e \in K^A$, and $sc^P(e) := \psi^A(sc^H(e))$, if $e \in H^A \setminus G^A$. tg^P is defined analogously. $\phi^* : K \hookrightarrow P$ is an inclusion graph homomorphism by construction and $\psi^* : H \to P$ is defined for $X \in \{A, N\}$ by $\psi^{*X}(v) := \psi^X(v)$, if $v \in G^X$ and $\psi^{*X}(v) := v$, if $v \in H^X \setminus G^X$.

The pair $G \setminus H := (H^N \setminus G^N, H^A \setminus G^A)$ of subsets of nodes and arrows of H is, in general, not establishing a subgraph of H. We will nevertheless use the notation $P = K + H \setminus G$ to indicate that P is constructed as described above. ψ^* can be described then as a sum of two parallel pairs of mappings

$$\psi^* = \psi + id_{H \setminus G} := (\psi^N + id_{H^N \setminus G^N}, \psi^A + id_{H^A \setminus G^A}) \qquad (4)$$

Pushout for Inclusion Chains with Equal Depth. We consider now the span

$$\mathcal{S}_{\downarrow f} \xleftarrow{\quad (\mu, id_{[n]}) \quad} \mathcal{L} \xhookrightarrow{\quad (\lambda, id_{[n]}) \quad} \mathcal{I}$$

of reduct morphisms in **Chain** (see Fig. 12). For each level $i \in [n]$ we construct the corresponding pushout of graph homomorphisms. This ensures, especially,

$$\lambda_i; \delta_i = \mu_i; \varsigma_{f(i)} \quad \text{for all} \quad i \in [n]. \qquad (5)$$

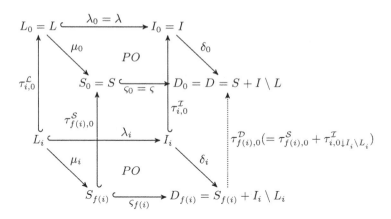

Fig. 13. Level-wise pushout construction

We look at an arbitrary level $n \geq i \geq 1$ together with the base level 0 (see Fig. 13). We get a cube where the top and bottom square are pushouts by construction. In addition, the left and back square are pullbacks since $(\mu, id_{[n]})$ and $(\lambda, id_{[n]})$, respectively, are reduct morphisms. We get a unique graph homomorphism $\tau^{\mathcal{D}}_{f(i),0} : D_{f(i)} \to D$ that makes the cube commute. By the uniqueness of mediating morphisms and the fact that the top pushout square has the Van Kampen property (see [12,25]), we can conclude that the front and the right square are pullbacks as well. That the back square is a pullback means nothing but $L_i = L \cap I_i$. This entails $I_i \setminus L_i \subseteq I \setminus L$ thus $\tau^{\mathcal{D}}_{f(i),0}$ turns out to be the sum of the two inclusions $\tau^{\mathcal{S}}_{f(i),0} : S_{f(i)} \hookrightarrow S$ and $\tau^{\mathcal{I}}_{i,0 \downarrow I_i \setminus L_i} : I_i \setminus L_i \hookrightarrow I \setminus L$ and is therefore an inclusion itself.

The sequence $[D_{f(n)}, D_{f(n-1)}, \ldots, D_{f(1)}, D_0]$ of subgraphs of $D = D_0$ defines the intended inclusion chain $\mathcal{D}_{\downarrow f}$. Since the front and right squares in Fig. 13 are pullbacks, Lemma 2 ensures that the family $\varsigma_{\downarrow f} = (\varsigma_{f(i)} : S_{f(i)} \hookrightarrow D_{f(i)} \mid i \in [n])$ of inclusion graph homomorphisms constitutes a reduct morphism $(\varsigma_{\downarrow f}, id_{[n]})$: $\mathcal{S}_{\downarrow f} \to \mathcal{D}_{\downarrow f}$ while the family $\delta = (\delta_i : I_i \to D_{f(i)} \mid i \in [n])$ of graph homomorphisms constitutes a reduct morphism $(\delta, id_{[n]}) : \mathcal{I} \to \mathcal{D}_{\downarrow f}$. Finally, Eq. 5 ensures that the resulting square (1) of reduct morphisms in Fig. 12 commutes. The proof that we have constructed a pushout in **Chain** is given in [26].

Remark 2 (Only one pushout). $\varsigma_{f(i)}$ and δ_i are jointly surjective for all $n \geq i \geq 1$ thus we can describe $D_{f(i)}$ as the union $D_{f(i)} = \varsigma(S_{f(i)}) \cup \delta(I_i)$. Hence in practice, there is no need for an explicit construction of pushouts at all the levels $n \geq i \geq 1$; these are all constructed implicitly by the pushout construction at level 0.

Pushout by Chain Extension. To obtain an inclusion chain $\mathcal{D}$ of length $m + 1$, we fill the gaps in $\mathcal{D}_{\downarrow f}$ by corresponding subgraphs of S: $D_a := D_a$ if $a \in f([n])$ and $D_a := S_a$ if $a \in [m] \setminus f([n])$ and obtain the intended inclusion chain $\mathcal{D} = (\overline{D}, m, \tau^{\mathcal{D}})$. The family $\overline{id}^{\mathcal{D}}_{\downarrow f} = (id_{D_{f(i)}} : D_{f(i)} \to D_{f(i)} \mid i \in [n])$ of identities defines trivially a reduct morphism $(\overline{Id}^{\mathcal{D}}_{\downarrow f}, f) : \mathcal{D}_{\downarrow f} \to \mathcal{D}$. One can show that the family $\varsigma = (\varsigma_a : S_a \to D_a \mid a \in [m])$ of graph homomorphisms defined by

$$\varsigma_a := \begin{cases} \varsigma_a : S_a \hookrightarrow D_a & \text{if } a \in f([n]) \\ id_{S_a} : S_a \to D_a = S_a & \text{if } a \in [m] \setminus f([n]) \end{cases}$$

establishes a reduct morphism $(\varsigma, id_{[m]}) : \mathcal{S} \to \mathcal{D}$. The two reduct morphisms $(\overline{id}^{\mathcal{D}}_{\downarrow f}, f) : \mathcal{D}_{\downarrow f} \to \mathcal{D}$ and $(\varsigma, id_{[m]}) : \mathcal{S} \to \mathcal{D}$ establish square (2) in Fig. 12 that commutes trivially. In [26] it is shown that square (2) is also a pushout in **Chain**.

Pullback Complement. We construct the reduct of $\mathcal{D} = (\overline{D}, m, \tau^{\mathcal{D}})$ along θ : $T \hookrightarrow D$ and $id_{[m]}$ by level-wise intersection (pullback) for all $n \geq i \geq 1$ (see the pullback square below). Due to Lemma 2, we obtain, in such a way, an inclusion chain $\mathcal{T} = (\overline{T}, m, \tau^{\mathcal{T}})$ together with a reduct morphism $(\theta, id_{[m]}) : \mathcal{T} \to \mathcal{D}$. The multilevel typing of $\mathcal{T}$ is simply borrowed from $\mathcal{D}$, that is, we define (see Fig. 11)

$$(\sigma^{\mathcal{T}}, id_{[m]}) := (\theta, id_{[m]}); (\sigma^{\mathcal{D}}, id_{[m]}) \tag{6}$$

and this gives us trivially the intended type compatibility of $(\theta, id_{[m]})$. The typing chain morphism $(\nu, f) : \mathcal{R} \to \mathcal{T}$ with $\nu = (\nu_i : R_i \to T_{f(i)} \mid i \in [n])$ such that

$$(\rho, id_{[n]}); (\delta, f) = (\nu, f); (\theta, id_{[m]}) \tag{7}$$

is simply given by pullback composition and decomposition in **Graph**: For each $n \geq i \geq 1$ we consider the following incomplete cube on the right-hand side:

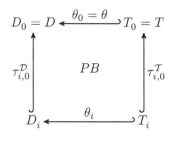

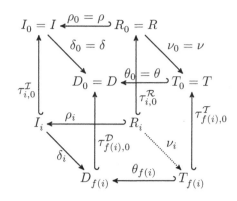

The back square, the left square as well as the front square are pullbacks since $(\rho, id_{[n]})$, (δ, f) and $(\theta, id_{[m]})$, respectively, are reduct morphisms. The top square is constructed as a pullback complement. The diagonal square from $\tau_{i,0}^{\mathcal{R}}$ to $\tau_{f(i),0}^{\mathcal{D}}$ is a pullback due to the composition of the back pullback and the left pullback. The decomposition of this diagonal pullback w.r.t. the front pullback gives us $\nu_i : R_i \to T_i$ making the cube, and especially the bottom square, commute and making the right square to a pullback as well.

According to Lemma 2 the family $\nu = (\nu_i : R_i \to T_{f(i)} \mid i \in [n])$ of graph homomorphisms defines a reduct morphism $(\nu, f) : \mathcal{R} \to \mathcal{T}$ where condition 7 is simply satisfied by construction. Finally, (ν, f) is also type compatible since conditions 6 and 7 ensure that the roof square in Fig. 11 commutes.

Example 4. To present a non-trivial rule application for our example, we discuss the undesired application of the plain version of rule *CreatePart* (see Fig. 14), mentioned in Example 3, for a state of the hammer configuration with only one node `ghead`, as shown in `hammer_config_0` in Fig. 15. So, we have $f : [1] \to [2]$, with $f(0) = 0, f(1) = 1$, and the "undesired match" of $MM_1 = ($ M1 $\xrightarrow{\text{cr}}$ P1 $)$ with $TG_1 = $ `generic_plant` $ = ($ Machine $\xrightarrow{\text{creates}}$ Part $)$ together with the trivial match of the left-hand side $L = ($ m1 $)$ of the rule with `hammer_config_0` $= ($ ghead $)$. The resulting inclusion chains $\mathcal{S}$, $\mathcal{L}$, $\mathcal{R}$ and two reduct morphisms between them are depicted in Fig. 14. Note, that the ellipse and cursive labels indicate here the corresponding typing chain morphisms $(\sigma^{\mathcal{S}}, id_{[2]})$, $(\sigma^{\mathcal{L}}, id_{[1]})$ and $(\sigma^{\mathcal{R}}, id_{[1]})$, respectively.

For the two levels in $f([1]) = \{0, 1\} \subset [2]$ we construct the pushouts D_0 and D_1 while D_2 is just taken as S_2. The lowest level in $\mathcal{D}$, where the new elements p1 and c appear, is level 1 thus the constructed direct types of p1 and c are Part and creates, resp., as shown in `hammer_config_1` in Fig. 15.

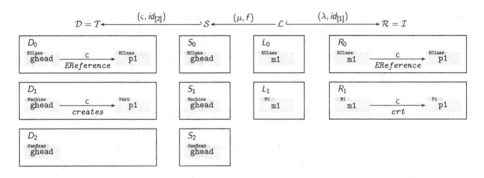

Fig. 14. Inclusion chains for the plain version of *CreatePart*

Fig. 15. Plain version of *CreatePart* and its application

4 Conclusions, Related and Future Work

Conclusion. Multilevel modeling offers more flexibility on top of traditional modeling techniques by supporting an unlimited number of abstraction levels. Our approach to multilevel modeling enhances reusability of concepts and their behaviour by allowing the definition of flexible transformation rules which are applicable to different hierarchies with a variable number of levels. In this paper, we have presented a formalization of these flexible and reusable transformation rules based on graph transformations. We represent multilevel models by multilevel typed graphs whose manipulation and transformation are carried out by multilevel typed graph transformation rules. These rules are cospans of three graphs and two inclusion graph homomorphisms where the three graphs are multilevel typed over a common typing chain. As these rules are represented as cospans, their application is carried out by a pushout and a final pullback complement construction for the underlying graphs in the category **Graph**. We have identified type compatibility conditions, for rules and their matches, which are crucial for rule applications. Moreover, we have shown that typed graph transformations can be generalized to multilevel typed graph transformations improving preciseness, flexibility and reusability of transformation rules.

Related work. The theory and practise of graph transformations are well-studied, and the concept of model transformations applied to MLM is not novel. Earlier works in the area have worked in the extension of pre-existing model transformation languages to be able to manipulate multilevel models and model hierarchies. In [3], the authors adapt ATL [15] to manipulate multilevel models built with the Melanee tool [2]. In a similar manner, [11] proposes the adaptation of ETL [16] and other languages from the Epsilon family [14] for the application of model transformation rules into multilevel hierarchies created with MetaDepth [8]. These works, however, tackle the problem from the practical point of view. That is, how to reuse mature off-the-shelf tools for model transformation in the context of MLM, via the manipulation of a "flattened" representation of the hierarchy to emulate multilevel transformations. Our approach, on the contrary, has been developed from scratch with a multilevel setting in mind, and we believe it can be further extended to tackle all scenarios considered by other approaches. Therefore, to the best of our knowledge, there are no formal treatments of multilevel typed graph transformations in the literature except for our previous works [19,20,26] (see Sect. 4 in [26]). Hence, we consider our approach the first approximation to formally address the challenges which come with multilevel modeling and multilevel model transformations.

Common for our work and [9] is that the concepts of typing chains, multilevel typed graphs and multilevel models originate in [22]. However, [9] presents partial morphisms as spans of total morphisms and does not use the composition of those spans explicitly. Wrt. typing chains, a multilevel model in [9] is a sequence of graphs $[G_n, G_{n-1}, \ldots, G_1, G_0]$ together with the subfamily $(\tau_{i+1,i}^{\mathcal{G}} : G_{i+1} \dashrightarrow G_i \mid n \geq i \geq 0)$ of typing morphisms.

Future work. Although it is trivial to see that the bottom square in the cube for the pullback complement step becomes a pullback for all $n \geq i \geq 1$, we leave it for future work to prove that we indeed have constructed a final pullback complement in **Chain**. A utilization of our theory to deal with coupled transformations [21] in the setting of multilevel typed modelling is also desirable. Furthermore, it would be interesting to investigate the category **Chain** for its own; e.g., study its monomorphisms and epimorphisms, possible factorization systems, and the conditions for existence of general pushouts and pullbacks.

References

1. Almeida, J.P.A., Frank, U., Kühne, T.: Multi-level modelling (Dagstuhl Seminar 17492). Dagstuhl Reports **7**(12), 18–49 (2018). https://doi.org/10.4230/DagRep. 7.12.18
2. Atkinson, C., Gerbig, R.: Flexible deep modeling with melanee. In: Stefanie Betz and Ulrich Reimer, editors, Modellierung 2016, 2.-4. März 2016, Karlsruhe - Workshopband, Modellierung 2016, Bonn, vol. 255, pp. 117–122 (2016). Gesellschaft für Informatik
3. Atkinson, C., Gerbig, R., Tunjic, C.V.: Enhancing classic transformation languages to support multi-level modeling. Software Syst. Model. **14**(2), 645–666 (2015). https://doi.org/10.1007/s10270-013-0384-y
4. Atkinson, C., Kühne, T.: The essence of multilevel metamodeling. In: Gogolla, M., Kobryn, C. (eds.) UML 2001. LNCS, vol. 2185, pp. 19–33. Springer, Heidelberg (2001). https://doi.org/10.1007/3-540-45441-1_3

5. Atkinson, C., Kühne, T.: Rearchitecting the UML infrastructure. ACM Trans. Model. Comput. Simul. **12**(4), 290–321 (2002). https://doi.org/10.1145/643120. 643123

6. Atkinson, C., Kühne, T.: Reducing accidental complexity in domain models. Software Syst. Model. **7**(3), 345–359 (2008). https://doi.org/10.1007/s10270-007-0061-0

7. Corradini, A., Heindel, T., Hermann, F., König, B.: Sesqui-pushout rewriting. In: Corradini, A., Ehrig, H., Montanari, U., Ribeiro, L., Rozenberg, G. (eds.) ICGT 2006. LNCS, vol. 4178, pp. 30–45. Springer, Heidelberg (2006). https://doi.org/10. 1007/11841883_4

8. de Lara, J., Guerra, E.: Deep meta-modelling with METADEPTH. In: Vitek, J. (ed.) TOOLS 2010. LNCS, vol. 6141, pp. 1–20. Springer, Heidelberg (2010). https://doi. org/10.1007/978-3-642-13953-6_1

9. de Lara, J., Guerra, E.: Multi-level model product lines. FASE 2020. LNCS, vol. 12076, pp. 161–181. Springer, Cham (2020). https://doi.org/10.1007/978-3-030-45234-6_8

10. de Lara, J., Guerra, E., Cuadrado, J.S.: When and how to use multilevel modelling. ACM Trans. Softw. Eng. Methodol. **24**(2), 12:1–12:46 (2014). https://doi.org/10. 1145/2685615

11. de Lara, J., Guerra, E., Cuadrado, J.S.: Model-driven engineering with domain-specific meta-modelling languages. Software Syst. Model. **14**(1), 429–459 (2015). https://doi.org/10.1007/s10270-013-0367-z

12. Implementation of typed attributed graph transformation by AGG. Fundamentals of Algebraic Graph Transformation. MTCSAES, pp. 305–323. Springer, Heidelberg (2006). https://doi.org/10.1007/3-540-31188-2_15

13. Ehrig, H., Hermann, F., Prange, U.: Cospan DPO approach: an alternative for DPO graph transformations. Bulletin EATCS **98**, 139–149 (2009)

14. The Eclipse Foundation. Epsilon (2012). http://www.eclipse.org/epsilon/

15. Jouault, F., Allilaire, F., Bézivin, J., Kurtev, I.: ATL: a model transformation tool. Sci. Comput. Program. **72**(1–2), 31–39 (2008). https://doi.org/10.1016/j. scico.2007.08.002

16. Kolovos, D.S., Paige, R.F., Polack, F.A.C.: The epsilon transformation language. In: Vallecillo, A., Gray, J., Pierantonio, A. (eds.) ICMT 2008. LNCS, vol. 5063, pp. 46–60. Springer, Heidelberg (2008). https://doi.org/10.1007/978-3-540-69927-9_4

17. Macías, F.: Multilevel modelling and domain-specific languages. Ph.D. thesis, Western Norway University of Applied Sciences and University of Oslo (2019)

18. Macías, F., Rutle, A., Stolz, V.: MultEcore: combining the best of fixed-level and multilevel metamodelling. In: MULTI, CEUR Workshop Proceedings, vol. 1722. CEUR-WS.org (2016)

19. Macías, F., Rutle, A., Stolz, V., Rodriguez-Echeverria, R., Wolter, U.: An approach to flexible multilevel modelling. Enterp. Model. Inf. Syst. Archit. **13**, 10:1–10:35 (2018). https://doi.org/10.18417/emisa.13.10

20. Macías, F., Wolter, U., Rutle, A., Durán, F., Rodriguez-Echeverria, R.: Multilevel coupled model transformations for precise and reusable definition of model behaviour. J. Log. Algebr. Meth. Program. **106**, 167–195 (2019). https://doi.org/10.1016/j.jlamp.2018.12.005

21. Mantz, F., Taentzer, G., Lamo, Y., Wolter, U.: Co-evolving meta-models and their instance models: a formal approach based on graph transformation. Sci. Comput. Program. **104**, 2–43 (2015). https://doi.org/10.1016/j.scico.2015.01.002

22. Rossini, A., de Lara, J., Guerra, E., Rutle, A., Wolter, U.: A formalisation of deep metamodelling. Formal Aspects Comput. **26**(6), 1115–1152 (2014). https://doi.org/10.1007/s00165-014-0307-x
23. Rutle, A., Rossini, A., Lamo, Y., Wolter, U.: A formal approach to the specification and transformation of constraints in MDE. J. Log. Algebr. Program. **81**(4), 422–457 (2012). https://doi.org/10.1016/j.jlap.2012.03.006
24. Steinberg, D., Budinsky, F., Paternostro, M., Merks, E.: EMF: Eclipse Modeling Framework, 2nd edn. Addison-Wesley Professional, Boston (2008)
25. Wolter, U., König, H.: Fibred amalgamation, descent data, and van kampen squares in topoi. Appl. Categor. Struct. **23**(3), 447–486 (2013). https://doi.org/10.1007/s10485-013-9339-2
26. Wolter, U., Macías, F., Rutle, A.: The category of typing Chains as a foundation of multilevel typed model transformations. Technical Report 2019–417, University of Bergen, Department of Informatics, November 2019

Application Domains

Rewriting Theory for the Life Sciences: A Unifying Theory of CTMC Semantics

Nicolas Behr[1]([⊠])(iD) and Jean Krivine[2](iD)

[1] Center for Research and Interdisciplinarity (CRI), Université de Paris,
INSERM U1284, 8-10 Rue Charles V, 75004 Paris, France
`nicolas.behr@cri-paris.org`
[2] Institut de Recherche en Informatique Fondamentale (IRIF), Université de Paris,
CNRS UMR 8243, 8 Place Aurélie Nemours, 75205 Paris Cedex 13, France
`jean.krivine@irif.fr`

Abstract. The Kappa biochemistry and the MØD organo-chemistry frameworks are amongst the most intensely developed applications of rewriting theoretical methods in the life sciences to date. A typical feature of these types of rewriting theories is the necessity to implement certain structural constraints on the objects to be rewritten (a protein is empirically found to have a certain signature of sites, a carbon atom can form at most four bonds, ...). In this paper, we contribute to the theoretical foundations of these types of rewriting theory a number of conceptual and technical developments that permit to implement a universal theory of continuous-time Markov chains (CTMCs) for stochastic rewriting systems. Our core mathematical concepts are a novel rule algebra construction for the relevant setting of rewriting rules with conditions, both in Double- and in Sesqui-Pushout semantics, augmented by a suitable stochastic mechanics formalism extension that permits to derive dynamical evolution equations for pattern-counting statistics.

Keywords: Double-pushout rewriting · Sesqui-pushout rewriting · Rule algebra · Stochastic mechanics · Biochemistry · Organic chemistry

1 Motivation

One of the key applications that rewriting theory may be considered for in the life sciences is the theory of continuous-time Markov chains (CTMCs) modeling complex systems. In fact, since Delbrück's seminal work on autocatalytic reaction systems in the 1940s [20], the mathematical theory of chemical reaction systems has effectively been formulated as a rewriting theory in disguise, namely via the rule algebra of discrete graph rewriting [11]. In the present paper, we provide the necessary technical constructions in order to consider the CTMCs and analysis methods of relevance for more general types of compositional rewriting

---

An extended version of this paper containing additional technical appendices is available online [9].

© Springer Nature Switzerland AG 2020
F. Gadducci and T. Kehrer (Eds.): ICGT 2020, LNCS 12150, pp. 185–202, 2020.
https://doi.org/10.1007/978-3-030-51372-6_11

theories with conditions, with key examples provided in the form of *biochemical graph rewriting* in the sense of the KAPPA framework (https://kappalanguage. org) [12], and *(organo-) chemical graph rewriting* in the sense of the MØD framework (https://cheminf.imada.sdu.dk/mod/) [1]. The present paper aims to serve two main purposes: the first consists in providing an extension of the existing category-theoretical rule-algebra frameworks [4,10,11] by the rewriting theoretical design feature of incorporating rules with conditions as well as constraints on objects (Sect. 3). Based upon these technical developments, we then investigate to which extent it is possible to utilize the rule-algebraic *stochastic mechanics frameworks* of the relevant types (Sect. 4) in order to compute evolution equations for the moments of pattern-count observables within the KAPPA and MØD frameworks (Sect. 5 and 6).

2   Compositional Rewriting Theories with Conditions

The well-established *Double-Pushout (DPO)* [21] and *Sesqui-Pushout (SqPO)* [13] frameworks for rewriting systems over categories with suitable adhesivity properties [23,24,26,30] provide a principled and very general foundation for rewriting theories. In practice, many applications require the rewriting of objects that are not part of an adhesive category themselves, but which may be obtained from a suitable "ambient" category via the notion of *conditions* on objects. Together with a corresponding notion of constraints on rewriting rules, this yields a versatile extension of rewriting theory. In the DPO setting, this modification had been well-known [21–23,27], while it has been only very recently introduced for the SqPO setting [8]. For the *rule algebra* constructions presented in the main part of this contribution, we require in addition a certain *compositionality* property of our rewriting theories (established for the DPO case in [10,11], for the SqPO case in [4], and for both settings augmented with conditions in [8]).

2.1   Category-Theoretical Prerequisites

Throughout this paper, we will make the following assumptions[1]:

Assumption 1. $\mathbf{C} \equiv (\mathbf{C}, \mathcal{M})$ *is a finitary $\mathcal{M}$-adhesive category with $\mathcal{M}$-initial object, $\mathcal{M}$-effective unions and epi-$\mathcal{M}$-factorization. In the setting of* Sesqui-Pushout (SqPO) *rewriting, we assume in addition that all final pullback complements (FPCs) along composable pairs of $\mathcal{M}$-morphisms exist, and that $\mathcal{M}$-morphisms are stable under FPCs.*

Both of the main application examples presented within this paper rely upon typed variants of undirected multigraphs.

---

[1] We review in Appendix A.1 of [9] (an extended version of the present paper) some of the salient background material on $\mathcal{M}$-adhesive categories and the relevant notational conventions.

Definition 1. *Let $\mathcal{P}^{(1,2)} :$ **Set** $\to$ **Set** be the restricted powerset functor (mapping a set S to the set of its subsets $P \subset S$ with $1 \leq |P| \leq 2$). The category* **uGraph** *[11] of finite undirected multigraphs is defined as the finitary restriction of the comma category $(ID_{\textbf{Set}}, \mathcal{P}^{(1,2)})$. Thus an undirected multigraph is specified as $G = (E_G, V_G, i_G)$, where E_G and V_G are (finite) sets of edges and vertices, respectively, and where $i_G : E_G \to \mathcal{P}^{(1,2)}(V_G)$ is the edge-incidence map.*

Theorem 1. **uGraph** *satisfies Assumption 1, both for the DPO- and for the extended SqPO-variant.*

Proof. As demonstrated in [11], **uGraph** is indeed a finitary $\mathcal{M}$-adhesive category with $\mathcal{M}$-initial object and $\mathcal{M}$-effective unions, for $\mathcal{M}$ the class of component-wise monic **uGraph**-morphisms. It thus remains to prove the existence of an epi-$\mathcal{M}$-factorization as well as the properties related to FPCs. To this end, utilizing the fact that the category **Set** upon which the comma category **uGraph** is based possesses an epi-mono-factorization, we may construct the following diagram from a **uGraph**-morphism $\varphi = (\varphi_E, \mathcal{P}^{(1,2)}(\varphi_V))$ (for component morphisms $\varphi_E : E \to E'$ and $\varphi_V : V \to V'$):

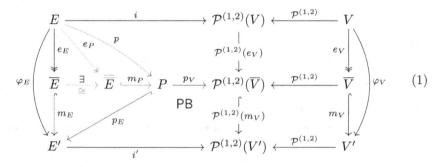

$$(1)$$

The diagram is constructed as follows:

1. Perform the epi-mono-factorizations $\varphi_E = m_E \circ e_E$ and $\varphi_V = m_V \circ e_V$, and apply the functor $\mathcal{P}^{(1,2)}$ in order to obtain the morphisms $\mathcal{P}^{(1,2)}(e_V)$ and $\mathcal{P}^{(1,2)}(m_V)$; since the functor $\mathcal{P}^{(1,2)}$ preserves monomorphisms [31], $\mathcal{P}^{(1,2)}(m_V) \in \mathsf{mono}(\textbf{Set})$.
2. Construct the pullback

$$(E' \leftarrow P \to \mathcal{P}^{(1,2)}(\overline{V})) := \mathsf{PB}(E' \to \mathcal{P}^{(1,2)}(V') \leftarrow \mathcal{P}^{(1,2)}(\overline{V})),$$

   Since monomorphisms are stable under pullback in **Set**, having proved that $\mathcal{P}^{(1,2)}(m_V) \in \mathsf{mono}(\textbf{Set})$ implies $(p_E : P \to E') \in \mathsf{mono}(\textbf{Set})$.
3. By the universal property of pullbacks, there exists a morphism $(p : E \to P)$. Let $p = m_P \circ e_P$ be the epi-mono-factorization of this morphism.
4. By stability of monomorphisms under composition in **Set**, we find that $p_E \circ m_P \in \mathsf{mono}(\textbf{Set})$, and consequently $\varphi_E = (p_E \circ m_P) \circ e_P$ yields an alternative epi-mono-factorization of φ_E. Then by uniqueness of epi-mono-factorizations up to isomorphism, there must exist an isomorphism $(\overline{E} \to \overline{\overline{E}}) \in \mathsf{iso}(\textbf{Set})$.

We have thus demonstrated that both $(e_E, \mathcal{P}^{(1,2)}(e_V))$ and $(m_E, \mathcal{P}^{(1,2)}(m_V))$ are morphisms in **uGraph**. Since morphisms in comma categories are mono-, epi- or iso-morphisms if they are so componentwise [21], we conclude that

$$(e_E, \mathcal{P}^{(1,2)}(e_V)) \in \mathsf{epi}(\mathbf{uGraph}), \quad (m_E, \mathcal{P}^{(1,2)}(m_V)) \in \mathsf{mono}(\mathbf{uGraph}),$$

which finally entails that we have explicitly constructed an epi-mono-factorization of the **uGraph**-morphism $(\varphi_E, \mathcal{P}^{(1,2)}(\varphi_V))$.

In order to demonstrate that FPCs along pairs of composable $\mathcal{M}$-morphisms $\varphi_A, \varphi_B \in \mathcal{M}$ in **uGraph** exist (for $\mathcal{M}$ the class of component-wise monomo-mophic **uGraph** morphisms), we provide the following explicit construction:

$$
\begin{array}{c}
\begin{array}{ccc}
A & \xrightarrow{\varphi_A} & B \\
{\scriptstyle\varphi_C}\big\uparrow & \text{FPC} & \big\downarrow{\scriptstyle\varphi_B} \\
C & \xrightarrow[\varphi_D]{} & D
\end{array}
\end{array}
\quad
\begin{array}{l}
V_C = V_D \setminus (V_B \setminus V_A) \\
E_C = \{e \in E_D \setminus (E_B \setminus E_A) \mid u_D(e) \in \mathcal{P}^{(1,2)}(V_C)\} \\
u_C = u_D|_{E_C} \\
\varphi_C = (E_A \hookrightarrow E_C, \mathcal{P}^{(1,2)}(V_A \hookrightarrow V_C)) \\
\varphi_D = (E_C \hookrightarrow E_D, \mathcal{P}^{(1,2)}(V_C \hookrightarrow V_D))
\end{array}
\tag{2}
$$

2.2 Conditions

Referring to [9, Appendix A.2] for further details and technical definitions, we will utilize as a **notational convention** the standard shorthand notations

$$\exists(X \hookrightarrow Y) := \exists(X \hookrightarrow Y, \mathsf{true}_Y), \quad \forall(X \hookrightarrow Y, \mathsf{c}_Y) := \neg\exists(X \hookrightarrow Y, \neg\mathsf{c}_Y). \tag{3}$$

For example, the constraints

$$\mathsf{c}_\varnothing^{(1)} = \exists(\varnothing \hookrightarrow \bullet \quad \bullet), \ \mathsf{c}_\varnothing^{(2)} = \nexists(\varnothing \hookrightarrow \bullet\!\!\leftrightarrows\!\!\bullet), \ \mathsf{c}_\varnothing^{(3)} = \forall(\varnothing \hookrightarrow \bullet\!\!\rightarrow\!\!\bullet, \exists(\bullet\!\!\rightarrow\!\!\bullet \hookrightarrow \bullet\!\!\leftrightarrows\!\!\bullet))$$

express for a given object $Z \in \mathsf{obj}(\mathbf{C})$ that Z contains at least two vertices (if $Z \vDash \mathsf{c}_\varnothing^{(1)}$), that Z does not contain parallel pairs of directed edges (if $Z \vDash \mathsf{c}_\varnothing^{(2)}$), and that for every directed edge in Z there also exists a directed edge between the same endpoints with opposite direction (if $Z \vDash \mathsf{c}_\varnothing^{(3)}$), respectively.

2.3 Compositional Rewriting with Conditions

Throughout this section, we assume that we are given a type $\mathbb{T} \in \{DPO, SqPO\}$ of rewriting semantics and an $\mathcal{M}$-adhesive category $\mathbf{C}$ satisfying the respective variant of Assumption 1. In categorical rewriting theories, the universal constructions utilized such as pushouts, pullbacks, pushout complements and final pullback complements are unique only up to universal isomorphisms. This motivates specifying a suitable notion of equivalence classes of rules with conditions:

Definition 2 (Rules with conditions). *Let* $\overline{\mathsf{Lin}}(\mathbf{C})$ *denote the class of* (linear) *rules with conditions, defined as*

$$\overline{\mathsf{Lin}}(\mathbf{C}) := \{(O \xleftarrow{o} K \xrightarrow{i} I; \mathsf{c}_I) \mid o, i \in \mathcal{M}, \ \mathsf{c}_I \in \mathsf{cond}(\mathbf{C})\}. \tag{4}$$

We define two rules with conditions $R_j = (r_j, c_{I_j})$ $(j = 1, 2)$ equivalent, denoted $R_2 \sim R_1$, iff $c_{I_1} \equiv c_{I_2}$ and if there exist isomorphisms $\omega, \kappa, \iota \in \mathsf{iso}(\mathbf{C})$ such that the diagram on the right commutes. We denote by $\overline{\mathsf{Lin}}(\mathbf{C})_\sim$ the set of equivalence classes under $\sim$ of rules with conditions.

$$
\begin{array}{ccccc}
O_1 & \longleftarrow & K_1 & \longrightarrow & I_1 \\
\cong \downarrow \omega & & \cong \downarrow \kappa & & \cong \downarrow \iota \\
O_2 & \longleftarrow & K_2 & \longrightarrow & I_2
\end{array}
\tag{5}
$$

Definition 3 (Direct derivations). *Let $r = (O \hookleftarrow K \hookrightarrow I) \in \mathsf{Lin}(\mathbf{C})$ and $c_I \in \mathsf{cond}(\mathbf{C})$ be concrete representatives of some equivalence class $R \in \overline{\mathsf{Lin}}(\mathbf{C})_\sim$, and let $X, Y \in \mathsf{obj}(\mathbf{C})$ be objects. Then a type $\mathbb{T}$ direct derivation is defined as a commutative diagram such as below right, where all morphism are in $\mathcal{M}$ (and with the left representation a shorthand notation)*

$$
\begin{array}{ccc}
O & \overset{r}{\longleftarrow\!\!\!-} & I \\
m^* \downarrow & \mathbb{T} & \downarrow m \\
Y & \longleftarrow\!\!\!- & X
\end{array}
\quad := \quad
\begin{array}{ccccc}
O & \longleftarrow & K & \longrightarrow & I \\
m^* \downarrow & (B) & \updownarrow k & (A) & \downarrow m \\
Y & \longleftarrow & \overline{K} & \longrightarrow & X
\end{array}
\quad .
\tag{6}
$$

with the following pieces of information required relative to the type:

1. *$\mathbb{T} = \mathbf{DPO}$: given $(m : I \hookrightarrow X) \in \mathcal{M}$, m is a DPO-admissible match of R into X, denoted $m \in \mathsf{M}_R^{DPO}(X)$, if $m \vDash c_I$ and (A) is constructable as a pushout complement, in which case (B) is constructed as a pushout.*
2. *$\mathbb{T} = \mathbf{SqPO}$: given $(m : I \hookrightarrow X) \in \mathcal{M}$, m is a SqPO-admissible match of R into X, denoted $m \in \mathsf{M}_R^{SqPO}(X)$, if $m \vDash c_I$, in which case (A) is constructed as a final pullback complement and (B) as a pushout.*
3. *$\mathbb{T} = \mathbf{DPO}^\dagger$: given just the "plain rule" r and $(m^* : O \hookrightarrow Y) \in \mathcal{M}$, m^* is a DPO†-admissible match of r into X, denoted $m \in \mathsf{M}_r^{DPO^\dagger}(Y)$, if (B) is constructable as a pushout complement, in which case (B) is constructed as a pushout.*

For types $\mathbb{T} \in \{DPO, SqPO\}$, we will sometimes employ the notation $R_m(X)$ for the object Y.

Note that at this point, we have not yet resolved a conceptual issue that arises from the non-uniqueness of a direct derivation given a rule and an admissible match. Concretely, the pushout complement, pushout and FPC constructions are only unique up to isomorphisms. This issue will ultimately be resolved as part of the rule algebraic theory. We next consider a certain *composition operation* on rules with conditions that is quintessential to our main constructions:

Definition 4 (Rule compositions). *Let $R_1, R_2 \in \overline{\mathsf{Lin}}(\mathbf{C})_\sim$ be two equivalence classes of rules with conditions, and let $r_j \in \mathsf{Lin}(\mathbf{C})$ and c_{I_j} be concrete representatives of R_j (for $j = 1, 2$). For $\mathbb{T} \in \{DPO, SqPO\}$, an $\mathcal{M}$-span $\mu = (I_2 \hookleftarrow M_{21} \hookrightarrow O_1)$ (i.e. with $(M_{21} \hookrightarrow O_1), (M_{21} \hookrightarrow I_2) \in \mathcal{M}$) is a $\mathbb{T}$-*

admissible match of R_2 into R_1 *if the diagram below is constructable (with N_{21} constructed by taking pushout)*

$$
\begin{array}{ccccccc}
O_2 & \xleftarrow{\;r_2\;} & I_2 & \longleftarrow & M_{21} & \longrightarrow & O_1 & \xleftarrow{\;r_1\;} & I_1 \\
\downarrow & \mathbb{T} & & \searrow & \text{PO} & \swarrow & & DPO^\dagger & \downarrow \\
O_{21} & & \longleftarrow & & N_{21} & \longleftarrow & & I_{21}
\end{array}
\tag{7}
$$

and if $c_{I_{21}} \not\equiv$ false. *Here, the condition* $c_{I_{21}}$ *is computed as*

$$
c_{I_{21}} := \mathsf{Shift}(I_1 \hookrightarrow I_{21}, c_{I_1}) \wedge \mathsf{Trans}(N_{21} \hookleftarrow I_{21}, \mathsf{Shift}(I_2 \hookrightarrow N_{21}, c_{I_2})).
\tag{8}
$$

In this case, we define the type $\mathbb{T}$ composition *of* R_2 *with* R_1 *along* μ, *denoted* $R_2{}^{\mu}\!\vartriangleleft_{\mathbb{T}} R_1$, *as*

$$
R_2{}^{\mu}\!\vartriangleleft_{\mathbb{T}} R_1 := [(O_{21} \hookleftarrow I_{21}; c_{I_{21}})]_\sim,
\tag{9}
$$

where $(O_{21} \hookleftarrow I_{21}) := (O_{21} \hookleftarrow N_{21}) \circ (N_{21} \hookleftarrow I_{21})$ *(with $\circ$ the* span composition *operation).*

We refer the interested readers to [9, Appendix A.3] for a summary of the relevant technical results on two variants of concurrency theorems following [8] (where the DPO-type concurrency theorem is of course classical, cf. e.g. [21]).

3   Rule Algebras for Compositional Rewriting with Conditions

The associativity property of rule compositions in both DPO- and SqPO-type semantics for rewriting with conditions as proved in [8] may be fruitfully exploited within rule algebra theory. One possibility to encode the non-determinism in sequential applications of rules to objects is given by lifting each possible configuration $X \in \mathrm{obj}(\mathbf{C})_{\cong}$ (i.e. isomorphism class of *objects*) to a basis vector $|X\rangle$ of a vector space $\hat{\mathbf{C}}$; then a rule r is lifted to a linear operator acting on $\hat{\mathbf{C}}$, with the idea that this operator acting upon a basis vector $|X\rangle$ should evaluate to the "sum over all possibilities to act with r on X". We will extend here the general rule algebra framework [4,6,10] to the present setting of rewriting rules with conditions.

We will first lift the notion of rule composition into the setting of a composition operation on a certain abstract vector space over rules, thus realizing the heuristic concept of "summing over all possibilities to compose rules".

Definition 5. *Let* $\mathbb{T} \in \{DPO, SqPO\}$ *be the rewriting type, and let* $\mathbf{C}$ *be a category satisfying the relevant variant of Assumption 1. Let* $\overline{\mathcal{R}}_{\mathbf{C}}$ *be an* $\mathbb{R}$-*vector space, defined via a bijection* $\delta : \overline{\mathsf{Lin}}(\mathbf{C})_\sim \xrightarrow{\;\cong\;} \mathrm{basis}(\overline{\mathcal{R}}_{\mathbf{C}})$ *from the set of equivalence classes of linear rules with conditions to the set of basis vectors of* $\overline{\mathcal{R}}_{\mathbf{C}}$. *Let* $\star_{\mathbb{T}}$ *denote the* type $\mathbb{T}$ rule algebra product, *a binary operation defined via its action on basis elements* $\delta(R_1), \delta(R_1) \in \overline{\mathcal{R}}_{\mathbf{C}}$ *(for* $R_1, R_2 \in \overline{\mathsf{Lin}}(\mathbf{C})_\sim$) *as*

$$
\delta(R_2) \star_{\mathbb{T}} \delta(R_1) := \sum_{\mu \in \mathcal{M}^{\mathbb{T}}_{R_2}(R_1)} \delta\left(R_2{}^{\mu}\!\vartriangleleft_{\mathbb{T}} R_1\right).
\tag{10}
$$

We refer to $\overline{\mathcal{R}}_{\mathbf{C}}^{\mathbb{T}} := (\mathcal{R}_{\mathbf{C}}, \star_{\mathbb{T}})$ *as the* $\mathbb{T}$-type rule algebra over $\mathbf{C}$.

Theorem 2. *For type* $\mathbb{T} \in \{DPO, SqPO\}$ *over a category* $\mathbf{C}$ *satisfying Assumption 1, the rule algebra* $\overline{\mathcal{R}}_{\mathbf{C}}^{\mathbb{T}}$ *is an associative unital algebra, with unit element* $\delta(R_\varnothing)$, *where* $R_\varnothing := (\varnothing \hookleftarrow \varnothing \hookrightarrow \varnothing; \mathsf{true})$.

Proof. Associativity follows from [9, Theorem 7], while *unitality*, i.e. that

$$\forall R \in \overline{\mathsf{Lin}}(\mathbf{C})_\sim : \quad \delta(R_\varnothing) \star_{\mathbb{T}} \delta(R) = \delta(R) \star_{\mathbb{T}} \delta(R_\varnothing) = \delta(R)$$

follows directly from an explicit computation of the relevant rule compositions. □

As alluded to in the introduction, the prototypical example of rule algebras are those of DPO- or (in this case equivalently) SqPO-type over discrete graphs, giving rise as a special case to the famous Heisenberg-Weyl algebra of key importance in mathematical chemistry, combinatorics and quantum physics (see [11] for further details). We will now illustrate the rule algebra concept in an example involving a more general base category.

Example 1. For the category **uGraph** and DPO-type rewriting semantics, consider as an example the following two rules with conditions:

$$R_C := \left(\overset{\bullet}{\underset{\bullet}{|}} \;\; \hookleftarrow \;\; \overset{\bullet}{\underset{\bullet}{\,}} \;\; \hookrightarrow \;\; \overset{\bullet}{\underset{\bullet}{\,}}; \neg \exists \left(\overset{\bullet}{\underset{\bullet}{\,}} \hookrightarrow \overset{\bullet}{\underset{\bullet}{|}} \right) \right), \quad R_V := (\bullet \hookleftarrow \varnothing \hookrightarrow \bullet; \mathsf{true}). \quad (11)$$

The first rule is a typical example of a rule with application conditions, i.e. here stating that the rule may only link two vertices if they were previously not already linked to each other. The second rule, owing to DPO semantics, can in effect only be applied to vertices without any incident edges. The utility of the rule-algebraic composition operation then consists in reasoning about sequential compositions of these rules, for example (letting $* := \star_{DPO}$):

$$\delta(R_C) * \delta(R_V) = \delta(R_C \uplus R_V) + 2\delta(R_C'), \quad R_C' := \left(\overset{\bullet}{\underset{\bullet}{|}} \;\; \hookleftarrow \;\; \overset{\bullet}{\,} \;\; \hookrightarrow \;\; \overset{\bullet}{\underset{\bullet}{\,}}; \mathsf{true} \right) \quad (12)$$

$$\delta(R_V) * \delta(R_C) = \delta(R_C \uplus R_V).$$

To provide some intuition: the first computation encodes the causal information that the two rules may either be composed along a trivial overlap, or rule R_C may overlap on one of the vertices in the output of R_V; in the latter case, any vertex to which first R_V and then R_C applies must not have had any incident edges, i.e. in particular no edge violating the constraint of R_C, which is why the composite rule R_C' does not feature any non-trivial constraint. In the other order of composition, the two vertices in the output of R_C are linked by an edge, so R_V cannot be applied to any of these two vertices (leaving just the trivial overlap contribution).

Just as the rule algebra construction encodes the compositional associativity property of rule compositions, the following *representation* construction encodes in a certain sense the properties described by the concurrency theorem:

Definition 6. *Let* $\mathbf{C}$ *be an* $\mathcal{M}$*-adhesive category satisfying Assumption 1. Let* $\hat{\mathbf{C}}$ *be defined as the* $\mathbb{R}$*-vector space whose set of basis vectors is isomorphic to the set[2] of iso-classes of objects of* $\mathbf{C}$ *via a bijection* $|.\rangle : \mathsf{obj}(\mathbf{C})_{\cong} \to \mathsf{basis}(\hat{\mathbf{C}})$*. Then the* $\mathbb{T}$*-type canonical representation of the* $\mathbb{T}$*-type rule algebra over* $\mathbf{C}$*, denoted* $\overline{\rho}_{\mathbf{C}}^{\mathbb{T}}$*, is defined as the morphism* $\overline{\rho}_{\mathbf{C}}^{\mathbb{T}} : \overline{\mathcal{R}}_{\mathbf{C}}^{\mathbb{T}} \to End_{\mathbb{R}}(\hat{\mathbf{C}})$ *specified via*

$$\forall R \in \overline{\mathsf{Lin}}(\mathbf{C})_{\sim}, X \in \mathsf{obj}(\mathbf{C})_{\cong} : \quad \overline{\rho}_{\mathbf{C}}^{\mathbb{T}}(\delta(R)) |X\rangle := \sum_{m \in \mathsf{M}_R^{\mathbb{T}}(X)} |R_m(X)\rangle . \quad (13)$$

Theorem 3. $\overline{\rho}_{\mathbf{C}}^{\mathbb{T}}$ *as defined above is an algebra homomorphism (and thus in particular a well defined representation).*

Proof. The proof that we provide in [9, Appendix B.1] is entirely analogous to the one for the case without application conditions [4,10]. □

4   Stochastic Mechanics Formalism

Referring to [6,7,11] for further details and derivations, suffice it here to highlight the key role played by the algebraic concept of *commutators* in stochastic mechanics. Let us first provide the constructions of continuous-time Markov chains (CTMCs) and observables in stochastic rewriting systems.

Definition 7. *Let* $\langle| : \hat{\mathbf{C}} \to \mathbb{R}$ *(referred to as* dual projection vector*) be defined via its action on basis vectors of* $\hat{\mathbf{C}}$ *as* $\langle|X\rangle := 1_{\mathbb{R}}$.

Theorem 4. *Let* $\mathbf{C}$ *be a category satisfying the relevant variant of Assumption 1, and let* $\overline{\mathcal{R}}_{\mathbf{C}}^{\mathbb{T}}$ *be the* $\mathbb{T}$*-type rule algebra of linear rules with conditions over* $\mathbf{C}$*. Let* $\rho \equiv \rho_{\mathbf{C}}^{\mathbb{T}}$ *denote the* $\mathbb{T}$*-type canonical representation of* $\mathcal{R}_{\mathbf{C}}^{\mathbb{T}}$*. Then the following results hold:*

1. *The basis elements of the space* $\mathsf{obs}(\mathbf{C})_{\mathbb{T}}$ *of* $\mathbb{T}$*-type observables, i.e. the diagonal linear operators that arise as (linear combinations of)* $\mathbb{T}$*-type canonical representations of rewriting rules with conditions, have the following structure (*$\hat{\mathcal{O}}_{P,q}^{c_P}$ *in the DPO case,* $\hat{\mathcal{O}}_{P}^{c_P}$ *in the SqPO case):*

$$\hat{\mathcal{O}}_{P,q}^{c_P} := \rho(\delta(P \xleftarrow{q} Q \xrightarrow{q} P; c_P)) \quad (P \in \mathsf{obj}(\mathbf{C})_{\cong}, q \in \mathcal{M}, c_P \in \mathsf{cond}(\mathbf{C})_{\sim})$$
$$\hat{\mathcal{O}}_{P}^{c_P} := \rho(\delta(P \xleftarrow{\cong} P \xrightarrow{\cong} P; c_P)) \quad (P \in \mathsf{obj}(\mathbf{C})_{\cong}, c_P \in \mathsf{cond}(\mathbf{C})_{\sim}).$$
$$(14)$$

---

[2] We assume here that the isomorphism classes of objects of $\mathbf{C}$ form a *set* (i.e. not a proper class).

2. **DPO-type jump closure property:** *for every linear rule with condition* $R \equiv (O \hookleftarrow K \hookrightarrow I, c_I) \in \overline{\mathsf{Lin}}(\mathbf{C})$, *we find that*

$$\langle\,|\,\rho(\delta(R)) = \langle\,|\,\hat{\mathbb{O}}(\delta(R)), \tag{15}$$

where $\hat{\mathbb{O}} : \overline{\mathcal{R}}_{\mathbf{C}}^{DPO} \rightarrow End_{\mathbb{R}}(\hat{\mathbf{C}})$ *is the homomorphism defined via its action on basis elements* $\delta(R)$ *for* $R = (O \hookleftarrow K \hookrightarrow I; c_I) \in \overline{\mathsf{Lin}}(\mathbf{C})_{\sim}$ *as*

$$\hat{\mathbb{O}}(\delta(R)) := \rho(\delta(I \hookleftarrow K \hookrightarrow I; c_I)) \in \mathsf{obs}(\mathbf{C}). \tag{16}$$

3. **SqPO-type jump closure property:** *for every linear rule with condition* $R \equiv (O \hookleftarrow K \hookrightarrow I, c_I) \in \overline{\mathsf{Lin}}(\mathbf{C})$, *we find that*

$$\langle\,|\,\rho(\delta(R)) = \langle\,|\,\hat{\mathbb{O}}(\delta(R)), \tag{17}$$

where[3] $\hat{\mathbb{O}} : \overline{\mathcal{R}}_{\mathbf{C}}^{SqPO} \rightarrow End_{\mathbb{R}}(\hat{\mathbf{C}})$ *is the homomorphism defined via*

$$\hat{\mathbb{O}}(\delta(R)) := \rho(\delta(I \xleftarrow{\cong} I \xrightarrow{\cong} I; c_I)) \in \mathsf{obs}(\mathbf{C}). \tag{18}$$

4. **CTMCs via stochastic rewriting systems:** *Let* $\mathsf{Prob}(\mathbf{C})$ *be the space of (sub-)probability distributions over* $\hat{\mathbf{C}}$ *(i.e.* $|\Psi\rangle = \sum_{X \in \mathsf{obj}(\mathbf{C})_{\cong}} \psi_X \,|X\rangle$*). Let* $\mathcal{T}$ *be a collection of* N *pairs of positive real-valued parameters* κ_j *(referred to as* base rates*) and linear rules* R_j *with application conditions,*

$$\mathcal{T} := \{(\kappa_j, R_j)\}_{1 \le j \le N} \qquad (\kappa_j \in \mathbb{R}_{\ge 0},\ R_j \equiv (r_j, c_{I_j}) \in \overline{\mathsf{Lin}}(\mathbf{C})). \tag{19}$$

Then given an initial state $|\Psi_0\rangle \in \mathsf{Prob}(\mathbf{C})$, *the* $\mathbb{T}$*-type stochastic rewriting system based upon the transitions* $\mathcal{T}$ *gives rise to the CTMC* $(\mathcal{H}, |\Psi(0)\rangle)$ *with time-dependent state* $|\Psi(t)\rangle \in \mathsf{Prob}(\mathbf{C})$ *(for* $t \ge 0$*) and evolution equation*

$$\forall t \ge 0: \quad \tfrac{d}{dt}|\Psi(t)\rangle = \mathcal{H}\,|\Psi(t)\rangle, \quad |\Psi(0)\rangle = |\Psi_0\rangle. \tag{20}$$

Here, the infinitesimal generator $\mathcal{H}$ *of the CTMC is given by*

$$\mathcal{H} = \hat{H} - \hat{\mathbb{O}}(\hat{H}), \quad \hat{H} = \sum_{j=1}^{N} \kappa_j\, \rho(\delta(R_j)). \tag{21}$$

Proof. See [9, Appendix B.2]. $\qquad\qquad\blacksquare$

Remark 1. The operation $\hat{\mathbb{O}}$ featuring in the DPO- and SqPO-type jump-closure properties has a very intuitive interpretation: given a linear rule with condition $R \equiv (r, c_I) \in \overline{\mathsf{Lin}}(\mathbf{C})$, the linear operator $\hat{\mathbb{O}}(\delta(R))$ is an observable that evaluates on a basis vector $|X\rangle \in \hat{\mathbf{C}}$ as $\hat{\mathbb{O}}(\delta(R))\,|X\rangle = (\#\ \text{of ways to apply } R \text{ to } X)\cdot |X\rangle$.

---

[3] Since in applications we will always fix the type of rewriting to either DPO or SqPO, we will use the same symbol for the jump-closure operator in both cases.

As for the concrete computational techniques offered by the stochastic mechanics formalism, one of the key advantages of this rule-algebraic framework is the possibility to reason about *expectation values* (and higher moments) of pattern-count observables in a principled and universal manner. The precise formulation is given by the following generalization of results from [7] to the setting of DPO- and SqPO-type rewriting for rules with conditions:

Theorem 5. *Given a CTMC* $(|\Psi_0\rangle, \mathcal{H})$ *with time-dependent state* $|\Psi(t)\rangle$ *(for* $t \geq 0$*), a set of observables* $O_1, \ldots O_n \in \mathsf{obs}(\mathbf{C})$ *and* n *formal variables* $\lambda_1, \ldots, \lambda_n$, *define the* exponential moment-generating function *(EMGF)* $M(t; \underline{\lambda})$ *as*

$$M(t; \underline{\lambda}) := \langle| e^{\underline{\lambda} \cdot \underline{O}} |\Psi(t)\rangle, \quad \underline{\lambda} \cdot \underline{O} := \sum_{j=1}^{n} \lambda_j O_j. \tag{22}$$

Then $M(t; \underline{\lambda})$ *satisfies the following formal evolution equation (for* $t \geq 0$*):*

$$\tfrac{d}{dt} M(t; \underline{\lambda}) = \sum_{q \geq 1} \tfrac{1}{q!} \langle| \left(ad^{\circ q}_{\underline{\lambda} \cdot \underline{O}}(\hat{H}) \right) e^{\underline{\lambda} \cdot \underline{O}} |\Psi(t)\rangle, \quad M(0; \underline{\lambda}) = \langle| e^{\underline{\lambda} \cdot \underline{O}} |\Psi_0\rangle. \tag{23}$$

Proof. In full analogy to the case of rules without conditions [7], the proof follows from the BCH formula $e^{\lambda A} B e^{-\lambda A} = e^{ad_{\lambda A}}(B)$ (for $A, B \in End_{\mathbb{R}}(\hat{\mathbf{C}})$). Here, $ad^{\circ 0}_A(B) := B$, $ad_A(B) := AB - BA$ (also referred to as the *commutator* $[A, B]$ of A and B), and $ad^{\circ(q+1)}_A(B) := ad_A(ad^{\circ q}_A(B))$ for $q \geq 1$. Finally, the $q = 0$ term in the above expression evaluates identically to 0 due to $\langle| \mathcal{H} = 0$.

Combining this theorem with the notion of $\mathbb{T}$-type jump-closure, one can in favorable cases express the EMGF evolution equation as a PDE on formal power series in $\lambda_1, \ldots, \lambda_n$ and with t-dependent real-valued coefficients. Referring the interested readers to [7] for further details on this technique, let us provide here a simple non-trivial example of such a calculation.

Example 2. Let us consider a stochastic rewriting system over the category $\mathbf{C} = \mathbf{uGraph}$ of finite undirected multigraphs, with objects further constrained by the structure constraint $c^S_{\varnothing} := \neg\exists(\varnothing \hookrightarrow \bullet\!\!\frown\!\!\bullet) \in \mathsf{cond}(\mathbf{uGraph})$ that prohibits

multiedges. Let us consider for type $\mathbb{T} = SqPO$ the four rules with conditions $R_{E_\pm}$ (edge-creation/-deletion) and $R_{V_\pm}$ (vertex creation/deletion), defined as

$$R_{E_+} := \tfrac{1}{2}\delta\left(\begin{smallmatrix}\bullet\\\bullet\end{smallmatrix} \hookleftarrow \begin{smallmatrix}\bullet\\\bullet\end{smallmatrix} \hookrightarrow \begin{smallmatrix}\bullet\\\bullet\end{smallmatrix}; \neg\exists\left(\begin{smallmatrix}\bullet\\\bullet\end{smallmatrix} \hookrightarrow \begin{smallmatrix}\bullet\\\bullet\end{smallmatrix}\right)\right), \quad R_{V_+} := \delta(\bullet \hookleftarrow \varnothing \hookrightarrow \varnothing; \mathsf{true})$$

$$R_{E_-} := \tfrac{1}{2}\delta\left(\begin{smallmatrix}\bullet\\\bullet\end{smallmatrix} \hookleftarrow \begin{smallmatrix}\bullet\\\bullet\end{smallmatrix} \hookrightarrow \begin{smallmatrix}\bullet\\\bullet\end{smallmatrix}; \mathsf{true}\right), \qquad\qquad R_{V_-} := \delta(\varnothing \hookleftarrow \varnothing \hookrightarrow \bullet; \mathsf{true}).$$

Here, the prefactors $\tfrac{1}{2}$ for $R_{E_\pm}$ are chosen purely for convenience. Note that R_{E_+} is the only rule requiring a non-trivial application condition, since linking two vertices with an edge might create a multiedge (precisely when the two vertices were already linked). Introducing base rates $\nu_\pm, \varepsilon_\pm \in \mathbb{R}_{>0}$ and letting $\hat{X} := \rho(R_X)$, we may assemble the infinitesimal generator $\mathcal{H}$ of a CTMC as

$$\mathcal{H} = \hat{H} + \hat{\mathbb{O}}(\hat{H}), \ \hat{H} := \nu_+ \hat{V}_+ + \nu_- \hat{V}_- + \varepsilon_+ \hat{E}_+ + \varepsilon_- \hat{E}_-. \tag{24}$$

One might now ask whether there is any interesting dynamical structure e.g. in the evolution of the moments of the observables that count the number of times each of the transitions of this system is applicable,

$$O_{\bullet|\bullet} := \hat{\mathbb{O}}(\delta(R_C)), \ O_{\bullet\!-\!\bullet} := \hat{\mathbb{O}}(\delta(R_D)), \ O_\bullet := \hat{\mathbb{O}}(\delta(R_{VD})). \tag{25}$$

The algebraic data necessary in order to formulate EMGF evolution equations are all **commutators** of the observables with the contributions $\hat{X} := \rho(\delta(R_X))$ to the "off-diagonal part" $\hat{H}$ of the infinitesimal generator $\mathcal{H}$. We will present here for brevity just those commutators necessary in order to compute the evolution equations for the averages of the three observables:

$$\begin{aligned}
[O_\bullet, \hat{V}_\pm] &= \pm \hat{V}_\pm, & [O_\bullet, \hat{E}_\pm] &= 0 \\
[O_{\bullet|\bullet}, \hat{V}_+] &= \hat{A}, & [O_{\bullet|\bullet}, \hat{V}_-] &= -\hat{B}, & [O_{\bullet|\bullet}, \hat{E}_\pm] &= \mp \hat{E}_\pm \\
[O_{\bullet\!-\!\bullet}, \hat{V}_+] &= 0, & [O_{\bullet\!-\!\bullet}, \hat{V}_-] &= -\hat{C}, & [O_{\bullet\!-\!\bullet}, \hat{E}_\pm] &= \pm \hat{E}_\pm
\end{aligned} \tag{26}$$

As typical in these types of commutator computations, we find a number of contributions (here $\hat{A}$, $\hat{B}$ and $\hat{C}$) that were not either observables or based upon rules of the SRS:

$$\hat{A} := \rho\left(\delta\left({\overset{\bullet}{\underset{\bullet}{}}} \hookleftarrow {\overset{\bullet}{}} \hookrightarrow {\overset{\bullet}{}}; \mathsf{true}\right)\right), \ \hat{B} := \rho\left(\delta\left({\overset{\bullet}{}} \hookleftarrow {\overset{\bullet}{}} \hookrightarrow {\overset{\bullet}{\underset{\bullet}{}}}; \neg\exists\left({\overset{\bullet}{\underset{\bullet}{}}} \hookrightarrow {\overset{\bullet}{\underset{\bullet}{\mathsf{I}}}}\right)\right)\right)$$

$$\hat{C} := \rho\left(\delta\left({\overset{\bullet}{\underset{\bullet}{\mathsf{I}}}} \hookleftarrow {\overset{\bullet}{\underset{\bullet}{}}} \hookrightarrow {\overset{\bullet}{\underset{\bullet}{}}}; \mathsf{true}\right)\right), \ \hat{\mathbb{O}}(\hat{A}) = O_\bullet, \ \hat{\mathbb{O}}(\hat{B}) = 2O_{\bullet|\bullet}, \ \hat{\mathbb{O}}(\hat{C}) = 2O_{\bullet\!-\!\bullet}$$

Picking for simplicity as an initial state $|\Psi(0)\rangle = |\varnothing\rangle$ just the empty graph, and invoking the SqPO-type jump-closure property (cf. Theorem 4) repeatedly in order to evaluate $\langle [O_P, \hat{H}] \rangle(t) = \langle \hat{\mathbb{O}}([O_P, \hat{H}]) \rangle(t)$, the moment EGF evolution equation (23) specializes to the following "Ehrenfest-like" [7] ODE system:

$$\begin{aligned}
\tfrac{d}{dt}\langle O_\bullet \rangle(t) &= \langle [O_\bullet, H] \rangle(t) = \nu_+ - \nu_- \langle O_\bullet \rangle(t) \\
\tfrac{d}{dt}\langle O_{\bullet|\bullet} \rangle(t) &= \langle [O_{\bullet|\bullet}, H] \rangle(t) = \nu_+ \langle O_\bullet \rangle(t) - (2\nu_- + \varepsilon_+)\langle O_{\bullet|\bullet} \rangle(t) + \varepsilon_- \langle O_{\bullet\!-\!\bullet} \rangle(t) \\
\tfrac{d}{dt}\langle O_{\bullet\!-\!\bullet} \rangle(t) &= \langle [\langle O_{\bullet\!-\!\bullet} \rangle(t), H] \rangle(t) = \varepsilon_+ \langle O_{\bullet|\bullet} \rangle(t) - (2\nu_- + \varepsilon_-)\langle O_{\bullet\!-\!\bullet} \rangle(t) \\
\langle O_\bullet \rangle(0) &= \langle O_{\bullet|\bullet} \rangle(t) = \langle O_{\bullet\!-\!\bullet} \rangle(t) = 0.
\end{aligned}$$

This ODE system may be solved exactly (see [9, Appendix C]). We depict in Fig. 1 two exemplary evolutions of the three average pattern counts for different choices of parameters. Since due to SqPO-semantics the vertex deletion and creation transitions are entirely independent of the edge creation and deletion transitions, the vertex counts stabilize on a Poisson distribution of parameter ν_+/ν_- (where we only present the average vertex count value here). As for

the non-linked vertex pair and edge patter counts, the precise average values
are sensitive to the parameter choices (i.e. whether or not vertices tend to be
linked by an edge or not may be freely tuned in this model via adjusting the
parameters).

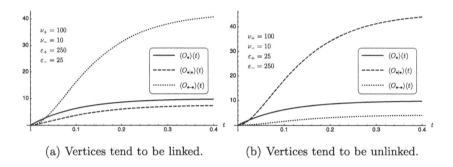

(a) Vertices tend to be linked.   (b) Vertices tend to be unlinked.

Fig. 1. Time-evolutions of pattern count observables for different parameter choices.

While the example presented was chosen mainly to illustrate the computational techniques, it highlights the typical feature of the emergence of contributions in the relevant (nested) commutator calculations that may not have
been included in the non-diagonal part $\hat{H}$ of the infinitesimal generator of the
CTMC. We refer the interested readers to [7] for an extended discussion of this
phenomenon, and for computation strategies for higher-order moment evolution
equations.

5   Application Scenario 1: Biochemistry with Kappa

The KAPPA platform [18,19] for rule-based modeling of biochemical reaction
systems is based upon the notion of so-called *site-graphs* that abstract proteins
and other complex macro-molecules into *agents* (with *sites* representing interaction capacities of the molecules). This open source platform offers a variety of
high-performance *simulation algorithms* (for CTMCs based upon KAPPA rewriting rules) as well as several variants of static analysis tools to analyze and verify
biochemical models [12]. In view of the present paper, it is interesting to note that
since the start of the KAPPA development, the simulation-based algorithms have
been augmented by *differential semantics* modules aimed at deriving ODE systems for the evolution of pattern-count observable average values [14,15,17,29].
In this section, we will experiment with a (re-)encoding of KAPPA in terms
of typed undirected graphs with certain structural constraints that permits to
express such moment statistics ODEs via our general rule-algebraic stochastic mechanics formalism. We will then provide an illustrative exemplary computation of ODEs in order to point out certain intrinsic intricacies (notably
non-closure properties) typical of such calculations. One of the key theoretical

features of KAPPA is its foundation upon the notion of *rigidity* [16]. In practice, the construction involves an *ambient category* $\mathbb{A}$ (which possesses suitable adhesivity properties), a *pattern category* $\mathbb{P}$ (obtained from $\mathbb{A}$ via certain *negative constraints*) and finally a *state category* $\mathbb{S}$ (obtained from $\mathbb{P}$ via additional *positive constraints*). We will now present one possible realization of KAPPA based upon the $\mathcal{M}$-adhesive category of typed undirected multigraphs:

Definition 8. *For a* KAPPA *model* K, *let* $\mathbb{A} = \mathbf{uGraph}/T_K$ *be the category of finite undirected multigraphs typed over* T_K, *where* T_K *distinguishes agent vertex types, site vertex types and three forms of edge types: agent-site, site-site and loops on sites. For each agent type vertex* $X \in \{A, B, \ldots\}$, *the type graph contains the site type vertices* $x_1 : X, \ldots, x_{n_X} : X$ *(incident to the* X-*type vertex via an edge, and where* $n_X < \infty$). T_K *also contains link type edges between sites that encode which sites can be linked, and loops on site type vertices that represent dynamic attributes, such as the phosphorylation state of a site. Indicating the three different edge types by wavy lines (agent-site), solid lines (site-site) and dotted lines (property loops), the agent vertices with filled circles* *and the site vertices by open circles* $\circledcirc$, *and using the placeholder* $\bullet$ *for a vertex and a dashed line for an edge of any type, we may introduce the* negative constraints *defining the* pattern category $\mathbb{P}_K$ *as* $\mathsf{c}_{\mathcal{N}_K} := \wedge_{N \in \mathcal{N}_K} \neg\exists(\varnothing \hookrightarrow N)$, *with the set* $\mathcal{N}_K$ *of "forbidden subgraphs" defined as*

$$\mathcal{N}_K := \{\bullet\!\!\!{:}\!\!{:}\!\!\bullet\} \cup \bigcup_x \left\{ \circledcirc \right\} \cup \bigcup_{X,x} \left\{ \circledcirc\!\!\wedge\!\!\!\text{X}\!\!\!\wedge\!\!\circledcirc, \text{X}\!\!\wedge\!\!\circledcirc\!\!\wedge\!\!\text{X} \right\}. \quad (27)$$

Finally, the state category $\mathbb{S}_K$ *is obtained from* $\mathbb{P}_K$ *via imposing a positive constraint* $\mathsf{c}_{\mathcal{P}_K}$ *that ensures that each agent* X *is linked to exactly one of each of its site vertices* $x : X$, *and if a site* $x : X$ *can carry a property or alternative variants thereof, it also carries a loop that signifies one of these properties (see the example below for further details). Moreover, a given site* $x : X$ *must be linked to an agent* X *(i.e. cannot occur in isolation).*

Example 3. Consider a simple KAPPA model with a type graph as below left that introduces two agent types K (for "kinase") and P (for "protein"), where K has a site $k : K$, and where P has sites $p_t, p_l, p_b : P$. Moreover, the sites p_t and p_b can carry properties u ("unphosphorylated") and p ("phosphorylated"), depicted as dotted loops in the type graph. Sites $k : K$ and $p_l : P$ can bind (as indicated by the solid line in the type graph).

As a prototypical example of a KAPPA stochastic rewriting system, consider a system based upon the rewriting rules $k_\pm$, $l_\pm$, $t_\pm$ and $b_\pm$. Here, for the rule l_+, we have indicated that it must be equipped with an application condition that ensures that the site of the K-type agent and the left site of the P-type agent must be *free* before binding. As common practice also in the standard KAPPA theory, we otherwise leave in the graphical depictions those application conditions necessary to ensure consistent matches implicit as much as possible. Consider then for a concrete computational example the time-evolution of the average count of the pattern described in the identity rule r_{obs_P}. As typical in KAPPA rule specifications r_{obs_P} as well as several of the other rules depicted only explicitly involve *patterns*, but not necessarily *states*, since e.g. in r_{obs_P} the left site of the P-type agent is not mentioned. In complete analogy to the computation presented in Example 2, let us first compute the commutators of the observable $O_K = \rho(\delta(r_{obs_K}; c_{obs_K}))$ with the operators $\hat{X} := \rho(\delta(r_X; c_X))$:

$$[O_K, \hat{K}_\pm] = \pm\hat{K}_\pm, \ [O_K, \hat{L}_\pm] = [O_K, \hat{T}_\pm] = [O_K, \hat{B}_\pm] = 0 \qquad (28)$$

However, letting $O_P^{(x,y)}$, $O_{link}^{(x,y)}$ and $O_{free}^{(x,y)}$ denote the observables for the patterns

one may easily demonstrate that even a comparatively simple observable such as $O_P^{(p,p)}$ already leads to an infinite cascade of contributions to the ODEs for the averages of pattern counts. As typical in these sorts of computations, the discovery of a new pattern observable via applying SqPO-type jump-closure (Theorem 4) to the commutator contributions to $\frac{d}{dt}\langle O_P^{(p,p)}\rangle(t)$ leads to the discovery of new pattern observables yet again, such as in

$$[O_P, \hat{T}_+] = \hat{T}_+^{(p)}, \ \hat{\mathbb{O}}(\hat{T}_+^{(p)}) = O_{link}^{(u,p)}, \ [O_{link}^{(u,p)}, \hat{L}_+] = \hat{L}^{(u,p)}, \ \hat{\mathbb{O}}(\hat{L}^{(u,p)}) = O_{free}^{(u,p)}.$$

In particular the last observable $O_{free}^{(u,p)}$ is found to lead to an infinite tower of other observables (i.e. "ODE non-closure"), starting from

This exemplary and preliminary analysis reveals that while the rule-algebraic CTMC implementation is in principle applicable to the formulation and analysis KAPPA systems, further algorithmic and theoretical developments will be necessary (including possibly ideas of *fragments* and *refinements* as in [14,15,29]) in order to obtain a computationally useful alternative rewriting-theoretic implementation of KAPPA.

6 Application Scenario 2: Organic Chemistry with MØD

The MØD platform [1] for organo-chemical reaction systems is a prominent example of a DPO-type rewriting theory of high relevance to the life sciences. From a theoretical perspective, MØD has been designed [3] as a rewriting system over so-called *chemical graphs*, a certain typed and undirected variant of the category **PLG** of partially labelled directed graphs. While the latter category had been introduced in [28] as a key example of an $\mathcal{M}$-$\mathcal{N}$-adhesive category, with the motivation of permitting label-changes in rewriting rules, it was also demonstrated in loc. cit. that **PLG** is *not* $\mathcal{M}$-adhesive. Since moreover no concrete construction of a tentative variant **uPLG** of **PLG** for undirected graphs, let alone results on the possible adhesivity properties of such a category are known in the literature, we propose here an alternative and equivalent encoding of chemical graphs. We mirror the constructions of [1,3] in that chemical graphs will be a certain typed variant of undirected graphs, with *vertex types* representing *atom types, edge types* ranging over the types $\{-,=,\#,:\}$ representing *single, double, triple and aromatic bonds*, respectively, and with the graphs being required to not contain multiedges. Inspired by the KAPPA constructions in the previous section, we opt to represent *properties* (such as e.g. *charges* on atoms) as *typed loop edges* on vertices representing atoms, whence the change of a property (which was the main motivation in [3] for utilizing a variant of **PLG**) may be encoded in a rewriting rule simply via deletion/creation of property-encoding loops. Unfortunately, while the heuristics presented thus far would suggest that chemical graphs in the alternative categorical setting should be just simple typed undirected graphs, the full specification of chemical graphs would also have to include additional, empirical information from the chemistry literature. Concretely, atoms such as e.g. carbon only support a limited variety of bond types and configurations of incident bonds (referred to as *valencies*), with additional complications such as poly-valencies possible for some types of atoms as illustrated by the following example.

Example 4. The *Meisenheimer-2-3-rearrangement* reaction [32] (cf. also [2]) constitutes an example[4] of a reaction where *polyvalence* is encountered:

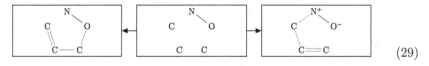

$$(29)$$

Upon matching this rule into a chemically valid mixture, the N atom on the input of the rule will have valence 5, while on the output it will have valence 3. This type of information is evidently in no way contained in the chemical graphs alone, and must therefore be encoded in terms of suitable additional typing on the graphs and application conditions.

---

[4] This example reaction was typeset directly via MØD (cf. [9, Appendix D]).

While thus at present no encoding of chemical graphs into a categorical framework with suitable adhesivity properties is available, we posit that it would be highly fruitful in light of the stochastic mechanics framework presented in this paper to develop such an encoding (joint work in progress with J.L. Andersen, W. Fontana and D. Merkle).

7   Conclusion and Outlook

Rewriting theories of DPO- and SqPO-type for rules with conditions over $\mathcal{M}$-adhesive categories are poised to provide a rich theoretical and algorithmic framework for modeling stochastic dynamical systems in the life sciences. The main result of the present paper consists in the introduction of a *rule algebra framework* that extends the pre-existing constructions [4,6,10] precisely via incorporating the notion of conditions. The sophisticated KAPPA [12] and MØD [1] bio-/organo-chemistry platforms and related developments have posed one of the main motivations for this work. For both of these platforms, we present a first analysis and stepping stones towards bridging category-theoretical rewriting theories and stochastic mechanics computations. Especially for the organo-chemistry setting, our work motivates the development of a full encoding of (at least a reasonable fragment of) organic chemistry in terms of *chemical graphs* and rewriting rules thereof, which to date is still unavailable. This encoding will be beneficial also in the development of tracelet-based techniques [5], and is current work in progress.

An intriguing perspective for future developments in categorical rewriting theory consists in developing a robust and versatile methodology for the analysis of ODE systems of pattern-counting observables in stochastic rewriting systems. While the results of this paper permit to formulate dynamical evolution equations for arbitrary higher moments of such observables, in general cases (as illustrated in Sect. 5) the non-closure of the resulting ODE systems remains a fundamental technical challenge. In the KAPPA literature, sophisticated conceptual and algorithmic approaches to tackle this problem have been developed such as refinements [14,16], model reduction techniques [15] and stochastic fragments [25] (see also [7] for an extended discussion). We envision that a detailed understanding of these approaches from within the setting of categorical rewriting and of rule algebra theory could provide a very fruitful enrichment of the methodology of rewriting theory.

References

1. Andersen, J.L., Flamm, C., Merkle, D., Stadler, P.F.: A software package for chemically inspired graph transformation. In: Echahed, R., Minas, M. (eds.) ICGT 2016. LNCS, vol. 9761, pp. 73–88. Springer, Cham (2016). https://doi.org/10.1007/978-3-319-40530-8_5

2. Andersen, J.L., Flamm, C., Merkle, D., Stadler, P.F.: An intermediate level of abstraction for computational systems chemistry. Philos. Trans. R. Soc. A: Math. Phys. Eng. Sci. **375**(2109), 20160354 (2017). https://doi.org/10.1098/rsta.2016.0354

3. Andersen, J.L., Flamm, C., Merkle, D., Stadler, P.F.: Rule composition in graph transformation models of chemical reactions. Match **80**(3), 661–704 (2018)

4. Behr, N.: Sesqui-pushout rewriting: concurrency, associativity and rule algebra framework. In: Echahed, R., Plump, D. (eds.) Proceedings of the Tenth International Workshop on Graph Computation Models (GCM 2019) in Eindhoven, The Netherlands. Electronic Proceedings in Theoretical Computer Science, vol. 309, pp. 23–52. Open Publishing Association (2019). https://doi.org/10.4204/eptcs.309.2

5. Behr, N.: Tracelets and tracelet analysis of compositional rewriting systems (accepted for ACT2019 in Oxford). arXiv:1904.12829 (2019)

6. Behr, N., Danos, V., Garnier, I.: Stochastic mechanics of graph rewriting. In: Proceedings of the 31st Annual ACM/IEEE Symposium on Logic in Computer Science - LICS 2016. ACM Press (2016). https://doi.org/10.1145/2933575.2934537

7. Behr, N., Danos, V., Garnier, I.: Combinatorial conversion and moment bisimulation for stochastic rewriting systems. arXiv:1904.07313 (2019)

8. Behr, N., Krivine, J.: Compositionality of rewriting rules with conditions. arXiv:1904.09322 (2019)

9. Behr, N., Krivine, J.: Rewriting theory for the life sciences: a unifying theory of CTMC semantics (extended version). arXiv:2003.09395 (2020)

10. Behr, N., Sobocinski, P.: Rule algebras for adhesive categories. In: Ghica, D., Jung, A. (eds.) 27th EACSL Annual Conference on Computer Science Logic (CSL 2018). Leibniz International Proceedings in Informatics (LIPIcs), vol. 119, pp. 11:1–11:21. Schloss Dagstuhl-Leibniz-Zentrum fuer Informatik, Dagstuhl, Germany (2018). https://doi.org/10.4230/LIPIcs.CSL.2018.11

11. Behr, N., Sobocinski, P.: Rule algebras for adhesive categories (invited extended journal version). arXiv:1807.00785 (2019)

12. Boutillier, P., et al.: The kappa platform for rule-based modeling. Bioinformatics **34**(13), i583–i592 (2018). https://doi.org/10.1093/bioinformatics/bty272

13. Corradini, A., Heindel, T., Hermann, F., König, B.: Sesqui-pushout rewriting. In: Corradini, A., Ehrig, H., Montanari, U., Ribeiro, L., Rozenberg, G. (eds.) Graph Transformations. Lecture Notes in Computer Science, vol. 4178, pp. 30–45. Springer, Berlin and Heidelberg (2006)

14. Danos, V., Feret, J., Fontana, W., Harmer, R., Krivine, J.: Rule-based modelling, symmetries, refinements. In: Fisher, J. (ed.) FMSB 2008. LNCS, vol. 5054, pp. 103–122. Springer, Heidelberg (2008). https://doi.org/10.1007/978-3-540-68413-8_8

15. Danos, V., Feret, J., Fontana, W., Harmer, R., Krivine, J.: Abstracting the differential semantics of rule-based models: exact and automated model reduction. In: 2010 25th Annual IEEE Symposium on Logic in Computer Science. IEEE (2010). https://doi.org/10.1109/lics.2010.44

16. Danos, V., Heckel, R., Sobocinski, P.: Transformation and refinement of rigid structures. In: Giese, H., König, B. (eds.) ICGT 2014. LNCS, vol. 8571, pp. 146–160. Springer, Cham (2014). https://doi.org/10.1007/978-3-319-09108-2_10

17. Danos, V., Heindel, T., Honorato-Zimmer, R., Stucki, S.: Moment semantics for reversible rule-based systems. In: Krivine, J., Stefani, J.-B. (eds.) RC 2015. LNCS, vol. 9138, pp. 3–26. Springer, Cham (2015). https://doi.org/10.1007/978-3-319-20860-2_1

18. Danos, V., Laneve, C.: Formal molecular biology. Theoret. Comput. Sci. **325**(1), 69–110 (2004). https://doi.org/10.1016/j.tcs.2004.03.065

19. Danos, V., Schachter, V. (eds.): CMSB 2004. LNCS, vol. 3082. Springer, Heidelberg (2005). https://doi.org/10.1007/b107287

20. Delbrück, M.: Statistical fluctuations in autocatalytic reactions. J. Chem. Phys. **8**(1), 120–124 (1940). https://doi.org/10.1063/1.1750549

21. Ehrig, H., Ehrig, K., Prange, U., Taentzer, G.: Fundamentals of Algebraic Graph Transformation. MTCSAES. Springer, Heidelberg (2006). https://doi.org/10.1007/3-540-31188-2

22. Ehrig, H., Golas, U., Habel, A., Lambers, L., Orejas, F.: $\mathcal{M}$-adhesive transformation systems with nested application conditions. Part 2: embedding, critical pairs and local confluence. Fundamenta Informaticae **118**(1–2), 35–63 (2012). https://doi.org/10.3233/FI-2012-705

23. Ehrig, H., Golas, U., Habel, A., Lambers, L., Orejas, F.: $\mathcal{M}$-adhesive transformation systems with nested application conditions. Part 1: parallelism, concurrency and amalgamation. Math. Struct. Comput. Sci. **24**(04) (2014). https://doi.org/10.1017/s0960129512000357

24. Ehrig, H., Habel, A., Padberg, J., Prange, U.: Adhesive high-level replacement categories and systems. In: Ehrig, H., Engels, G., Parisi-Presicce, F., Rozenberg, G. (eds.) ICGT 2004. LNCS, vol. 3256, pp. 144–160. Springer, Heidelberg (2004). https://doi.org/10.1007/978-3-540-30203-2_12

25. Feret, J., Koeppl, H., Petrov, T.: Stochastic fragments: a framework for the exact reduction of the stochastic semantics of rule-based models. Int. J. Softw. Inf. (IJSI) **7**(4), 527–604 (2014)

26. Gabriel, K., Braatz, B., Ehrig, H., Golas, U.: Finitary $\mathcal{M}$-adhesive categories. Math. Struct. Comput. Sci. **24**(04) (2014). https://doi.org/10.1017/S0960129512000321

27. Habel, A., Pennemann, K.H.: Correctness of high-level transformation systems relative to nested conditions. Math. Struct. Comput. Sci. **19**(02), 245 (2009). https://doi.org/10.1017/s0960129508007202

28. Habel, A., Plump, D.: $\mathcal{M},\mathcal{N}$-adhesive transformation systems. In: Ehrig, H., Engels, G., Kreowski, H.-J., Rozenberg, G. (eds.) ICGT 2012. LNCS, vol. 7562, pp. 218–233. Springer, Heidelberg (2012). https://doi.org/10.1007/978-3-642-33654-6_15

29. Harmer, R., Danos, V., Feret, J., Krivine, J., Fontana, W.: Intrinsic information carriers in combinatorial dynamical systems. Chaos: Interdisc. J. Nonlinear Sci. **20**(3), 037108 (2010). https://doi.org/10.1063/1.3491100

30. Lack, S., Sobociński, P.: Adhesive and quasiadhesive categories. RAIRO - Theoret. Inf. Appl. **39**(3), 511–545 (2005). https://doi.org/10.1051/ita:2005028

31. Padberg, J.: Towards M-adhesive categories based on coalgebras and comma categories. arXiv:1702.04650 (2017)

32. Smith, M.B., March, J.: March's Advanced Organic Chemistry. Wiley, Hoboken (2006). https://doi.org/10.1002/0470084960

Algebras for Tree Decomposable Graphs

Roberto Bruni[1]($\boxtimes$) (iD), Ugo Montanari[1] (iD), and Matteo Sammartino[2,3] (iD)

[1] Dipartimento di Informatica, Università di Pisa, Pisa, Italy
{bruni,ugo}@di.unipi.it
[2] Royal Holloway University of London, Egham, UK
matteo.sammartino@rhul.ac.uk
[3] University College London, London, UK

Abstract. Complex problems can be sometimes solved efficiently via recursive decomposition strategies. In this line, the *tree decomposition* approach equips problems modelled as graphs with tree-like parsing structures. Following Milner's *flowgraph algebra*, in a previous paper two of the authors introduced a *strong network* algebra to represent open graphs (up to isomorphism), so that homomorphic properties of open graphs can be computed via structural recursion. This paper extends this graphical-algebraic foundation to tree decomposable graphs. The correspondence is shown: (i) on the algebraic side by a *loose network* algebra, which relaxes the restriction reordering and scope extension axioms of the strong one; and (ii) on the graphical side by Milner's *binding bigraphs*, and *elementary* tree decompositions. Conveniently, an interpreted loose algebra gives the evaluation complexity of each graph decomposition. As a key contribution, we apply our results to dynamic programming (DP). The initial statement of the problem is transformed into a term (this is the secondary optimisation problem of DP). Noting that when the scope extension axiom is applied to reduce the scope of the restriction, then also the complexity is reduced (or not changed), only so-called canonical terms (in the loose algebra) are considered. Then, the canonical term is evaluated obtaining a solution which is locally optimal for complexity. Finding a global optimum remains an NP-hard problem.

Keywords: Tree decomposition · Bigraphs · Graph algebras

1 Introduction

Many quite relevant complex graph problems become easy for specific classes of graphs. Usually these graphs are equipped with a suitable recursive structure which allows to compute the solution by problem reduction. The typical structure studied in the literature is *tree decomposition* [3,4,7,11,14,18]. Another suggestive approach is to consider a hyperedge replacement grammar [5], where the structure of a derived graph is its derivation tree.

Research supported by MIUR PRIN 201784YSZ5 *ASPRA*, by Univ. of Pisa PRA_2018_66 *DECLWARE*, and by EPSRC Grant *CLeVer* (EP/S028641/1).

F. Gadducci and T. Kehrer (Eds.): ICGT 2020, LNCS 12150, pp. 203–220, 2020.
https://doi.org/10.1007/978-3-030-51372-6_12

In [17], a *strong network* algebra, called *Soft Constraint Evaluation Problems* (SCEP), is introduced (see also [12]). The algebra has operations of parallel composition, node restriction and permutation and, in particular, has the axioms of restriction reordering and of scope extension. Equivalence classes of strong terms correspond exactly to open graphs up to isomorphism, which thus can be seen as standard representatives of the classes. Consequently, homomorphic properties of open graphs can be conveniently computed via structural recursion. While two strongly equivalent graphs evaluate to the same result by construction, it may happen that the complexities of their evaluations be vastly different. To represent explicitly similar additional information, it is convenient to introduce a finer graphical-algebraic initial pair of models. For instance, an algebra for the computational complexity of problems should fit the new axiomatisation.

In this paper, we choose *elementary* tree decompositions, a simple variant of the classical tree decomposition approach, as the reference graphical model. Interestingly, for obtaining an adequate algebraic model it is enough to eliminate the axioms of restriction reordering and of scope extension from the strong specification. The resulting specification is called *loose*. In [17], an alternative version of algebraic model was chosen, by eliminating only the axiom of scope extension, and called *weak*. The present axiomatisation is needed in order to achieve a tighter correspondence with tree decompositions.

Here we also consider Milner's *bigraphs* [16], a widely studied graphical model for process calculi. Once equipped with a suitable signature, bigraphs can be put into bijective correspondence with equivalence classes of loose terms: the *link* graph represents the variables, and the *place* graph the nesting of restrictions.

As in the strong case, the existence of graphical standard representatives for the initial algebra makes it easy to define interesting algebras of the class. For instance, the evaluation complexity of a term can be easily computed by its interpretation within a simple loose algebra. Notably, the reverse application of the scope extension axiom (i.e. aimed to reduce the scope of the restriction) reduces, or does not change, the evaluation complexity of a term. Thus minimal complexity must be achieved by terms which are fully reduced with respect to the extension axiom (they are called *canonical*) and search for optimal evaluations can be restricted to canonical terms. To take advantage of this property, we define a type system where only canonical terms are typeable.

An algebra of graphs of special interest naturally arises in the case of dynamic programming (DP) [1]. DP usually consists of minimising a cost function F of variables X while keeping variables Y as parameters, i.e. $F(Y) = \min_X F(X, Y)$. Typically, function F has the form

$$F(X, Y) = F_1(X_1, Y_1) + ... + F_n(X_n, Y_n).$$

where each function $F_1, ..., F_n$ is dependent only on a few variables. The key issue is how the variables in X and Y are used in $F_1, ..., F_n$. The sharing structure can be conveniently represented by a hypergraph F where nodes are variables and hyperedges are labelled by functions F_i, represented as multidimensional tables.

The evaluation procedure corresponds to compute by structural recursion the optimal cost of F, where values are multidimensional tables representing intermediate functions, parallel composition is sum, constants are hyperedge labels and restriction with respect to a variable x, $[\![(x)F(X,Y)]\!]$, corresponds to eliminate variable x in table $F(X,Y)$: $\min_x[\![F(X,Y)]\!]$. In conclusion, the solution of an optimisation problem via dynamic programming consists of two steps: (i) find a canonical loose term of low complexity for F; and (ii) evaluate the term.

To compute the complexity, it is enough to define a loose homomorphism where parallel composition $F_1|F_2$ is the max of the complexities of F_1 and F_2 and of the number of free variables of $F_1|F_2$, a constant is the number of variables in the corresponding hyperedge, and restriction is the identity. Notice that since table handling is typically of exponential complexity with respects to dimension, the complexity of a sequence of steps is assumed to be just the max.

The computational cost of (ii) typically depends on the chosen term, but not on the values stored in the tables. Thus the chosen term determines the complexity of step (ii). This property allows to separate the two optimisation procedures: the first step is called the *secondary optimisation problem* of DP [2]. Unfortunately, the secondary problem is NP hard [21], thus typically it is convenient to solve it exactly only if the evaluation must be executed many times.

Structure of the Paper. In Sect. 2 we briefly recall some basic notions and some results from [17]. The original contribution starts in Sect. 3, where we present the loose network specification and draw the graphical-algebraic correspondence with binding bigraphs. Section 4 focuses on dynamic programming, tree decomposition and canonical form, showing how to move from one to the other. There we also define the (loose) algebra we introduced above for computing the evaluation complexity. A simple type system characterises canonical forms. Finally, it is shown that all and only canonical forms are computed by an algorithm based on bucket elimination. Concluding remarks are in Sect. 5.

2   Background

Notation. Given a set V we denote by $V^\star$ the set of (finite) sequences over V and we let $|\cdot|$ return the length of a sequence. Given a function $f\colon V_1 \to V_2$ we overload the symbol f to denote also its lifting $f\colon V_1^\star \to V_2^\star$, defined elementwise.

Hypergraphs. A *ranked alphabet* $\mathcal{E}$ is a set where each element $e \in \mathcal{E}$ has an arity $\mathsf{ar}(e) \in \mathbb{N}$. A *labelled hypergraph* over a ranked alphabet $\mathcal{E}$ is a tuple $G = (V_G, E_G, a_G, lab_G)$, where: V_G is the set of vertices (also called nodes); E_G is the set of (hyper)edges; $a_G\colon E_G \to V_G^\star$ assigns to each hyperedge e the sequence of nodes attached to it; $lab_G\colon E_G \to \mathcal{E}$ is a labeling function, assigning a label to each hyperedge e such that $|a_G(e)| = \mathsf{ar}(lab_G(e))$.

Given two hypergraphs G_1 and G_2 over $\mathcal{E}$, a *homomorphism* between them is a pair of functions $h = (h_V\colon V_{G_1} \to V_{G_2}, h_E\colon E_{G_1} \to E_{G_2})$ preserving connectivity and labels, namely: $h_V \circ a_{G_1} = a_{G_2} \circ h_E$ and $lab_{G_2} \circ h_E = lab_{G_1}$. We say that G_1 and G_2 are *isomorphic*, denoted $G_1 \cong G_2$, whenever there exists a homomorphism between them which is a component-wise isomorphism. We write $G_1 \uplus G_2$ for the component-wise disjoint union of G_1 and G_2.

Permutation Algebras. Given a countable set of variables $\mathbb{V}$, we write $Perm(\mathbb{V})$ for the set of finite permutations over $\mathbb{V}$, i.e., bijective functions $\pi\colon \mathbb{V} \to \mathbb{V}$. A *permutation algebra* is an algebra for the signature comprising all finite permutations and the formal equations $x \text{ id} = x$ and $(x\, \pi_1)\, \pi_2 = x\, (\pi_2 \circ \pi_1)$.

2.1   Strong Network Algebras

In [15] Milner introduced an *algebra of flowgraphs* defined by simple axioms. Here we introduce an *algebra of networks* that has essentially the same axioms, but that exploits a nominal structure for nodes. Hereafter we fix a ranked alphabet $\mathcal{E}$ and a countable set of variables $\mathbb{V}$. We also assume functions $\text{var}\colon \mathcal{E} \to \mathbb{V}^\star$ (with $\text{ar}(A) = |\text{var}(A)|$, for all $A \in \mathcal{E}$), assigning a tuple of *distinct* canonical variables to each symbol of the alphabet. We require $\text{var}(A) \cap \text{var}(B) = \emptyset$ whenever $A \neq B$.

We explicitly equip hypergraphs with an interface, specifying which nodes allow them to interact when composed. We call these hypergraphs *networks*.

Definition 1 (Concrete network). *A concrete network is a pair $I \blacktriangleright G$ of a hypergraph G without isolated nodes[1] such that $V_G \subseteq \mathbb{V}$, and a set $I \subseteq V_G$.*

Every edge e in a network can be connected to the same node multiple times. This can be understood as having a variable substitution σ mapping the tuple of canonical variables $\text{var}(lab_G(e))$ to the actual variables $a_G(e)$ to which e is connected.

Two networks $I_1 \blacktriangleright G_1$ and $I_2 \blacktriangleright G_2$ are *isomorphic* whenever $I_1 = I_2$ and there exists an isomorphism $\iota\colon G_1 \to G_2$ such that $\iota|_{I_1} = \text{id}_{I_1}$.

Definition 2 (Abstract network). *An* abstract *network is an isomorphism-class of a concrete network.*

Intuitively, abstract networks are taken up to α-conversion of non-interface nodes. We write $I \triangleright G$ to denote the abstract network that corresponds to the equivalence class of the concrete network $I \blacktriangleright G$.

Concrete networks can be seen as terms of an algebraic specification which we call *strong network specification*, where free variables correspond to the interface, and variable restriction (written $(x)P$) is used to declare local (non-interface) nodes (x is local to P in $(x)P$). Following well-known algebraic descriptions of nominal calculi [10], terms will carry a permutation algebra structure. This enables an algebraic treatment of variable binding, together with associated notions of scope, free/bound variables and α-conversion.

The syntax of networks is given by the following grammar:

$$P, Q := A(\tilde{x}) \mid P|Q \mid (x)P \mid P\pi \mid \text{nil}$$

where $A \in \mathcal{E}$, $\pi \in Perm(\mathbb{V})$, $x \in \mathbb{V}$, $\tilde{x} \in \mathbb{V}^\star$ and $|\tilde{x}| = \text{ar}(A)$. The free variables $\text{fv}(P)$ of P are the unrestricted ones, and are defined by recursion as expected.

---

[1] In the network specification below, nodes are introduced as support of the permutation algebra of hyperarcs. Isolated nodes would require additional items with singleton support of little use in our model.

(AX$_|$)

$$P|Q \equiv_s Q|P \qquad (P|Q)|R \equiv_s P|(Q|R) \qquad P|\text{nil} \equiv_s P$$

(AX$_{(x)}$)

$$(x)(y)P \equiv_s (y)(x)P \quad (x)\text{nil} \equiv_s \text{nil}$$

(AX$_\alpha$)

$$(x)P \equiv_s (y)P[x \mapsto y] \qquad (y \notin \text{fv}(P))$$

(AX$_{SE}$)

$$(x)(P|Q) \equiv_s (x)P \mid Q \qquad (x \notin \text{fv}(Q))$$

(AX$_\pi$)

$$P \text{ id} \equiv_s P \qquad (P\pi')\pi \equiv_s P(\pi \circ \pi')$$

(AX$_\pi^p$)

$$A(x_1,\ldots,x_n)\pi \equiv_s A(\pi(x_1),\ldots,\pi(x_n)) \qquad \text{nil } \pi \equiv_s \text{nil} \qquad (P|Q)\pi \equiv_s P\pi \mid Q\pi$$

$$((x)P)\pi \equiv_s (\pi(x))(P\pi)$$

Fig. 1. Axioms of strong networks.

The *atom* $A(\tilde{x})$ represents an A-labelled hyperedge, connecting the nodes $\tilde{x}$, possibly with repetitions. The *parallel composition* $P|Q$ represents the union of networks P and Q, possibly sharing some nodes. The *restriction* $(x)P$ represents a network where x is local, and hence cannot be shared. The *permutation* $P\pi$ is P where its free variables have been renamed according to π. As usual in permutation algebras, π is not a capture-avoiding substitution, but just a renaming of all global and local names that appear in the term. The constant nil represents the empty graph. We say that a term P is nil-free if nil is not a subterm of P.

We now introduce a *strong* network specification, which, as opposed to the loose one, shown later, identifies more terms.

Definition 3 (Strong network specification). *The* strong network specification *consists of the syntax given above, subject to the axioms of Fig. 1.*

The operator $|$ forms a commutative monoid **(AX$_|$)**. Restrictions can be α-converted **(AX$_\alpha$)**, reordered and removed whenever their scope is nil **(AX$_{(x)}$)**. The scope of restricted variables can be narrowed to terms where they occur free by the scope extension axiom **(AX$_{SE}$)**. Axioms for permutations say that identity and composition behave as expected **(AX$_\pi$)** and that permutations distribute over syntactic operators **(AX$_\pi^p$)**. Permutations replace all names bijectively, including the bound ones.

Example 1. Consider the terms

$$(x)(y)(z)(A(x,y) \mid B(y,z)) \qquad \text{and} \qquad (y)((x)A(x,y) \mid (z)B(y,z))$$

They are proved to be (strong) equivalent by exploiting **(AX$_{(x)}$)** to switch the order of restrictions on x and y and then **(AX$_{SE}$)** (twice) to move the restrictions on x and z inside parallel composition.

An *s-algebra* $\mathcal{A}$ is a set together with an interpretation $\text{op}^{\mathcal{A}}$ of each operator op. The set of freely generated terms modulo the axioms of Fig. 1 is an *initial* algebra. By initiality, for any such term P there is a unique interpretation $[\![P]\!]^{\mathcal{A}}$ of P as an element of $\mathcal{A}$.

In [17] we show that abstract networks form an initial s-algebra. Hence, we have a unique evaluation of abstracts networks into any other s-algebra.

Definition 4 (Initial s-algebra). *The initial s-algebra $\mathcal{N}$ consists of abstract networks, and the following interpretation of operations:*

$$A^{\mathcal{N}}(x_1, x_2, \ldots, x_n) = \boxed{A} \qquad \text{nil}^{\mathcal{N}} = \emptyset \triangleright 0_G$$
$$\underbrace{}_{\boxed{x_1}\ \boxed{x_2}\ \ldots\ \boxed{x_n}}$$

$$(I \triangleright G)\pi^{\mathcal{N}} = \pi(I) \triangleright G_\pi \qquad (x)^{\mathcal{N}}(I \triangleright G) = I \setminus \{x\} \triangleright G$$
$$I_1 \triangleright G_1|^{\mathcal{N}}I_2 \triangleright G_2 = I_1 \cup I_2 \triangleright G_1 \uplus_{I_1, I_2} G_2$$

where: G_π is G where each node v is replaced with $\pi(v)$; $G_1 \uplus_{I_1, I_2} G_2$ is the disjoint union of G_1 and G_2 where nodes in $I_1 \cup I_2$ with the same name are identified; and 1_G is the empty hypergraph.

Permutations in the specification allow computing the set of free variables, called *support*, in any s-algebra.

Definition 5 (Support). *Let $\mathcal{A}$ be an s-algebra. We say that a finite $X \subset \mathbb{V}$ supports $P \in \mathcal{A}$ whenever $P\pi = P$, for all permutations π such that $\pi|_X = \text{id}_X$. The (minimal) support $\text{supp}(P)$ is the intersection of all sets supporting P.*

It is important to note that $\text{supp}(I \triangleright G) = I$.

2.2   Tree Decomposition

A decomposition of a graph can be represented as a *tree decomposition* [3,4,7, 11,14,18], i.e., a tree where each node is a piece of the graph. Following [17], we introduce a notion of *rooted tree decomposition*. Recall that a *rooted tree* $T = (V_T, E_T)$ is a set of nodes V_T and a set of edges $E_T \subseteq V_T \times V_T$, such that there is a *root*, i.e. a node $r \in V_T$:

- with no ingoing edges: there are no edges (v, r) in E_T;
- such that, for every $v \in V_T$, $v \neq r$, there is a unique path from r to v.

Definition 6 (Rooted tree decomposition). *A rooted tree decomposition of a hypergraph G is a pair $\mathcal{T} = (T, X)$, where T is a rooted tree and $X = \{X_t\}_{t \in V_T}$ is a family of subsets of V_G, one for each node of T, such that:*

1. *for each node $v \in V_G$, there exists a node t of T such that $v \in X_t$;*
2. *for each hyperedge $e \in E_G$, there is a node t of T such that $a_G(e) \subseteq X_t$;*
3. *for each node $v \in V_G$, let $S_v = \{t \mid v \in X_t\}$, and $E_v = \{(x, y) \in E_T \mid x, y \in S_v\}$; then (S_v, E_v) is a rooted tree.*

We gave a slightly different definition of tree decomposition: the original one refers to a non-rooted, undirected tree. In our dynamic programming application it is convenient to model hierarchical problem reductions as rooted structures. All tree decompositions in this paper are rooted, so we will just call them tree decompositions, omitting "rooted". Of course the above definition includes trivial decompositions, like the one with a single node, or the one where $X_t = V_G$ for every node t. They will be ruled out by the notion of *elementary* tree decomposition on which our contribution is centred (see Sect. 4.3).

Tree decompositions are suited to decompose networks: we require that interface variables are located at the root.

Definition 7 (Decomposition of a network). *The decomposition of a network $I \triangleright G$ is a decomposition of G rooted in r, such that $I \subseteq X_r$.*

3   Loose Specification

In strong network specifications the order and positions of restrictions are immaterial. However, the order and positions in which restrictions appear in a term provide some sort of parsing structure for the underlying network. In this section we relax some axioms of the strong algebra to make explicit the hierarchical structure in the network. This is achieved by showing a tight correspondence between terms of the relaxed algebra and Milner's binding bigraphs [13]. In the next sections we will show that the same correspondence can be extended to characterise some special kinds of tree decompositions, called elementary, and also the output produced by an algorithm based on bucket elimination [9].

Definition 8 (Loose network specification). *The* loose network specification *is the strong one without axioms* $(\mathbf{AX}_{SE})$ *and* $(\mathbf{AX}_{(x)})$.

The removal of axioms $(\mathbf{AX}_{(x)})$ means that the order in which restrictions are applied is recorded in each equivalence class. The removal of axiom $(\mathbf{AX}_{SE})$ means that the hierarchy imposed by a restriction on name x is not permeable to all hyperarcs, even those that are not attached to x, in the sense that the axioms do not allow to freely move them down and sideways to the restriction on x. We write $P \equiv_l Q$ if P and Q are in the same loose equivalence class.

3.1   Initial Loose Algebra of Binding Bigraphs

Our first result shows that, in the same way as the algebra of strong terms offers a syntax for open graphs, the algebra of loose terms offers a syntax for a well-known model of structured hypergraphs, called binding bigraphs [8,13].

We recall that bigraphs are structures where two dimensions coexist: one related to a tree of nested components, mimicking the structure of a term; and another related to the sharing of names. The first is called place graph, the second link graph. The nodes of the graph are labelled by so-called controls that

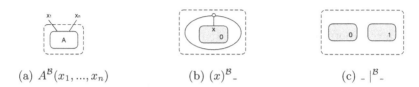

(a) $A^B(x_1, ..., x_n)$ (b) $(x)^B$_   (c) _|^B_

Fig. 2. The loose algebra of binding bigraphs

fix their type and arity. A set of controls gives the signature of a class of bigraphs. See [8] for the exact definitions.

In binding bigraphs, places can act as binders for names and the scope rule guarantees that whenever a name is linked to some component then it is either a free name or one bound to some parent of the (place of the) component.

Graphically a bigraph consists of some *roots* (dashed boxes) where *nodes* (solid boxes) can be nested inside (according to the place graph). Each node is labelled by a control that indicates the number of ports for linkage of the node. Binding ports are denoted by circular attachments. Bigraphs can also contain *sites* (grey boxes) that represent some holes where other bigraphs can be plugged in. Sites are numbered, starting from 0. Bigraphs have also *names* (denoted by $x, y, z, ...$) that are local if introduced by some binding port or global otherwise. Ports and names are linked by lines, (according to the link graph).

The tensor product $A \otimes B$ of two bigraphs corresponds to put them side by side, while the composition $A \circ B$ of two bigraphs is defined when the number of holes in A matches with the number of roots in B and it corresponds to plug each root of B in the corresponding hole of A.

In the following we mostly consider *ground* bigraphs, i.e., without sites, with a unique root and with just global outer names. As explained in the Appendix, we take (lean support) equivalence classes of concrete bigraphs, up-to graph isomorphism, renaming of local names and presence of unused names.

The correspondence between loose terms and binding bigraphs is obtained by taking one control A (drawn as a rounded box) for each constant $A(\tilde{x})$ with binding arity $\mathsf{ar}_b(A) = 0$ and free arity $\mathsf{ar}_f(A) = |\tilde{x}|$ and one control ν (drawn as an oval) for each binding expression (x), with $\mathsf{ar}_b(\nu) = 1$ and $\mathsf{ar}_f(\nu) = 0$.

Definition 9 (The loose algebra of binding bigraphs). *The l-algebra of binding bigraphs consists of (lean-support equivalence classes of) ground binding bigraphs (i.e., with one root and no sites). The symbols of the signature are mapped to the binding bigraphs as shown in the table in Fig. 2 (permutations are just applied to rename the global ports of the graph) and term substitution corresponds to bigraph composition.*

Some examples of interpretations are in Fig. 3.

Binding bigraphs offer a convenient model for the loose network specification.

Proposition 1. *Binding bigraphs form an initial loose network algebra.*

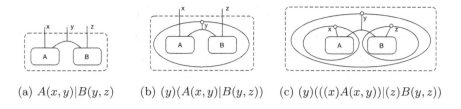

(a) $A(x,y)|B(y,z)$ (b) $(y)(A(x,y)|B(y,z))$ (c) $(y)(((x)A(x,y))|(z)B(y,z))$

Fig. 3. Some examples of bigraphs

4   Dynamic Programming

The structure of a number of optimisation problems can be conveniently represented as dynamic programming (DP) networks, where hyperedges correspond to atomic subproblems and nodes to (possibly shared) variables. Costs of subproblems are summed up and restricted variables are assigned optimal values.

In this section we will first show how optimisation problems are represented in our framework. We will then turn our attention to the *secondary optimisation problem*, i.e., the problem of finding a decomposition into subproblems of minimal complexity. This is a problem of paramount practical importance for DP.

We have so far given two equivalent ways of decomposing a network: l-terms and binding bigraphs. We will introduce a notion of evaluation complexity for l-terms, and we will characterise local optima as *l-canonical* terms. Then we will establish a formal connection with another way of decomposing DP problems, namely (*elementary*) tree decompositions.

Finally, we will show that the well-known bucket elimination algorithm (see, e.g., [19, 5.2.4]) precisely corresponds to computing the l-canonical form of a term w.r.t. a given ordering on restricted variables. Leveraging the algebraic representation of networks as terms, and the correspondence between l-terms and tree decompositions, this result provides us with a way to compute a locally optimal decomposition of a network in three equivalent ways: as a l-canonical term, a binding bigraph, or a tree decomposition.

4.1   Networks as Optimisation Problems

We now introduce an s-algebra of *cost functions*, where networks are evaluated to solutions of the corresponding optimisation problem. We fix a domain $\mathbb{D}$ of values for variables. Then an element of the s-algebra is a cost function $\varphi \colon (\mathbb{V} \to \mathbb{D}) \to \mathbb{R}_\infty$ that given an assignment of values to variables returns its cost. To interpret all terms we assume that an interpretation is given of each symbol $A \in \mathcal{E}$ as a cost function $\mathsf{func}_A \colon (\mathbb{V} \to \mathbb{D}) \to \mathbb{R}_\infty$ such that, for any $\rho, \rho' \colon \mathbb{V} \to \mathbb{D}$, $\rho_{|\mathsf{var}(A)} = \rho'_{|\mathsf{var}(A)}$ implies $\mathsf{func}_A\,\rho = \mathsf{func}_A\,\rho'$, i.e., func_A only depends on the canonical variables of A.

Definition 10 (S-algebra of cost functions). *The s-algebra $\mathcal{V}$ consists of cost functions $\varphi \colon (\mathbb{V} \to \mathbb{D}) \to \mathbb{R}_\infty$ and the following interpretation of operations, for any $\rho \colon \mathbb{V} \to \mathbb{D}$:*

$$A^\mathcal{V}(\tilde{x})\rho = \mathsf{func}_A(\rho \circ \sigma) \qquad \mathsf{nil}^\mathcal{V}\rho = 1$$

$$((x)^\mathcal{V}\phi)\rho = \min\{\phi\rho[x \mapsto d]\}_{d \in \mathbb{D}} \qquad (\phi\pi^\mathcal{V})\rho = \phi(\rho \circ \pi) \qquad (\phi_1|^\mathcal{V}\phi_2)\rho = \phi_1\rho + \phi_2\rho$$

where $\sigma \colon \mathbb{V} \to \mathbb{V}$ *is a substitution mapping* $\mathsf{var}(A)$ *to* $\tilde{x}$ *component-wise and expressing the connection between the canonical vertices of an* A-*labelled hyperedge and the actual nodes of the graph it is connected to.*

Example 2. Consider the term $P = (y)(x)(z)(\; A(x,y) \mid B(y,z)\;)$. This is evaluated as the optimisation problem consisting in minimising the sum of the cost functions for A and B w.r.t. to all the variables. Explicitly:

$$\llbracket P \rrbracket^\mathcal{V} = \lambda\rho.\min\;\{\mathsf{func}_A(\rho[x \mapsto d_1, y \mapsto d_2]) + \mathsf{func}_B(\rho[y \mapsto d_2, z \mapsto d_3])\}_{d_1,d_2,d_3 \in \mathbb{D}}$$

which, since func_A and func_B only depends on $\{x,y\}$ and $\{y,z\}$, respectively, is a single value that does not depend on ρ.

Although all terms for the same network have the same evaluation in any algebra, different ways of computing such an evaluation, represented as different terms, may have different computational costs.

We make this precise by introducing a notion of evaluation complexity.

4.2   Evaluation Complexity

We define the complexity of a term P as the maximum "size" of elements of an algebra $\mathcal{A}$ computed while inductively constructing $\llbracket P \rrbracket^\mathcal{A}$, the size being given by the number of variables in the support. Intuitively, a step of DP consists of solving a subproblem parametrically with respect to a number n of variables. Thus if the number of possible values of a variable is $|\mathbb{D}|$, then the number of cases to consider for solving the subproblem is $|\mathbb{D}|^n$, namely it is exponential with the number n of parameters of the subproblem. As a consequence, we can approximate the complexity of the whole problem with the complexity of the hardest subproblem.

In our algebraic representation, DP problems are terms P of the strong algebra interpreted as functions $\varphi \colon (\mathbb{V} \to \mathbb{D}) \to \mathbb{R}_\infty$ and their cost is just the evaluation of P. Solving a subproblem corresponds to evaluating a restriction operator $(x)P$ of a term P, while the parameters are the variables in its support. On the other hand, the cost of a single case is again proportional to $|\mathbb{D}|$: we fix the parameters and we optimize with respect to the values of the restricted variable. Thus the complexity of evaluating $(x)P$ is $|\mathsf{fv}(P)|$. Notice that it represents the space and time complexity of the problem. In fact, it correctly coincides with the dimension of the matrix needed to represent the function $\varphi \colon (\mathbb{V} \to \mathbb{D}) \to \mathbb{R}_\infty$ corresponding to $\llbracket P \rrbracket^\mathcal{A}$. The key observation is that if we take two strong equivalent terms P_1 and P_2, they will necessarily evaluate to the same cost function in $(\mathbb{V} \to \mathbb{D}) \to \mathbb{R}_\infty$, but they will not have necessarily the same complexity.

Example 3. Consider the following terms:

$$P = (y)(x)(z)(\; A(x,y) \mid B(y,z)\;) \qquad Q = (y)(\; (x)A(x,y) \mid (z)B(y,z)\;).$$

Although $P \equiv_s Q$, these terms have different complexities. The term P has complexity 3 because, when evaluating it in any algebra, one has to evaluate $A(x,y)|B(y,z)$, and then solve it w.r.t. all its variables. Intuitively, $A(x,y)|B(y,z)$ is the most complex subproblem one considers in P, with 3 variables. Instead, the complexity of Q is 2, because its evaluation requires solving $A(x,y)$ and $B(y,z)$ w.r.t. x and z, which are problems with 2 variables.

Given a term in the strong algebra, the problem of finding a (syntactical) term with minimal complexity corresponds to the secondary optimisation problem of DP. In [17] we have inductively defined a complexity function for terms.

Definition 11. *Given a term P, its complexity $\langle\langle P \rangle\rangle$ is defined as follows:*

$$\langle\langle P|Q \rangle\rangle = \max\{\langle\langle P \rangle\rangle, \langle\langle Q \rangle\rangle, |\mathsf{fv}(P|Q)|\} \qquad \langle\langle (x)P \rangle\rangle = \langle\langle P \rangle\rangle \qquad \langle\langle P\pi \rangle\rangle = \langle\langle P \rangle\rangle$$
$$\langle\langle A(\tilde{x}) \rangle\rangle = |\mathsf{set}(\tilde{x})| \qquad \langle\langle \mathsf{nil} \rangle\rangle = 0$$

Complexity is well-defined for loose terms but not for strong terms, as applying $(\mathbf{AX}_{SE})$ may change the complexity (see [17]).

Lemma 1. *Given $(x)(P|Q)$, with $x \notin \mathsf{fv}(Q)$, we have $\langle\langle (x)P|Q \rangle\rangle \leq \langle\langle (x)(P|Q) \rangle\rangle$.*

We now classify l-terms according to their complexity. We say an l-term is *pure* if every subterm $(x)P$ is such that $x \in \mathsf{fv}(P)$.

Definition 12 (L-normal and l-canonical forms). *An l-term is in* l-normal *form whenever it is of the form*

$$(\tilde{x})(\ A_1(\tilde{x}_1)\ |\ A_2(\tilde{x}_2)\ |\ \ldots\ |\ A_n(\tilde{x}_n)\ )$$

A l-term is in l-canonical *form whenever the directed form of $(\mathbf{AX}_{SE})$*

$$(x)(P|Q) \to (x)P\ |\ Q \qquad (x \notin \mathsf{fv}(Q))$$

cannot be applied to it. For both forms, we assume that they are pure and that nil *sub-terms are removed via $(\mathbf{AX}_|)$.*

L-normal forms have *maximal* complexity. A term can be s-equivalent ($\equiv_s$) to several l-normal forms, all with the same complexity: they differ for the order in which restrictions are applied.

Example 4. The l-term $(z)(x)(y)(A(x,y)|B(y,z))$ is in l-canonical form, while the l-term $(x)(y)(z)(A(x,y)|B(y,z))$ is not, because the axiom $(\mathbf{AX}_{SE})$ is applicable to restrict the scope of x as in $(x)(y)(A(x,y)|(z)B(y,z))$.

Due to Lemma 1, l-canonical forms are local minima of complexity w.r.t. the application of strong axioms minus $(\mathbf{AX}_{(x)})$. In fact, $(\mathbf{AX}_{(x)})$ may enable further applications of $(\mathbf{AX}_{SE})$, and lead to a further complexity reduction, as shown in [17] for weak terms. This phenomenon is exemplified in Example 4, where bringing (z) closer to the parallel via restriction swaps makes the term suboptimal. By forbidding $(\mathbf{AX}_{(x)})$, we considerably simplify the algorithm for computing local complexity optima, and we recover a full correspondence with the bucket elimination algorithm [9], as we shall see later.

4.3   Elementary Tree Decompositions

In this section we establish a correspondence between two ways of decomposing problems that admit a graph model: loose terms and tree decompositions. We first introduce the novel notion of *elementary* tree decomposition (e.t.d.).

Definition 13 (Elementary tree decomposition). *A tree decomposition (T, X) for a network $I \rhd G$ is* elementary *whenever $|X_r \setminus I| \leq 1$ and, for all non-root nodes t of T, $|X_t \setminus X_{\mathsf{parent}(t)}| = 1$.*

We can now adapt the translation function from tree decompositions to terms from [17] to elementary tree decomposition (*e.t.d.* for short).

Definition 14 (From elementary tree decompositions to l-terms). *Let $\mathcal{T} = (T, X)$ be an e.t.d. for $I \rhd G$. For each node t of T, let $E(t) \subseteq E_G$ and $V(t) \subseteq V_G$ be sets of nodes and edges of G such that $e \in E(t)$ (resp. $v \in V(t)$) if and only if t is the closest node to the root of T such that $\alpha_G(e) \subseteq X_t$ (resp. $v \in X_t$). Let $\tau(t)$ be recursively defined on nodes of T as follows:*

$$\tau(t) = (x)(A_1(\tilde{x}_1)|A_2(\tilde{x}_2)|\ldots|A_n(\tilde{x}_n)|\tau(t_1)|\tau(t_2)|\ldots|\tau(t_k))$$

where $E(t) = \{e_1, e_2, \ldots, e_n\}$, $x \in V(t)\setminus I$ (we drop the restriction if $V(t)\setminus I = \emptyset$), $\mathsf{lab}_G(e_i) = A_i$, $\alpha_G(e_i) = \tilde{x}_i$, and $t_1, t_2, \ldots, t_k$ are the children of t in T. Then we define $\mathsf{term}(\mathcal{T}) := \tau(r)$, where r is the root of $\mathcal{T}$.

It is immediate to observe that each e.t.d. is mapped to an l-term. In fact, at each node we add at most one restriction, which results in an ordering over restricted names. In general the mapping is not injective, as the following example shows.

Example 5. Consider the l-term $(x)(A(x, y)|(z)B(y, z))$. Then $\mathsf{term}(\_)$ maps both the elementary tree decompositions depicted in Fig. 4a to it.

We shall define a converse mapping, from pure l-terms to e.t.d.s. In the following we exploit the fact that every pure l-term is congruent to the form

$$(R)\,(A_1(\tilde{x}_1)|\ldots|A_1(\tilde{x}_n)|P_1|\ldots|P_k)$$

where $|R| \leq 1$ and it occurs free in the rest of the term if non-empty, $n, k \geq 0$, and P_i are l-terms of the same form with a top level restriction.

Definition 15 (From pure l-terms to elementary tree decompositions). *Given a pure l-term P such that*

$$P \equiv_l (R)\,(A_1(\tilde{x}_1)|\ldots|A_n(\tilde{x}_n)|P_1|\ldots|P_k)$$

the corresponding e.t.d. $\mathsf{etd}(P)$ is recursively defined as follows:

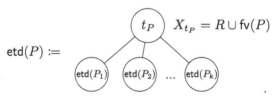

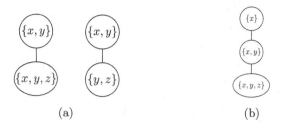

Fig. 4. E.t.d.s for Examples 5(a) and 6(b). For simplicity, we have annotated each node t with the corresponding set X_t.

$$\frac{}{A(\tilde{x}) : \mathsf{set}(\tilde{x})} \qquad \frac{P : X}{P\pi : \pi(X)} \qquad \frac{P : X \cup \{z\}}{(z)P : \mathsf{fv}((z)P)} \qquad \frac{P_1 : X_1 \quad P_2 : X_2}{P_1|P_2 : X_1 \cap X_2}$$

Fig. 5. Type rules for l-canonical terms

The place graph of the binding bigraph $P^{\mathcal{B}}$ is tightly connected to $\mathsf{etd}(P)$. In fact, if we remove from the place graph all leaves whose controls are atoms A we get the tree structure of (controls that are) restrictions. Then each node of the graph can be tagged with the names of the ports that are used by nested controls and we get $\mathsf{etd}(P)$. The free variables of P will also appear in the root.

Lemma 2. *We have that* $\mathsf{etd}(P)$ *is an elementary tree decomposition of* $[\![P]\!]^{\mathcal{N}}$.

As for $\mathsf{term}(\_)$, this translation is not injective (see the following example).

Example 6. Consider the l-terms $P = (x)(y)(A(x,y)|(z)B(x,y,z))$ and $Q = (x)(y)(z)(A(x,y)|B(x,y,z))$. Then $\mathsf{etd}(P) = \mathsf{etd}(Q)$ is the e.t.d. in Fig. 4b.

The translation is injective if we restrict the domain to l-canonical terms.

Proposition 2. *Different l-canonical terms give rise to different e.t.d.s.*

We conjecture that $\mathsf{term}(\_)$ is left-inverse to $\mathsf{etd}(\_)$ on l-canonical terms.

4.4   A Type System for l-canonical Terms

We use a type system to characterise l-canonical forms. Type judgements are of the form $P : X$ where X denotes the set of names of P that can be restricted immediately on top of P. The typing rules are in Fig. 5. Those for atoms and for permutations are trivial.

The rule for restriction deserves some explanation. Let P be a term such that the names in $X \cup \{z\}$ can be restricted, then for the term $(z)P$ all remaining names $\mathsf{fv}((z)P)$ can also be restricted. This is because taken any name $x \in \mathsf{fv}((z)P) \setminus X$, in the l-term $(x)(z)P$ we cannot swap x with z in order to apply the axiom $(\mathbf{AX}_{SE})$. The fact that $(\mathbf{AX}_{SE})$ cannot be applied to z is guaranteed by z being one of the names that can be restricted on top of P (see premise).

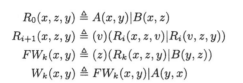

$$\frac{\begin{array}{cc} & \dfrac{B(y,z):\{y,z\}}{(z)B(y,z):\{y\}} \\[4pt] \dfrac{\quad}{A(x,y):\{x,y\}} & \end{array}}{}$$

Fig. 6. Two type derivations

$$R_0(x,z,y) \triangleq A(x,y)|B(x,z)$$
$$R_{i+1}(x,z,y) \triangleq (v)(R_i(x,z,v)|R_i(v,z,y))$$
$$FW_k(x,y) \triangleq (z)(R_k(x,z,y)|B(y,z))$$
$$W_k(x,y) \triangleq FW_k(x,y)|A(y,x)$$

Fig. 7. The wheel network $W_2(v,x)$

The rule for parallel composition follows a simple criterion: only names that are in common between all constituents can be restricted. If a name z appears in P_1 but not in P_2, then obviously the term $(z)(P_1|P_2)$ is not in l-canonical form.

Typing is preserved by the axioms of the loose network specification, in the sense that for all nil-free terms $P \equiv_l Q$ and type X, if $P : X$, then $Q : X$.

Proposition 3. *For any P and type X, if $P : X$ then P is in l-canonical form.*

Proposition 4. *If $P \neq$ nil is in l-canonical form, then $P : X$ for some X.*

Example 7. The term $(x)(y)(A(x,y) \mid (z)B(y,z))$ is in l-canonical form as witnessed by the derivation in Fig. 6, left. Also the term $(y)((x)A(x,y) \mid (z)B(y,z))$ is in l-canonical form (see Fig. 6, right). The term $(x)(y)(z)(A(x,y)|B(y,z))$ is not typeable, because $A(x,y)|B(y,z) : \{y\}$ and $z \notin \{y\}$ thus the subterm $(z)(A(x,y)|B(y,z))$ is not typeable. In fact it is not in l-canonical form because the axiom $(\mathbf{AX}_{SE})$ can be applied to restrict the scope of z to $B(y,z)$.

Example 8. Consider the wheel example from [17] in Fig. 7 (see [20] for a graph grammar presentation). Then $R_0(x,z,y) : \{x\}$, but $R_1(x,z,y) = (v)(R_0(x,z,v)|R_0(v,z,y))$ is not typeable, as $R_0(x,z,v)|R_0(v,z,y) : \emptyset$. We can give an alternative typeable definition of wheels where all terms are l-canonical:

$$R_0(x,z,y) \triangleq (v)(A(x,v)|B(v,z)|A(v,y)) : \{x,z,y\}$$
$$R_{i+1}(x,z,y) \triangleq (v)(R_i(x,z,v)|R_i(v,z,y)|B(v,z)) : \{x,z,y\}$$
$$FW_k(x,y) \triangleq (z)(R_k(x,z,y)|B(x,z)|B(y,z)) : \{x,y\}$$
$$W_k(x,y) \triangleq FW_k(x,y)|A(y,x) : \{x,y\}$$

Input: l-term $(R)M$ in normal form.
Output: l-term P in canonical form.

1: $P \leftarrow (R)M$
2: **while** $R \neq \emptyset$ **do**
3:     $x \leftarrow \max R$
4:     take largest $M' \subseteq M$ such that x occurs free in all elements of M'
5:     $P' \leftarrow (x)M'$
6:     $M \leftarrow M \setminus M' \cup \{P'\}$
7:     $R \leftarrow (R \setminus \{x\})$
8: **return** M

Fig. 8. Bucket elimination algorithm for l-terms.

4.5   Computing l-canonical Forms

We now give an algorithm to compute an l-canonical form of a term. This is based on *bucket elimination* (see, e.g., [19, 5.2.4]), also known as adaptive consistency.

We briefly recall the bucket elimination algorithm. Given a network representing an optimisation problem, and a total order over its variables, sub-problems are partitioned into *buckets*: each sub-problem is placed into the bucket of its last variable in the ordering. At each step, the bucket of a variable x is eliminated by creating a new sub-problem involving all the variables in the bucket different from x. This new problem is put into the bucket of its last variable, and the process is iterated.

In [17] we have extended the algorithm to modify the ordering of variables in the attempt of reducing the size of subproblems. This required an additional backtracking step. Here we show that l-canonical forms can be computed via the ordinary bucket elimination algorithm, suitably adapted to l-terms.

Our algorithm is shown in Fig. 8. Here putting a constraint in the bucket of its last variable amounts to applying the scope extension axiom, and eliminating a variable amounts to restricting it. The algorithm takes as input an l-term in normal form $(R)M$, represented by a totally ordered set of variables R (recall that restricted variables are assumed to be distinct), and a multiset of atomic terms M. The algorithm first picks the max variable in the total order (line 3), then it partitions the input l-term into subterms according to whether x occurs free or not (line 4), and from the former it creates a new term P' where x is restricted. It then adds P' to the remaining terms of P, removes x from the total order R, and iterates the process for the resulting term.

Example 9. We now show an example of execution. We will run the algorithm with the following l-normal term as input:

$$(a < b < c < d) \, \{A_1(c,d), A_2(d,b), A_3(d,a), A_4(b,a)\}$$

On the first iteration, line 3 picks d, and line 4 will select $M' = \{A_1(c,d), A_2(d,b), A_3(d,a)\}$. Then R and M in line 6 and 7 are

$$(a < b < c) \qquad \{A_4(b,a), \, (d)\{A_1(c,d), A_2(d,b), A_3(d,a)\} \, \}$$

At the next iteration, line 3 picks c, and line 6 and 7 give the following:

$$(a < b) \qquad \{A_4(b, a),\ (c)(d)\{A_1(c, d), A_2(d, b), A_3(d, a)\}\}$$

Despite c occurring only in $A_1(c, d)$, the presence of (d), which is greater in the ordering, prevents (c) from permeating the parallel composition. After two more iterations, where the variables b and a are processed, the output is:

$$(a)(b)(A_4(b, a) \mid (c)(d)(A_1(c, d) \mid A_2(d, b) \mid A_3(d, a)))$$

The algorithm outputs all and only the l-canonical forms for a given term.

Proposition 5. *Given a term P, a term $C \equiv_s P$ is l-canonical if and only if there is an l-normal form for P which, if provided as input to the bucket elimination algorithm, outputs C.*

5   Conclusion

Along the graphical-algebraic correspondence introduced by Milner [15] for networks and flow algebras, in [17] two of the authors have studied the connections between tree decompositions and certain weak network algebras. In this paper the correspondence of the two former models and, in addition, of a version of Milner's bigraphs [16], has been fully formalised introducing a small variation of weak network algebras, called loose network algebras. Milner's flow algebras also fit in the schema as strong algebras. The algebraic treatment is instrumental for conveniently expressing important computational properties of the complex problem at hand. In this line, we examine the very relevant case of graphical optimisation problems solved via dynamic programming. We show that the solution of a problem corresponds to the evaluation of a strong term, while the cost of such a computation is obtained by evaluating the same term in a loose algebra. It is also shown that reducing a loose term w.r.t. the axiom of scope expansion produces a modified loose term of better or equal complexity. The notion of l-canonical form for loose terms captures local minima of complexity and is tightly related to particular kinds of tree decompositions, called elementary. We also define a type system for checking if a term is in l-canonical form. Finally, when a total ordering is imposed on the variables of a term, a term in l-canonical form can be uniquely derived by applying Dechter's bucket algorithm [9].

The problem of how to represent parsing trees for (hyper)graphs has been studied in depth in the literature. We mention the work of Courcelle on graph algebras [6]. Here tree decompositions are used to bound the complexity of checking graph properties, and a term encoding allows employing automata-theoretic tools. The focus is different than ours, and investigating connections is left for future work. Other approaches based on tree decompositions are [3,4,7,11,14,18]. Here we provide an algebraic characterisation of tree decompositions with locally minimal complexity. The application of bigraphs to dynamic programming is also novel. We plan to study the relation between loose network specifications and graph grammars for hyperedge replacement [5], where the structure of a derived graph is its derivation tree: it seems that our approach has a simpler compositional structure, and an up-to-date foundation for name handling.

References

1. Bellman, R.: Some applications of the theory of dynamic programming - a review. Oper. Res. **2**(3), 275–288 (1954)
2. Bertelè, U., Brioschi, F.: On non-serial dynamic programming. J. Comb. Theory Ser. A **14**(2), 137–148 (1973)
3. Blume, C., Bruggink, H.J.S., Friedrich, M., König, B.: Treewidth, pathwidth and cospan decompositions with applications to graph-accepting tree automata. J. Vis. Lang. Comput. **24**(3), 192–206 (2013)
4. Bodlaender, H.L., Koster, A.M.C.A.: Combinatorial optimization on graphs of bounded treewidth. Comput. J. **51**(3), 255–269 (2008)
5. Chiang, D., Andreas, J., Bauer, D., Hermann, K.M., Jones, B.K., Knight, K.: Parsing graphs with hyperedge replacement grammars. In: Proceedings of ACL 2013, Long Papers, vol. 1, pp. 924–932. The Association for Computer Linguistics (2013)
6. Courcelle, B., Engelfriet, J.: Graph Structure and Monadic Second-Order Logic - A Language-Theoretic Approach, Encyclopedia of Mathematics and Its Applications, vol. 138. Cambridge University Press, Cambridge (2012)
7. Courcelle, B., Mosbah, M.: Monadic second-order evaluations on tree-decomposable graphs. Theoret. Comput. Sci. **109**(1&2), 49–82 (1993)
8. Damgaard, T.C., Birkedal, L.: Axiomatizing binding bigraphs. Nord. J. Comput. **13**(1–2), 58–77 (2006)
9. Dechter, R.: Constraint Processing. Elsevier Morgan Kaufmann, Amsterdam (2003)
10. Gadducci, F., Miculan, M., Montanari, U.: About permutation algebras, (pre)sheaves and named sets. High.-Order Symbolic Comput. **19**(2), 283–304 (2006)
11. Gogate, V., Dechter, R.: A complete anytime algorithm for treewidth. In: Proceedings of UAI 2004, pp. 201–208. AUAI Press (2004)
12. Hoch, N., Montanari, U., Sammartino, M.: Dynamic programming on nominal graphs. In: Proceedings of GaM@ETAPS 2015, EPTCS, vol. 181, pp. 80–96 (2015)
13. Jensen, O.H., Milner, R.: Bigraphs and mobile processes (revised). Technical report 580, University of Cambridge (2004)
14. Kloks, T. (ed.): Treewidth. LNCS, vol. 842. Springer, Heidelberg (1994). https://doi.org/10.1007/BFb0045375
15. Milner, R.: Flowgraphs and flow algebras. J. ACM **26**(4), 794–818 (1979)
16. Milner, R.: Bigraphical reactive systems. In: Larsen, K.G., Nielsen, M. (eds.) CONCUR 2001. LNCS, vol. 2154, pp. 16–35. Springer, Heidelberg (2001). https://doi.org/10.1007/3-540-44685-0_2
17. Montanari, U., Sammartino, M., Tcheukam Siwe, A.: Decomposition structures for soft constraint evaluation problems: an algebraic approach. In: Graph Transformation, Specifications, and Nets - In Memory of H. Ehrig, pp. 179–200 (2018)
18. Robertson, N., Seymour, P.D.: Graph minors. III. Planar tree-width. J. Comb. Theory Ser. B **36**(1), 49–64 (1984)

19. Rossi, F., van Beek, P., Walsh, T. (eds.): Handbook of Constraint Programming, Foundations of Artificial Intelligence, vol. 2. Elsevier, Amsterdam (2006)
20. Rozenberg, G. (ed.): Handbook of Graph Grammars and Computing by Graph Transformations: Foundations, vol. 1. World Scientific, Singapore (1997)
21. Yannakakis, M.: Computing the minimum fill-in is NP-complete. SIAM J. Algebraic Discrete Methods **2**(1), 77–79 (1981)

Graph Parsing as Graph Transformation
Correctness of Predictive Top-Down Parsers

Frank Drewes[1], Berthold Hoffmann[2]([✉]), and Mark Minas[3]

[1] Umeå Universitet, Umeå, Sweden
drewes@cs.umu.se
[2] Universität Bremen, Bremen, Germany
hof@uni-bremen.de
[3] Universität der Bundeswehr München, Neubiberg, Germany
mark.minas@unibw.de

Abstract. Hyperedge replacement (HR) allows to define context-free graph languages, but parsing is NP-hard in the general case. Predictive top-down (PTD) is an efficient, backtrack-free parsing algorithm for subclasses of HR and contextual HR grammars, which has been described and implemented in earlier work, based on a representation of graphs and grammar productions as strings. In this paper, we define PTD parsers for HR grammars by graph transformation rules and prove that they are correct.

Keywords: Graph transformation · Hyperedge replacement · Parsing · Correctness

1 Introduction

Hyperedge replacement (HR, [8]) is one of the best-studied mechanisms for generating graphs. Being context-free, HR grammars inherit most of the favorable structural and computational properties of context-free string grammars. Unfortunately, simplicity of parsing is not one of these, as there are NP-complete HR languages [1, 14]. Hence, efficient parsing can only be done for suitable subclasses. The authors have devised predictive top-down (PTD, [4]) and predictive shift-reduce (PSR, [6]) parsing for subclasses of HR grammars and, in fact, for subclasses of contextual HR grammars (CHR grammars, [2,3]), which are a modest extension of HR grammars that allows to overcome some of the structural limitations of HR languages.

Although the concepts and implementation of PTD parsers have been described at depth in [4], their correctness has not yet been formally established. We show in this paper how PTD parsing can be defined by graph transformation rules and use this in order to prove the correctness of PTD parsers. Our experience with the correctness proof for PSR parsing in [6] seems to indicate that a graph- and rule-based definition of parsers can make this task easier.

Related work on using graph transformation for defining parsers has dealt with LR string grammars [11] and two-level string grammars [12]. For a broader

© Springer Nature Switzerland AG 2020
F. Gadducci and T. Kehrer (Eds.): ICGT 2020, LNCS 12150, pp. 221–238, 2020.
https://doi.org/10.1007/978-3-030-51372-6_13

discussion of related work on parsing algorithms for graph grammars in general we refer to [6, Sect. 10.1].

The paper is structured as follows. After recalling graph transformation concepts (Sect. 2) and HR grammars (Sect. 3), we introduce threaded HR grammars (Sect. 4), which impose a total order on the edges of their derived graphs, which in turn induces a dependency relation on their nodes. In Sect. 5, we define a general top-down parser for HR grammars that respects edge order and node dependencies, and prove it correct. Since this parser is nondeterministic and hence inefficient, we introduce properties that make the parser predictive, and backtrack-free (Sect. 6) and show that this yields correct parsers that terminate for grammars without left recursion.[1] We conclude the paper by indicating some future work (Sect. 7).

2   Preliminaries

In this paper, $\mathbb{N}$ denotes the set of non-negative integers and $[n]$ denotes $\{1, \ldots, n\}$ for all $n \in \mathbb{N}$. A^* denotes the set of all finite sequences over a set A; the empty sequence is denoted by ε, and the length of a sequence α by $|\alpha|$. As usual, $\to^+$ and $\to^*$ denote the transitive and the transitive reflexive closure of a binary relation $\to$. For a function $f \colon A \to B$, its extension $f^* \colon A^* \to B^*$ to sequences is defined by $f^*(a_1 \cdots a_n) = f(a_1) \cdots f(a_n)$, for all $n \in \mathbb{N}$ and $a_1, \ldots, a_n \in A$. The *composition* of functions $f \colon A \to B$ and $g \colon B \to C$ is denoted as $g \circ f$ and defined by $(g \circ f)(x) = g(f(x))$ for $x \in A$. The restriction of f to some subset $X \subseteq A$ is denoted as $f|_X$.

Definition 1 (Hypergraph). An *alphabet* Σ is a finite set of symbols that comes with an *arity function* $arity \colon \Sigma \to \mathbb{N}$. A *hypergraph* (over Σ) is a tuple $G = (\dot{G}, \bar{G}, att, lab)$, where $\dot{G}$ and $\bar{G}$ are finite sets of *nodes* and *hyperedges*, respectively, the function $att \colon \bar{G} \to \dot{G}^*$ attaches hyperedges to sequences of nodes, and the function $lab \colon \bar{G} \to \Sigma$ labels hyperedges so that $|att(e)| = arity(lab(e))$ for every $e \in \bar{G}$, i.e., the number of attached nodes of hyperedges is dictated by the arity of their labels.

$\mathcal{G}_\Sigma$ denotes the class of hypergraphs over Σ; $\langle\rangle$ denotes the *empty hypergraph*, with empty sets of nodes and hyperedges. A set of hyperedges $E \subseteq \bar{G}$ induces the subgraph consisting of these hyperedges and their attached nodes.

For brevity, we omit the prefix "hyper" in the sequel. Instead of "$x \in \dot{G}$ or $x \in \bar{G}$", we often write "$x \in G$". We often refer to the functions of a graph G by att_G and lab_G. An edge carrying a label in an alphabet Σ is also called a Σ-edge. And a node is called *isolated* if no edge is attached to it.

Definition 2 (Graph Morphism). Given graphs G and H, a *graph morphism* (morphism, for short) $m \colon G \to H$ is a pair $m = (\dot{m}, \bar{m})$ of functions $\dot{m} \colon \dot{G} \to \dot{H}$

---

[1] Since this paper is dedicated to proving the correctness of PTD parsers, and it has been established in [4] that they run in quadratic time at worst, we shall not dwell on issues of efficiency here.

and $\bar{m}\colon \bar{G} \to \bar{H}$ that preserve attachments and labels, i.e., $att_H \circ \bar{m} = \dot{m}^* \circ att_G$ and $lab_H \circ \bar{m} = lab_G$. The morphism is *injective* or *surjective* if both $\dot{m}$ and $\bar{m}$ are, and a *subgraph inclusion* of G in H if $m(x) = x$ for every $x \in G$; then we write $G \subseteq H$. If m is surjective and injective, it is called an *isomorphism*, and G and H are called *isomorphic*, written as $G \cong H$.

For transforming graphs, we use the classical approach of [7], with injective matching and non-injective rules [9], but without rules that delete nodes.

Definition 3 (Rule). A *graph transformation rule* $r = (P, R, r^\circ)$ consists of a *pattern graph* P, a *replacement graph* R, and a mapping $r^\circ\colon \dot{P} \to \dot{R}$. [2] We briefly call r a *rule* and denote it as $r\colon P \rightarrowtail R$. An injective morphism $m\colon P \to G$ into a graph G is a *match* of r, and r *transforms* G at m to a graph H as follows:

- Remove all edges $m(e)$, $e \in \bar{P}$, from G to obtain a graph K.
- Construct H from the disjoint union of K and R by identifying $m(x)$ with $r^\circ(x)$ for every $x \in \dot{P}$.

Then we write $G \Rightarrow_r^m H$, but may omit m if it is irrelevant, and write $G \Rightarrow_{\mathcal{R}} H$ if $\mathcal{R}$ is a set of rules such that $G \Rightarrow_r H$ for some $r \in \mathcal{R}$.

Sometimes it is necessary to restrict the application of a rule by requiring the existence or non-existence of certain graphs in the context of its match. Our definition of application conditions is based on [10].

Definition 4 (Conditional Rule). For a graph P, the set of *conditions over* P is defined inductively as follows: (*i*) a subgraph relation $P \subseteq C$ defines a *basic condition* $\exists C$ over P. (*ii*) if c, c' are conditions over P, then $\neg c$, $(c \wedge c')$, and $(c \vee c')$ are conditions over P. [3]

An injective morphism $m\colon P \to G$ *satisfies* a condition c, written $m \vDash c$, if

- $c = \exists C$ and there is an injective morphism $m'\colon C \to G$ so that $m'|_P = m$;
- $c = \neg c'$ and $m \nvDash c'$;
- $c = (c' \wedge c'')$ and both $m \vDash c'$ and $m \vDash c''$;
- $c = (c' \vee c'')$ and $m \vDash c'$ or $m \vDash c''$.

A *conditional rule* $r' = (r, c)$ consists of a rule $r\colon P \rightarrowtail R$ and a condition c over P, and is denoted as $r'\colon c[\!] P \rightarrowtail R$. We let $G \Rightarrow_{r'}^m H$ if $m \vDash c$ and $G \Rightarrow_r^m H$. Note that each rule $P \rightarrowtail R$ without a condition can also be seen as a conditional rule $\exists P[\!] P \rightarrowtail R$. If $\mathcal{C}$ is a finite set of conditional rules, $\Rightarrow_{\mathcal{C}}$ denotes the conditional transformation relation using these rules.

Examples of graphs and rules, with and without conditions, will be shown below.

---

[2] This corresponds to a DPO rule $P \supseteq I \to R$, where the interface I is the discrete graph with nodes $\dot{P}$, and the morphism $I \to R$ is given by $(r^\circ, \varnothing)$.

[3] We omit *nested conditions* like "$\forall(C, \exists C' \wedge \neg \exists C'')$" since we do not need them.

3 Hyperedge Replacement Graph Grammars

We recall graph grammars based on hyperedge replacement [8].[4]

Definition 5 (Hyperedge Replacement Grammar). Consider a finite alphabet Σ and a subset $\mathcal{N} \subseteq \Sigma$ of *nonterminals*. Edges with labels in $\mathcal{N}$ are accordingly *nonterminal edges*; those with labels in $\Sigma \setminus \mathcal{N}$ are *terminal edges*.

A rule $p\colon P \rightarrowtail R$ is a *hyperedge replacement production* (*production*, for short) over Σ if the pattern P consists of a single edge e and its attached nodes, where $lab_P(e) \in \mathcal{N}$, and the mapping $p^\circ\colon \dot{P} \to \dot{R}$ is injective.

A *hyperedge-replacement grammar* (*HR grammar*) $\Gamma = \langle \Sigma, \mathcal{N}, \mathcal{P}, Z \rangle$ consists of Σ and $\mathcal{N} \subseteq \Sigma$ as above, a finite set $\mathcal{P}$ of productions over Σ, and a start graph $Z \in \mathcal{G}_\Sigma$.

The *language* generated by Γ is given by $\mathcal{L}(\Gamma) = \{G \in \mathcal{G}_{\Sigma \setminus \mathcal{N}} \mid Z \Rightarrow^*_{\mathcal{P}} G\}$.

Example 1 (HR Grammars for Trees). As a running example for the constructions in this paper, we use the productions in Fig. 1. They derive n-ary trees like the one in Fig. 2, if the pattern of production s is the start graph. We draw nodes as circles, and nonterminal edges as boxes that contain their labels. Edges are connected to their attached nodes by lines, called tentacles. Tentacles are ordered counter-clockwise around the edge, starting in the north.

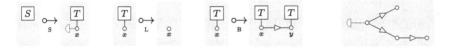

Fig. 1. HR productions for trees   **Fig. 2.** A tree

For the purpose of this paper, we restrict ourselves to this simple example because illustrations would otherwise become too complex. Further examples of well-known HR languages for which PTD parsers can be built include string graph languages such as palindromes, non-context-free ones like $a^n b^n c^n$, arithmetic expressions, and Nassi-Shneiderman diagrams.

In our running example, edges of shape ◁ with $arity(◁) = 1$ designate root nodes, whereas edges of shape ▷ with $arity(▷) = 2$ connect parent nodes to their children.

In productions (and later in other rules), nodes of the pattern P have the same identifier ascribed in P as their images in R under p°, like x in our example. In the following, the letters s, l, and b under the arrows in Fig. 1 are used as identifiers that refer to the corresponding production.

---

[4] In contrast to [8] and [4], "merging rules", with a non-injective node mapping, are prohibited here as they complicate the following formal discussion considerably.

Assumption 1. Throughout the remainder of this paper, we consider only HR grammars $\Gamma = \langle \Sigma, \mathcal{N}, \mathcal{P}, Z \rangle$ that satisfy the following conditions:

1. Z consists of a single edge e of arity 0.
2. $\mathcal{L}(\Gamma)$ does not contain graphs with isolated nodes.

These assumptions imply no loss of generality: a new initial nonterminal with a single start production according to Assumption 1 can be added easily. A grammar that violates Assumption 1 and produces isolated nodes can be transformed easily into an equivalent grammar that attaches virtual unary edges to those nodes.

4 Threaded HR Grammars

We now prepare HR grammars for parsing. The edges in graphs, productions and derivations will be ordered linearly with the idea that the parser is instructed to process the symbols of a grammar in this order when it attempts to construct a derivation for a given input graph. The edge order induces a dependency relation between nodes of a graph as follows: for an edge, an attached node is "known" if it is also attached to some preceeding edge, which will be processed earlier by the parser; it is "unknown" otherwise. This defines what we call the profile of an edge: a node is classified as incoming if it is known, and as outgoing otherwise.

Technically, edge order and profiles are represented by extending the structure and labels of a graph: Every edge is equipped with two additional tentacles by which edges are connected to a thread, and the label ℓ of an edge is equipped with a profile $\nu \subseteq \mathbb{N}$ indicating the positions of its incoming nodes. Unary hyperedges labeled with a fresh symbol distinguish thread nodes from kernel nodes of a graph.

Definition 6 (Threaded Graph). The *profiled* alphabet of an alphabet Σ is $\tilde{\Sigma} = \{ \ell^\nu \mid \ell \in \Sigma, \nu \subseteq [arity(\ell)] \} \cup \{ \bullet \}$ with $arity(\ell^\nu) = arity(\ell) + 2$ and $arity(\bullet) = 1$. The *profile* of an edge labelled by ℓ^ν is ν.

Let $G \in \mathcal{G}_{\tilde{\Sigma}}$. A node $v \in \dot{G}$ is called a *thread* node if a $\bullet$-edge is attached to it and a *kernel* node otherwise. $\underline{\dot{G}}$ and $\tilde{\dot{G}}$ denote the sets of all kernel nodes and thread nodes of G, respectively. An edge $e \in \bar{G}$ is a *profiled edge* if $lab_G(e) \neq \bullet$. The set of all profiled edges of G is denoted by $pe(G)$. The profile ν divides the set of attached kernel nodes of e into sets $in_G(e) = \{ v_i \mid i \in \nu \}$ and $out_G(e) = \{ v_i \mid i \in [arity(lab_G(e))] \setminus \nu \}$ of *incoming* and *outgoing* nodes, respectively.

A graph $G \in \mathcal{G}_{\tilde{\Sigma}}$ is *threaded* if the following hold:

1. Each node of G has at most one attached $\bullet$-edge.
2. For every $e \in pe(G)$ with $lab_G(e) = \ell^\nu$ and $att_G(e) = v_1 \ldots v_k v_{k+1} v_{k+2}$, the nodes $v_1, \ldots, v_k$ are kernel nodes of G and v_{k+1}, v_{k+2} are thread nodes of G. (Hence, $in_G(e)$ and $out_G(e)$ partition the kernel nodes of e into incoming and outgoing nodes.)

3. The profiled edges and thread nodes of G can be ordered as $pe(G) = \{e_1, \ldots, e_n\}$ and $\tilde{G} = \{v_0, \ldots, v_n\}$ so that, for $i \in [n]$,
   (a) $att_G(e_i)$ ends in $v_{i-1}v_i$ and
   (b) no edge e_j with $j \in [i-1]$ is attached to any node in $out_G(e_i)$.

We call v_0 the *first* and v_n the *last* thread node of G, and define furthermore $in(G) = \dot{G} \setminus \bigcup_{i \in [n]} out_G(e_i)$.

The *kernel graph* of G is the graph $\underline{G} \in \mathcal{G}_\Sigma$ obtained by removing the profiles of edge labels, the ●-edges, the thread nodes and their attached tentacles. $\tilde{\mathcal{G}}_{\tilde{\Sigma}}$ denotes the set of threaded graphs over $\tilde{\Sigma}$; $\langle \bullet \rangle$ denotes the *empty threaded graph* that consists of a single thread node with its attached ●-edge.

Remark 1 It is important to note that the profiles of the (profiled) edges of a threaded graph G are uniquely determined by $in(G)$ and the structure of G. To see this, let $pe(G) = \{e_1, \ldots, e_n\}$, threaded in this order. For every $v \in \dot{G}$, let

$$first(v) = \begin{cases} 0 \text{ if } v \in in(G) \\ i \text{ if } v \notin in(G) \text{ and } i = \min\{j \in [n] \mid att_G(e_j) \text{ contains } v\}. \end{cases}$$

Then $v \in in_G(e_i)$ if $v \in att_G(e_i)$ and $first(v) < i$.

Let the *concatenation* $H = G \circ G'$ of two threaded graphs G and G' with $\bar{G} \cap \bar{G}' = \dot{G} \cap \dot{G}' = \varnothing$ be the threaded graph H that is constructed from the union of G and G' by identifying the last thread node of G with the first thread node of G' (and removing one of their attached ●-edges). Note that kernel nodes of G may also occur in G'.

Definition 7 (Threaded Production and HR grammar). A rule $p: P \circ\!\!\rightarrow R$ is a *threaded production* if P and R are threaded and the following conditions are satisfied:

1. the rule $\underline{p}: \underline{P} \circ\!\!\rightarrow \underline{R}$, where $\underline{p}^\circ$ is the restriction of p° to $\underline{P}$, is a production, called *kernel production* of p,
2. p° maps the first and last thread nodes of P onto the first and last thread nodes of R, respectively, and
3. $p^\circ(in(P)) = in(R)$.

An application $G \Rightarrow_p^m H$ of a threaded production p to a threaded graph G is called *leftmost*, written $G \Rightarrow_{\mathrm{lm}\,p}^m H$, if it replaces the first nonterminal on the thread of G.

A HR grammar $\tilde{\Gamma} = \langle \tilde{\Sigma}, \tilde{\mathcal{N}}, \tilde{\mathcal{P}}, \tilde{Z} \rangle$ over a profiled alphabet $\tilde{\Sigma}$ is *threaded* if all its productions are threaded.

As in the case of context-free string grammars, the context-freeness of hyperedge replacement implies that derivations can be restricted to leftmost ones:

Fact 1. For every threaded HR grammar $\tilde{\Gamma} = \langle \tilde{\Sigma}, \tilde{\mathcal{N}}, \tilde{\mathcal{P}}, \tilde{Z} \rangle$ and every $G \in \mathcal{L}(\tilde{\Gamma})$, there is a leftmost derivation $\tilde{Z} \Rightarrow_{\mathrm{lm}\,\tilde{\mathcal{P}}}^* G$, i.e., a derivation in which all applications of productions are leftmost.

This fact will be important, as top-down parsers for HR grammars attempt to construct leftmost derivations of a graph.

It follows from Remark 1 and condition 3 of Definition 7 that the profiles of edges in the replacement graph of a threaded production are uniquely determined by the profile of the pattern. Hence, given a HR grammar $\Gamma = \langle \Sigma, \mathcal{N}, \mathcal{P}, Z \rangle$ and an order on $\bar{R}$ for each of its productions $p \colon P \circ\!\!\rightarrow R$, a unique threaded version $\tilde{\Gamma}$ of Γ is obtained as follows:

1. The threaded start graph $\tilde{Z}$ of $\tilde{\Gamma}$ is given by $\underline{\tilde{Z}} = Z$ (recall that $arity(Z) = 0$).
2. Every production $p \colon P \circ\!\!\rightarrow R$ of Γ is turned into all threaded productions $\tilde{p} \colon \tilde{P} \circ\!\!\rightarrow \tilde{R}$ where $\underline{\tilde{P}} = P$, $\underline{\tilde{R}} = R$, and the edges of $\tilde{R}$ are threaded according to the chosen order on $\bar{R}$ (which defines the profiles of edges in $\tilde{R}$ uniquely).

While the procedure above creates an exponential number of profiles and thus productions, in most cases many of them will be useless. A more efficient way of constructing $\tilde{\Gamma}$ is thus to choose the threading order and then construct the useful threaded productions inductively. The procedure would initially construct the threaded start production (in which $in(P) = \varnothing$) and then, as long as a replacement graph of one of the constructed productions contains a hitherto unseen profiled nonterminal, continue by constructing the threaded productions for this nonterminal. This leads to the following definition:

Definition 8 (Threaded Version of a HR Grammar). Let $\Gamma = \langle \Sigma, \mathcal{N}, \mathcal{P}, Z \rangle$ be a HR grammar. A *threaded version of* Γ is a threaded grammar $\tilde{\Gamma} = \langle \tilde{\Sigma}, \tilde{\mathcal{N}}, \tilde{\mathcal{P}}, \tilde{Z} \rangle$, such that

1. $\mathcal{P} = \{\underline{p} \mid p \in \tilde{\mathcal{P}}\}$ and $Z = \underline{\tilde{Z}}$,
2. all threaded productions with the same kernel production $p \colon P \circ\!\!\rightarrow R$ order the edges of R identically, and
3. $\tilde{\Gamma}$ is reduced, i.e., every production $p \in \tilde{\mathcal{P}}$ can participate in the generation of a graph in $\mathcal{L}(\tilde{\Gamma})$: there is a derivation $\tilde{Z} \overset{*}{\underset{\tilde{\mathcal{P}}}{\Rightarrow}} G \underset{p}{\Rightarrow} G' \overset{*}{\underset{\tilde{\mathcal{P}}}{\Rightarrow}} H$ such that $H \in \mathcal{L}(\tilde{\Gamma})$.

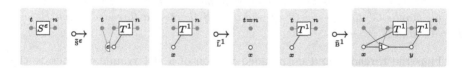

Fig. 3. Threaded tree productions

Example 2 (Threaded Tree Grammar). We consider a threaded version of the tree grammar, given by the threaded productions in Fig. 3. In examples such as this one we draw thread nodes in gray and omit the attached ●-edges, and we write profiles as ascending sequences of numbers rather than as sets. The profiles of profiled terminal edges are inscribed into the label symbols, i.e., $\rhd$ for $\rhd^1$ and

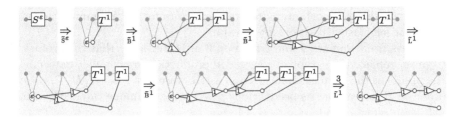

Fig. 4. A leftmost threaded derivation of the tree in Fig. 2

ⓔ for ⓒ$^{\varepsilon}$ Moreover, we distinguish threaded productions with the same kernel productions by the profile of the (unique edge in the) pattern in the production name. The profiled symbols T^{ε}, ▷$^{\varepsilon}$, ▷^{2}, ▷^{12}, and ⓒ^{1} do not appear as they occur only in useless productions.

It is worthwhile to note that production $\tilde{L}^{1}$ merges thread nodes t and n, which we indicate in the drawing by annotating the corresponding node in the replacement graph with "$t{=}n$".

We arrange thread nodes from left to right and draw thread tentacles in gray so that the kernel graph can be better identified. To make it easier to distinguish incoming from outgoing attached nodes, we draw the former to the left of an edge and the latter to the right of it.

In production $\tilde{B}^{1}$, left-recursion was avoided by choosing the terminal edge to be the first one on the thread. Figure 4 shows a threaded derivation of the tree in Fig. 2, which is leftmost.

Threaded productions derive threaded graphs to threaded graphs.

Fact 2. If $G \Rightarrow_{\tilde{p}} H$ and G is a threaded graph, H is a threaded graph as well, and $in(H) = in(G)$.

Threaded derivations and unthreaded ones correspond to each other.

Lemma 1. *Let $\Gamma = \langle \Sigma, \mathcal{N}, \mathcal{P}, Z \rangle$ be a HR grammar, $\tilde{\Gamma} = \langle \tilde{\Sigma}, \tilde{\mathcal{N}}, \tilde{\mathcal{P}}, \tilde{Z} \rangle$ a threaded version of Γ, and G a threaded graph such that $\tilde{Z} \Rightarrow_{\tilde{\mathcal{P}}}^{*} G$. Then it holds for all graphs G' that $\underline{G} \Rightarrow_{\mathcal{P}} G'$ if and only if there is a threaded graph H with $\underline{H} = G'$ and $G \Rightarrow_{\tilde{p}} H$.*

Thus the threaded and unthreaded version of a HR grammar generate the same language of kernel graphs.

Theorem 1. *If $\Gamma = \langle \Sigma, \mathcal{N}, \mathcal{P}, Z \rangle$ is a HR grammar and $\tilde{\Gamma} = \langle \tilde{\Sigma}, \tilde{\mathcal{N}}, \tilde{\mathcal{P}}, \tilde{Z} \rangle$ is a threaded version of Γ, then $\mathcal{L}(\Gamma) = \{\underline{G} \mid G \in \mathcal{L}(\tilde{\Gamma})\}$.*

Proof. Easy induction on the length of derivations, using Lemma 1.   □

5 General Top-Down Parsing for HR Grammars

We define top-down parsers for HR grammars as stack automata, which perform transitions of configurations that represent the input graph and a stack. Configurations are graphs, and transitions are described by graph transformation rules. This definition is more precise than the original definition of PTD parsing in [4], but avoids the technical complications occuring in the precise definition of PSR parsing for HR grammars [6], where graphs are represented textually as sequences of literals, and transitions are defined by the transformation of literal sequences, involving substitution and renaming operations on node identifiers. The use of graph transformation and graph morphisms avoids the explicit handling of these technical issues.

A configuration consists of a threaded graph as in Definition 6, which represents its stack and its read input, edges without profile that induce its unread input, and further edges that serve as flags, distinguishing different types of nodes.

Definition 9 (Configuration). Given a HR grammar $\Gamma = \langle \Sigma, \mathcal{N}, \mathcal{P}, Z \rangle$ and its profiled alphabet $\tilde{\Sigma}$, let ■, $\otimes$, and $\bigcirc$ be fresh symbols of arity 1. A graph G without isolated nodes is a *configuration* (of Γ) if the following hold:

- The subgraph *thread*(G) induced by its $\tilde{\Sigma}$-edges is a threaded graph.
- Exactly one thread node h of *thread*(G) is attached to a ■-edge, representing the top of the stack.
- Every kernel node of every profiled edge between the start node of the thread and h is attached to a $\bigcirc$-edge, marking it as read.
- Every node of every Σ-edge that is not attached to a profiled edge at the same time is attached to a $\otimes$-edge, marking it as unread.
- No node is attached to several edges with labels in $\{■, \otimes, \bigcirc\}$.

We let *read*(G), the *read input*, denote the subgraph of *thread*(G) induced by the profiled edges between the first thread node and h (including the ●-edges attached to those nodes). The (threaded) subgraph of *thread*(G) induced by the profiled edges between h and the last node of the thread (again including the ●-edges attached to those nodes) represents the *stack stack*(G), and the subgraph *unread*(G) induced by the Σ-edges represents the *unread input*. The union of *unread*(G) and the kernel of *read*(G) is the *input* represented by G, denoted by *input*(G).

A configuration G is

- *initial* if *read*(G) = $\langle \bullet \rangle$ and *stack*(G) = $\tilde{Z}$, and
- *accepting* if *stack*(G) = $\langle \bullet \rangle$ and *unread*(G) = $\langle \rangle$.

Definition 10 (Top-Down Parser). Let Γ be a HR grammar and $\mathcal{R}$ a set of conditional rules. A derivation $G \Rightarrow^*_{\mathcal{R}} H$ is a *parse* if G is an initial configuration. A parse $G \Rightarrow^*_{\mathcal{R}} H$ is *successful* if H is an accepting configuration. A configuration G is *promising* (with respect to $\mathcal{R}$) if there is an accepting configuration H so that $G \Rightarrow^*_{\mathcal{R}} H$. $\mathcal{R}$ is a *top-down parser* for Γ if, for each initial configuration G, $unread(G) \in \mathcal{L}(\Gamma)$ if and only if G is promising. $\mathcal{R}$ *terminates* if there is no infinite parse.

Consider in the following a threaded version $\tilde{\Gamma} = \langle \tilde{\Sigma}, \tilde{\mathcal{N}}, \tilde{\mathcal{P}}, \tilde{Z} \rangle$ of a HR grammar $\Gamma = \langle \Sigma, \mathcal{N}, \mathcal{P}, Z \rangle$. We define two types of general top-down parsing rules, called match and expand rules.

Definition 11 (Match and Expand Rules). For every terminal symbol $a^\nu \in \tilde{\Sigma} \setminus \tilde{\mathcal{N}}$, the *match rule* $t_{a^\nu} : P \circ\!\!\!\rightarrow R$ is given as follows:

- The pattern P is a configuration where
    - $read(P) = \langle \bullet \rangle$,
    - $unread(P)$ consists of one a-edge $\underline{e}$ with $a \in \Sigma \setminus \mathcal{N}$ and $att_P(\underline{e}) = \underline{v}_1 \ldots \underline{v}_k$ (where $arity(a) = k$), with a $\bigcirc$-edge attached to every $\underline{v}_i$ with $i \in \nu$ and a $\otimes$-edge attached to every $\underline{v}_i$ with $i \notin \nu$, and
    - $stack(P)$ consists of one a^ν-edge e with $att_P(e) = v_1 \ldots v_k v_{k+1} v_{k+2}$ such that $v_i = \underline{v}_i$ if $i \in \nu$. If $i \notin \nu$, then v_i is not attached to $\underline{e}$.
- The replacement R is a configuration where
    - $read(R) = stack(P)$, with a $\bigcirc$-edge attached to every v_i, for $i \in [k]$,
    - $stack(R) = \langle \bullet \rangle$,
    - $unread(R) = \langle \rangle$.
- The mapping $t^\circ_{a^\nu}$ identifies node v_i with $\underline{v}_i$ if and only if $i \notin \nu$.

For each of the threaded productions $p : \tilde{P} \circ\!\!\!\rightarrow \tilde{R}$ in $\tilde{\mathcal{P}}$, the *expand rule* $t_p : P \circ\!\!\!\rightarrow R$ is given as follows:

- $read(P) = read(R) = \langle \bullet \rangle$,
- $unread(P) = unread(R) = \langle \rangle$,
- $stack(P) = \tilde{P}$ and $stack(R) = \tilde{R}$,
- the mapping t°_p is the same as in p;

We let $\mathcal{R}^M_{\tilde{\Gamma}}$ denote the set of all match rules for terminal symbols, and $\mathcal{R}^E_{\tilde{\Gamma}}$ the set of all expand rules for productions of $\tilde{\Gamma}$. In the following, we will show that $\mathcal{R}_{\tilde{\Gamma}} = \mathcal{R}^M_{\tilde{\Gamma}} \cup \mathcal{R}^E_{\tilde{\Gamma}}$ is in fact a top-down parser for Γ, hence we call $\mathcal{R}_{\tilde{\Gamma}}$ a *general top-down parser of $\tilde{\Gamma}$ (for Γ)*.

Example 3 (General Top-Down Parser for Trees). The expand rules of the general top-down parser for trees in Fig. 5 differ from the threaded productions only in the ■-edge marking the top of the stack. (We draw ■- and $\bigcirc$-edges *around* the nodes to which they are attached, so that they look like distinguished kinds of nodes. Nodes with an attached $\otimes$-edge are drawn as $\otimes$, omitting the attached edge in the drawing.) The match rules for the two edge patterns needed are shown in Fig. 6.

Figure 7 shows snapshots of a successful parse of the tree in Fig. 2 with these rules, where five configurations are omitted for brevity. The parse constructs the leftmost derivation in Fig. 4.

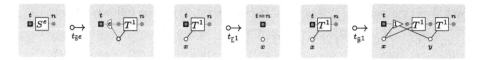

Fig. 5. Expand rules of the general top-down parser for trees

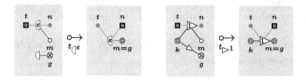

Fig. 6. Two match rules of the general top-down parser for trees

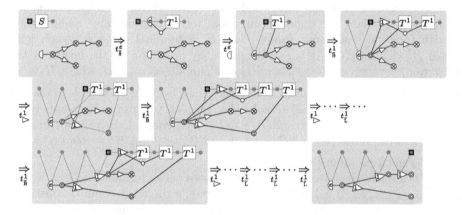

Fig. 7. A top-down parse of the tree in Fig. 2

Note that match rules do not change the thread, but just "move" the matched terminal from the unread to the read subgraph of the configuration. In contrast, expand rules do not modify the unread or read subgraphs of the configuration, but just replace the first nonterminal on the thread by the replacement graph of a threaded production for this nonterminal. We can summarize these observations in the following fact:

Fact 3. For a parse $G \Rightarrow^*_{\mathcal{R}_{\tilde{\Gamma}}} G' \Rightarrow_r H$ (where $r \in \mathcal{R}_{\tilde{\Gamma}}$), the following hold:

1. $input(G) \cong input(G') \cong input(H)$;
2. if $r = t_{a^\nu}$ is a match for some $a \in \Sigma \setminus \mathcal{N}$, then $thread(G') \cong thread(H)$;
3. if $r = t_p$ for some threaded production $p \in \tilde{\mathcal{P}}$, then $thread(G') \overset{\text{lm}}{\Rightarrow}_p thread(H)$.

Thus $\mathcal{R}_{\tilde{\Gamma}}$ constitutes a top-down parser: there is a successful parse if and only if its input graph is in the language of the grammar.

Theorem 2. *For every HR grammar Γ and each threaded version $\tilde{\Gamma}$ of Γ, $\mathcal{R}_{\tilde{\Gamma}}$ is a top-down parser for Γ.*

Proof Sketch. Let $G \Rightarrow^*_{\mathcal{R}_{\tilde{\Gamma}}} H$ be a successful parse. $\tilde{Z} = thread(G) \Rightarrow^*_{\tilde{\mathcal{P}}} thread(H)$ and $unread(G) = input(G) \cong input(H)$ hold by Fact 3; $input(H)$ is the kernel of $thread(H)$ because H is accepting, and hence $unread(G) \in \mathcal{L}(\Gamma)$ by Lemma 1.

In order to show the opposite direction, let us consider any configuration G with terminal read input $read(G)$ and H' a terminal threaded graph with kernel $\underline{H}' = unread(G)$. It is easy to prove, by induction on the length of the derivation, that $thread(G) \overset{\text{lm}}{\Rightarrow}^*_{\tilde{\mathcal{P}}} read(G) \circ H'$ implies $G \Rightarrow^*_{\mathcal{R}_{\tilde{\Gamma}}} H$ where H is an accepting configuration obtained from $read(G) \circ H'$ by adding a ■-edge to the last thread node and ○-edges to all kernel nodes, i.e., G is promising. Now let G be an initial configuration with $unread(G) \in \mathcal{L}(\Gamma)$. By Lemma 1, there is a threaded graph H' with kernel $unread(G)$ and $thread(G) = \tilde{Z} \overset{\text{lm}}{\Rightarrow}^*_{\tilde{\mathcal{P}}} H' = read(G) \circ H'$. Hence, G must be promising. □

If $\tilde{\Gamma}$ is not left-recursive, the general top-down parser terminates. Here, we say that $\tilde{\Gamma} = \langle \tilde{\Sigma}, \tilde{\mathcal{N}}, \tilde{\mathcal{P}}, \tilde{Z} \rangle$ is left-recursive if there is a threaded graph G consisting of a single nonterminal edge labeled A (for some nonterminal A) and there is a derivation $G \Rightarrow^+_{\tilde{\mathcal{P}}} H$ for some graph H such that the first profiled edge of H is also labeled with A.

Theorem 3 (Termination). *Let $\tilde{\Gamma}$ be a threaded version of a HR grammar. The general top-down parser $\mathcal{R}_{\tilde{\Gamma}}$ terminates unless $\tilde{\Gamma}$ is left-recursive.*

Proof Assume that there is an infinite parse $G \Rightarrow_{t_1} G_1 \Rightarrow_{t_2} G_2 \Rightarrow_{t_3} \cdots$ with $t_i \in \mathcal{R}_{\tilde{\Gamma}}$ for $i \in \mathbb{N}$. Since $unread(G)$ is finite and each match operation "removes" an unread edge, there must be a $k \in \mathbb{N}$ such that t_i is an expand rule for all $i > k$. As their number is finite, there must be numbers i and j, $k < i < j$, such that $stack(G_i)$ and $stack(G_j)$ start with edges labeled with the same nonterminal. By Fact 3, $thread(G_i) \overset{\text{lm}}{\Rightarrow}^+_{\tilde{\mathcal{P}}} thread(G_j)$, which proves that $\tilde{\Gamma}$ is left-recursive. □

Inconveniently, the steps of the general top-down parser are nondeterministic:

1. The *expansion* of a nonterminal A^ν may choose any of its productions.
2. The *match* of an edge a^ν may choose any unread edge fitting the profile ν.

We consider a parse $G \Rightarrow^*_{\mathcal{R}_{\tilde{\Gamma}}} H$ as a *blind alley* if the configuration H is not accepting, but does not allow further steps (using $\mathcal{R}_{\tilde{\Gamma}}$). This is the case if

- $stack(H)$ starts with an edge a^ν, but t_{a^ν} does not apply (*edge mismatch*), or
- $stack(H) = \langle \bullet \rangle$ but $unread(H) \neq \langle \rangle$ (*input too big*).

Due to nondeterminism, a successful parse may nevertheless exist in such a situation. Exploring the entire search space of parses to determine whether a successful one exists is very inefficient.

6 Predictive Top-Down Parsing for HR Grammars

The aim of predictive top-down parsing for threaded HR grammars is to avoid backtracking, the major source of inefficiency of a straightforward implementation of the general top-down parser. So we have to cope with the nondeterminism identified in the previous section. In every configuration of a parse, it must efficiently be possible to predict which choices of moves are *wrong* in the sense that they lead into a blind alley, whereas other moves could still lead to a successful parse if there is any. However, this is most likely not achievable for every threaded HR grammar $\tilde{\Gamma}$ because Theorem 2 in combination with the known NP-completeness of some HR languages would otherwise imply that P=NP. For such a grammar, certain configurations will allow more than one expansion, and it may be the case that any of them is promising, or just some of them (or none).

Thus backtrack-free parsing only seems to be possible for HR grammars that make correct moves of their top-down parsers predictable.

Let us first define *predictive expand rules* that will prevent a parser from running into blind alleys by additionally checking so-called *lookahead conditions*. Henceforth, given a rule $r\colon P \rightarrowtail R$ and a condition c over P, we denote the conditional rule $r'\colon c[\!] P \rightarrowtail R$ by $r[c]$.

Definition 12 (Predictive expand rules). Let Γ be a HR grammar, $\tilde{\Gamma}$ a threaded version of Γ, and $\mathcal{R}_{\tilde{\Gamma}} = \mathcal{R}^{\mathrm{M}}_{\tilde{\Gamma}} \cup \mathcal{R}^{\mathrm{E}}_{\tilde{\Gamma}}$ its general top-down parser. For an expand rule $t_{p^\nu}\colon P \rightarrowtail R \in \mathcal{R}^{\mathrm{E}}_{\tilde{\Gamma}}$, a condition c over P is a *lookahead condition for t_{p^ν}* if the following holds:

For every derivation $G \Rightarrow^*_{\mathcal{R}_{\tilde{\Gamma}}} H \Rightarrow^m_{t_{p^\nu}} H'$ where G is an initial configuration and H is promising,[5] if $m \vDash c$ then H' is promising.

A set $\mathcal{R} = \{t_{p^\nu}[c_{p^\nu}] \mid t_{p^\nu} \in \mathcal{R}^{\mathrm{E}}_{\tilde{\Gamma}}\}$ of conditional rules is a set of *predictive expand rules* for $\tilde{\Gamma}$ if c_{p^ν} is a lookahead condition for every $t_{p^\nu} \in \mathcal{R}^{\mathrm{E}}_{\tilde{\Gamma}}$.

In the following, we briefly describe a simple way to check whether a set of predictive expand rules can be obtained from $\mathcal{R}^{\mathrm{E}}_{\tilde{\Gamma}}$. For this purpose, let G be any initial configuration and $t_{p^\nu}\colon P \rightarrowtail R$ any expand rule so that $G \Rightarrow^*_{\mathcal{R}_{\tilde{\Gamma}}} H \Rightarrow^m_{t_{p^\nu}} H'$ where H' is promising, i.e., there is an accepting configuration F such that

---

[5] From now on, we call a configuration promising if it is in fact promising with respect to $\mathcal{R}_{\tilde{\Gamma}}$.

$$\text{either} \quad H \Rightarrow^m_{t_{p^\nu}} H' \Rightarrow^*_{\mathcal{R}^E_{\tilde{\Gamma}}} K \Rightarrow_{\mathcal{R}^M_{\tilde{\Gamma}}} K' \Rightarrow^*_{\mathcal{R}_{\tilde{\Gamma}}} F \tag{1}$$

$$\text{or} \quad H \Rightarrow^m_{t_{p^\nu}} H' \Rightarrow^*_{\mathcal{R}^E_{\tilde{\Gamma}}} F \tag{2}$$

Consider case (1) first. There is an isomorphism $iso : unread(K) \rightarrow unread(H)$ because K is obtained from H by expand rules only. Let e be the edge of $unread(K)$ that is read by the match operation $K \Rightarrow_{\mathcal{R}^M_{\tilde{\Gamma}}} K'$ and E the subgraph of K induced by e. Clearly, $m(P)$ as well as $iso(E)$ are both subgraphs of H. Now select a graph C and an injective morphism m' so that $P \subseteq C$, $m = m'|_P$, and $m'(C) = m(P) \cup iso(E)$. By definition, $m \models \exists C$. In case (2), $unread(H)$ is empty and $m \models \exists P$.

We can make use of this as follows. For an expand rule t_{p^ν}, performing the above analysis for all derivations of types (1) and (2) yields only finitely many distinct graphs C (up to isomorphism). These graphs $C_1, \ldots, C_n$ can be computed by procedures similar to the construction of FIRST and FOLLOW sets for LL(k) parsing [15, Sect. 5.5]. Defining $\hat{c}_{p^\nu} = \exists C_1 \vee \exists C_2 \vee \cdots \vee \exists C_n$ we thus obtain for all promising graphs H, H' that $H \Rightarrow^m_{t_{p^\nu}} H'$ implies $m \models \hat{c}_{p^\nu}$. Thus, by contraposition, if H is promising and $H \Rightarrow^m_{t_{p^\nu}} H'$ but $m \not\models \hat{c}_{p^\nu}$, then H' cannot be promising.

Note, however, that $m \models \hat{c}_{p^\nu}$ does not necessarily imply that H' is promising if $H \Rightarrow^m_{t_{p^\nu}} H'$ and H is promising. Therefore, $\hat{c}_{p^\nu}$ can in general not directly serve as a lookahead condition. To solve this problem, we define a relation $\sqsubset$ on expand rules. For this purpose, let us consider two different expand rules $t_{p^\nu_a}, t_{p^\nu_b} \in \mathcal{R}^E_{\tilde{\Gamma}}$ with isomorphic left-hand sides. Without loss of generality, we assume that the left-hand sides are identical. We define $t_{p^\nu_a} \sqsubset t_{p^\nu_b}$ if there is an initial configuration G and a derivation $G \Rightarrow^*_{\mathcal{R}_{\tilde{\Gamma}}} H \Rightarrow^m_{t_{p^\nu_a}} H'$ where H' is promising and $m \models \hat{c}_{p^\nu_b}$. In fact, relation $\sqsubset$ can be defined while conditions $\hat{c}_{p^\nu_i}$ are constructed.[6]

Note that $\sqsubset$ is in general not an ordering and that it may even contain cycles $t_{p^\nu_a} \sqsubset t_{p^\nu_b} \sqsubset \cdots \sqsubset t_{p^\nu_a}$. But if there are no such cycles, one can create (by topological sorting) a linear ordering $\prec$ on all expand rules with isomorphic left-hand sides (where we again assume that they have in fact identical left-hand sides) so that $t_{p^\nu_a} \sqsubset t_{p^\nu_b}$ always implies $t_{p^\nu_a} \prec t_{p^\nu_b}$. We then define, for each expand rule t_{p^ν}, the condition $c_{p^\nu} \equiv \hat{c}_{p^\nu} \wedge \neg c_1 \wedge \neg c_2 \wedge \cdots \wedge \neg c_n$ where $\{c_1, c_2, \ldots c_n\} = \{\hat{c}_{\bar{p}^\nu} \mid t_{\bar{p}^\nu} \prec t_{p^\nu}\}$. The following lemma states that these conditions can serve as lookahead conditions for predictive expand rules:

Lemma 2. *Let Γ be a HR grammar, $\tilde{\Gamma}$ a threaded version of Γ, and $\mathcal{R}_{\tilde{\Gamma}} = \mathcal{R}^M_{\tilde{\Gamma}} \cup \mathcal{R}^E_{\tilde{\Gamma}}$ its general top-down parser. If $\sqsubset$ is acyclic and the condition c_{p^ν} is defined as above for each expand rule $t_{p^\nu} \in \mathcal{R}^E_{\tilde{\Gamma}}$, then $\{t_{p^\nu}[c_{p^\nu}] \mid t_{p^\nu} \in \mathcal{R}^E_{\tilde{\Gamma}}\}$ is a set of predictive expand rules for $\tilde{\Gamma}$.*

---

[6] $\hat{c}_{p^\nu_i}$ identifies edges that must occur in H if $H \Rightarrow^m_{t_{p^\nu_b}} H''$ where H'' is promising. And if these edges may also occur in H if H' is promising, we define $t_{p^\nu_a} \sqsubset t_{p^\nu_b}$.

Proof. Consider any derivation $G \Rightarrow^*_{\mathcal{R}_{\tilde{\Gamma}}} H \Rightarrow_{\mathcal{R}_{\tilde{\Gamma}}^{\mathrm{E}}} H'$ where G is an initial configuration, and H is promising. Then there is an expand rule t_{p^ν} so that $H \Rightarrow^m_{t_{p^\nu}} K$ and K is promising. By construction, $m \vDash \hat{c}_{p^\nu}$. If there were a smaller expand rule $t_{\bar{p}^\nu} \prec t_{p^\nu}$ with $m \vDash \hat{c}_{\bar{p}^\nu}$, then this would imply $t_{p^\nu} \sqsubset t_{\bar{p}^\nu}$ because K is promising, and therefore, $t_{p^\nu} \prec t_{\bar{p}^\nu}$, contradicting the linearity of $\prec$. Therefore, $m \vDash \neg \hat{c}_{\bar{p}^\nu}$ for $t_{\bar{p}^\nu} \prec t_{p^\nu}$ and $m \vDash \hat{c}_{p^\nu}$, i.e., t_{p^ν} is the only expand rule that satisfies its lookahead condition for H, i.e., $m \vDash c_{p^\nu}$. □

The proof shows that these lookahead conditions always select a unique expand rule. Clearly, this cannot succeed for situations where expand rules can turn a promising configuration into two or more promising successor configurations.

However, the existence of a set of predictive expand rules is not sufficient for obtaining a predictive top-down parser. The threaded HR grammar must satisfy the following property as well:

Definition 13 (Free edge choice property). Let Γ be a HR grammar, $\tilde{\Gamma}$ a threaded version of Γ, and $\mathcal{R}_{\tilde{\Gamma}} = \mathcal{R}_{\tilde{\Gamma}}^{\mathrm{M}} \cup \mathcal{R}_{\tilde{\Gamma}}^{\mathrm{E}}$ its general top-down parser. $\tilde{\Gamma}$ is said to possess the *free edge choice property* if, for every derivation $G \Rightarrow^*_{\mathcal{R}_{\tilde{\Gamma}}} H \Rightarrow_{\mathcal{R}_{\tilde{\Gamma}}^{\mathrm{M}}} H'$ where G is an initial configuration and H is promising, H' is promising as well.

Theorem 4. *Let Γ be a HR grammar, $\tilde{\Gamma}$ a threaded version of Γ without left-recursion, and $\mathcal{R}_{\tilde{\Gamma}} = \mathcal{R}_{\tilde{\Gamma}}^{\mathrm{M}} \cup \mathcal{R}_{\tilde{\Gamma}}^{\mathrm{E}}$ its general top-down parser. $\mathcal{R}^{\mathrm{ptd}} = \mathcal{R}_{\tilde{\Gamma}}^{\mathrm{M}} \cup \mathcal{R}$ is a terminating top-down parser for Γ that cannot run into blind alleys if $\mathcal{R}$ is a set of predictive expand rules for $\tilde{\Gamma}$ and $\tilde{\Gamma}$ has the free edge choice property.*

Proof. Let Γ, $\tilde{\Gamma}$, and $\mathcal{R}^{\mathrm{ptd}}$ be as in the theorem. Moreover, let $\tilde{\Gamma}$ satisfy the free edge choice property, and let $\mathcal{R}$ be a set of predictive expand rules for $\tilde{\Gamma}$. Each derivation $G \Rightarrow^*_{\mathcal{R}^{\mathrm{ptd}}} H$ where G and H are initial and accepting configurations, resp., is also a successful parse in $\mathcal{R}_{\tilde{\Gamma}}$, i.e., $unread(G) \in \mathcal{L}(\Gamma)$ by Theorem 2.

Now let G be any initial configuration with $unread(G) \in \mathcal{L}(\Gamma)$, i.e., G is promising. Any infinite derivation $G \Rightarrow_{\mathcal{R}^{\mathrm{ptd}}} H_1 \Rightarrow_{\mathcal{R}^{\mathrm{ptd}}} H_2 \Rightarrow_{\mathcal{R}^{\mathrm{ptd}}} \cdots$ would also be an infinite parse $G \Rightarrow_{\mathcal{R}_{\tilde{\Gamma}}} H_1 \Rightarrow_{\mathcal{R}_{\tilde{\Gamma}}} H_2 \Rightarrow_{\mathcal{R}_{\tilde{\Gamma}}} \cdots$, contradicting Theorem 3.

Finally assume that $\mathcal{R}^{\mathrm{ptd}}$ runs into a blind alley starting at G, i.e., there is a derivation $G \Rightarrow^*_{\mathcal{R}^{\mathrm{ptd}}} H$, H is not accepting, and there is no configuration H' so that $H \Rightarrow_{\mathcal{R}^{\mathrm{ptd}}} H'$. By the free edge choice property and $\mathcal{R}$ being a set of predictive expand rules, H must be promising, i.e., there is a configuration H'' so that $H \Rightarrow_{\mathcal{R}_{\tilde{\Gamma}}^{\mathrm{M}}} H''$ or $H \Rightarrow_{\mathcal{R}_{\tilde{\Gamma}}^{\mathrm{E}}} H''$. In either case, there is a configuration H' so that $H \Rightarrow_{\mathcal{R}^{\mathrm{ptd}}} H'$, contradicting the assumption. □

This theorem justifies to call a threaded HR grammar $\tilde{\Gamma}$ *predictively top-down parsable* (PTD for short) if $\tilde{\Gamma}$ satisfies the free edge choice property and there is a set of predictive expand rules for $\tilde{\Gamma}$.

Example 4 (A Predictive Top-Down Tree Parser). The threaded tree grammar in Example 2 is PTD. To see this, let us construct lookahead conditions for expand rule $t_{\bar{B}^1}$ and $t_{\bar{L}^1}$ as described above.

Inspection of expand rule $t_{\tilde{B}^1}$ shows that choosing this rule cannot produce a promising configuration if the unread part of the input does not contain a $\triangleright$-edge starting at node x. The existence of this edge is hence requested by the graph condition $\hat{c}_{\tilde{B}^1} \equiv \exists C_{\tilde{B}^1}$, defined by the supergraph $C_{\tilde{B}^1}$ of the pattern of $t_{\tilde{B}^1}$ (see Fig. 8). No such edge can be requested for expand rule $t_{\tilde{L}^1}$; each match of $t_{\tilde{L}^1}$ satisfies $\hat{c}_{\tilde{L}^1} \equiv \exists C_{\tilde{L}^1}$ since $C_{\tilde{L}^1}$ is just the pattern of $t_{\tilde{L}^1}$. Condition $\hat{c}_{\tilde{L}^1}$ is in particular satisfied if choosing $t_{\tilde{B}^1}$ produces a promising configuration, and therefore $t_{\tilde{B}^1} \sqsubseteq t_{\tilde{L}^1}$. By Lemma 2, we can choose lookahead conditions $c_{\tilde{B}^1} \equiv \hat{c}_{\tilde{B}^1} \equiv \exists C_{\tilde{B}^1}$ and $c_{\tilde{L}^1} \equiv \hat{c}_{\tilde{L}^1} \wedge \neg c_{\tilde{B}^1} \equiv \neg \exists C_{\tilde{B}^1}$.

$$C_{\tilde{B}^1}: \qquad\qquad C_{\tilde{L}^1}:$$

Fig. 8. Graphs defining $\hat{c}_{\tilde{B}^1} \equiv \exists C_{\tilde{B}^1}$ and $\hat{c}_{\tilde{L}^1} \equiv \exists C_{\tilde{L}^1}$ for expand rule $t_{\tilde{B}^1}$ and $t_{\tilde{L}^1}$, resp.

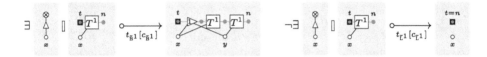

Fig. 9. Predictive expand operations of the tree parser

Figure 9 shows the resulting predictive expand rules for the nonterminal T of the tree parser. For brevity, lookahead conditions show only those subgraphs that must or must not exist in order to apply $t_{\tilde{B}^1}$ or $t_{\tilde{L}^1}$. The match rules and the expand rule $t_{\tilde{S}^\varepsilon}$ for the start production remain the same as in Example 3. Moreover, it is easy to see that match rule $t_{\triangleright^1}$ produces a promising configuration for each of its matches, i.e., the threaded tree grammar has the free edge choice property. With these modified expand rules, the predictive parser can select the same parse as in Fig. 7. As mentioned earlier, other well-known examples that allow for predictive parsing include palindromes, $a^n b^n c^n$, arithmetic expressions, and Nassi-Shneiderman diagrams.

7   Conclusions

In this paper, we have defined PTD parsers for HR grammars by graph transformation rules, and shown their correctness. The definition is consistent with the implementation of PTD parsers in the *graph parser distiller* GRAPPA[7] described

[7] Available at www.unibw.de/inf2/grappa. GRAPPA also distills PSR and generalized PSR parsers for CHR grammars [5,13].

in [4], but some features are still missing: First, productions that *merge* nodes of the left-hand side have been omitted. Such productions may occur when a HR grammar is "left-factorized" in order to allow for predictive expansion. (This corresponds to left-factorization of CF string grammars for LL-parsing.) Second, PTD parsing for *contextual* HR grammars [2,3] has not been considered. Finally, a more sophisticated way of calculating lookahead conditions, by approximating Parikh images, has been ignored.

So our next step will be to extend our definitions and proofs to cover these concepts as well. Our ultimate goal ist to use this definition to relate the power of PTD parsing to that of PSR parsing, probably by using a definition of PSR parsing that is based on graph transformation as well.

Acknowledgements. The authors thank Annegret Habel for her valuable suggestions in several stages of this work.

References

1. Aalbersberg, I., Ehrenfeucht, A., Rozenberg, G.: On the membership problem for regular DNLC grammars. Discrete Appl. Math. **13**, 79–85 (1986)
2. Drewes, F., Hoffmann, B.: Contextual hyperedge replacement. Acta Informatica **52**(6), 497–524 (2015). https://doi.org/10.1007/s00236-015-0223-4
3. Drewes, F., Hoffmann, B., Minas, M.: Contextual hyperedge replacement. In: Schürr, A., Varró, D., Varró, G. (eds.) AGTIVE 2011. LNCS, vol. 7233, pp. 182–197. Springer, Heidelberg (2012). https://doi.org/10.1007/978-3-642-34176-2_16
4. Drewes, F., Hoffmann, B., Minas, M.: Predictive top-down parsing for hyperedge replacement grammars. In: Parisi-Presicce, F., Westfechtel, B. (eds.) ICGT 2015. LNCS, vol. 9151, pp. 19–34. Springer, Cham (2015). https://doi.org/10.1007/978-3-319-21145-9_2
5. Drewes, F., Hoffmann, B., Minas, M.: Extending predictive shift-reduce parsing to contextual hyperedge replacement grammars. In: Guerra, E., Orejas, F. (eds.) ICGT 2019. LNCS, vol. 11629, pp. 55–72. Springer, Cham (2019). https://doi.org/10.1007/978-3-030-23611-3_4
6. Drewes, F., Hoffmann, B., Minas, M.: Formalization and correctness of predictive shift-reduce parsers for graph grammars based on hyperedge replacement. J. Locical Algebraic Methods Programm. (JLAMP) **104**, 303–341 (2019). https://doi.org/10.1016/j.jlamp.2018.12.006
7. Ehrig, H., Ehrig, K., Prange, U., Taentzer, G.: Implementation of typed attributed graph transformation by AGG. In: Fundamentals of Algebraic Graph Transformation. An EATCS Series, pp. 305–323. Springer, Heidelberg (2006). https://doi.org/10.1007/3-540-31188-2_15
8. Habel, A.: Hyperedge Replacement: Grammars and Languages. LNCS, vol. 643. Springer, Heidelberg (1992). https://doi.org/10.1007/BFb0013875
9. Habel, A., Müller, J., Plump, D.: Double-pushout graph transformation revisited. Math. Struct. Comput. Sci. **11**(5), 633–688 (2001)
10. Habel, A., Pennemann, K.H.: Correctness of high-level transformation systems relative to nested conditions. Math. Struct. Comput. Sci. **19**(2), 245–296 (2009)
11. Heilbrunner, S.: A parsing automata approach to LR theory. Theor. Comput. Sci. **15**, 117–157 (1981). https://doi.org/10.1016/0304-3975(81)90067-0

12. Hoffmann, B.: Modelling compiler generation by graph grammars. In: Ehrig, H., Nagl, M., Rozenberg, G. (eds.) Graph Grammars 1982. LNCS, vol. 153, pp. 159–171. Springer, Heidelberg (1983). https://doi.org/10.1007/BFb0000105

13. Hoffmann, B., Minas, M.: Generalized predictive shift-reduce parsing for hyperedge replacement graph grammars. In: Martín-Vide, C., Okhotin, A., Shapira, D. (eds.) LATA 2019. LNCS, vol. 11417, pp. 233–245. Springer, Cham (2019). https://doi.org/10.1007/978-3-030-13435-8_17

14. Lange, K.J., Welzl, E.: String grammars with disconnecting or a basic root of the difficulty in graph grammar parsing. Discrete Appl. Math. **16**, 17–30 (1987)

15. Sippu, S., Soisalon-Soininen, E.: Parsing Theroy I: Languages and Parsing. EATCS Monographs in Theoretical Computer Science, vol. 15. Springer, Heidelberg (1988). https://doi.org/10.1007/978-3-642-61345-6

Graph Consistency as a Graduated Property
Consistency-Sustaining and -Improving Graph Transformations

Jens Kosiol[1]([✉])[iD], Daniel Strüber[2][iD], Gabriele Taentzer[1][iD],
and Steffen Zschaler[3][iD]

[1] Philipps-Universität Marburg, Marburg, Germany
{kosiolje,taentzer}@mathematik.uni-marburg.de
[2] Radboud University, Nijmegen, The Netherlands
d.strueber@cs.ru.nl
[3] King's College London, London, UK
szschaler@acm.org

Abstract. Where graphs are used for modelling and specifying systems, consistency is an important concern. To be a valid model of a system, the graph structure must satisfy a number of constraints. To date, consistency has primarily been viewed as a binary property: a graph either is or is not consistent with respect to a set of graph constraints. This has enabled the definition of notions such as constraint-preserving and constraint-guaranteeing graph transformations. Many practical applications—for example model repair or evolutionary search—implicitly assume a more graduated notion of consistency, but without an explicit formalisation only limited analysis of these applications is possible. In this paper, we introduce an explicit notion of consistency as a graduated property, depending on the number of constraint violations in a graph. We present two new characterisations of transformations (and transformation rules) enabling reasoning about the gradual introduction of consistency: while consistency-sustaining transformations do not decrease the consistency level, consistency-improving transformations strictly reduce the number of constraint violations. We show how these new definitions refine the existing concepts of constraint-preserving and constraint-guaranteeing transformations. To support a static analysis based on our characterisations, we present criteria for deciding which form of consistency ensuring transformations is induced by the application of a transformation rule. We illustrate our contributions in the context of an example from search-based model engineering.

Keywords: Graph consistency · Graph transformation systems · Evolutionary search · Graph repair

1 Introduction

Graphs and graph transformations [8] are a good means for system modelling and specification. Graph structures naturally relate to the structures typically

© Springer Nature Switzerland AG 2020
F. Gadducci and T. Kehrer (Eds.): ICGT 2020, LNCS 12150, pp. 239–256, 2020.
https://doi.org/10.1007/978-3-030-51372-6_14

found in many (computer) systems and graph transformations provide intuitive tools to specify the semantics of a model or implement refinement and analysis techniques for specifications.

In all of these scenarios, it is important that the graphs used are consistent; that is, that their structures satisfy a set of constraints. Some constraints can be captured by typing graphs over so-called type graphs [8]—these allow capturing basic structural constraints such as which kinds of nodes may be connected to each other. To allow the expression of further constraints, the theory of nested graph constraints has been introduced [11]. A graph is considered consistent if it is correctly typed and satisfies all given constraints. Note that this notion of consistency is binary: a graph either is consistent or it is not consistent. It is impossible to distinguish different degrees of consistency.

In software engineering practice, it is often necessary to live with, and manage, a degree of inconsistency [23]. This requires tools and techniques for identifying, measuring, and correcting inconsistencies. In the field of graph-based specifications, this has led to many practical applications, where a more fine-grained notion of graph consistency is implicitly applied. For example, research in model repair has aimed to automatically produce graph-transformation rules that will gradually improve the consistency of a given graph. Such a rule may not make a graph completely consistent in one transformation step, but performing a sequence of such transformations will eventually produce a consistent graph (*e.g.,* [12,21,22,25]). In the area of search-based model engineering (*e.g.,* [5,9]), rules are required to be applicable to inconsistent graphs and, at least, not to produce new inconsistencies. In earlier work, we have shown how such rules can be generated at least with regard to multiplicity constraints [5]. However, in all of these works, the notion of "partial" graph consistency remains implicit. Without explicitly formalising this notion, it becomes difficult to reason about the validity of the rules generated or the correctness of the algorithm by which these rules were produced.

In this paper, we introduce a new notion of graph consistency as a graduated property. A graph can be consistent *to a degree,* depending on the number of constraint violations that occur in the graph. This conceptualisation allows us to introduce two new characterisations of graph transformations: a *consistency-sustaining* transformation does not decrease the overall consistency level, while a *consistency-improving* transformation strictly decreases the number of violations in a graph. We lift these characterisations to the level of graph transformation rules, allowing rules to be characterised as consistency sustaining and consistency improving, respectively. We show how these definitions fit with the already established terminology of constraint-preserving and constraint-guaranteeing transformations/rules. Finally, we introduce formal criteria that allow checking whether a given graph-transformation rule is consistency sustaining or consistency improving w.r.t. constraints in specific forms.

Thus, the contributions of our paper are:

1. We present the first formalisation of graph consistency as a graduated property of graphs;

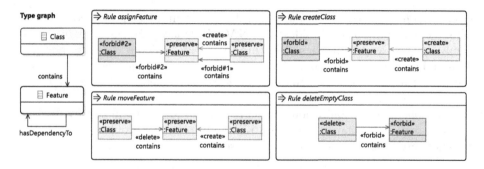

Fig. 1. Type graph and four mutation rules for the CRA problem.

2. We present two novel characterisations of graph transformations and transformation rules with regard to this new definition of graph consistency and show how these refine the existing terminology;
3. We present static analysis techniques for checking whether a graph-transformation rule is consistency sustaining or improving.

The remainder of this paper is structured as follows: We introduce a running example in Sect. 2 before outlining some foundation terminology in Sect. 3. Section 4 introduces our new concepts and Sect. 5 discusses how graph-transformation rules can be statically analysed for these properties. A discussion of related work in Sect. 6 concludes the paper. All proofs are provided in an extended version of this paper [16].

2 Example

Consider *class responsibility assignment* (CRA, [4]), a standard problem in object-oriented software analysis. Given is a set of features (methods, fields) with dependencies between them. The goal is to create a set of classes and assign the features to classes so that a certain *fitness function* is maximized. The fitness function rewards the assignment of dependent features to the same class (cohesion), while punishing dependencies that run between classes (coupling) and solutions with too few classes. Solutions can be expressed as instances of the type graph shown in the left of Fig. 1. For realistic problem instances, an exhaustive enumeration of all solutions to find the optimal one is not feasible.

Recently, a number of works have addressed the CRA problem via a combination of graph transformation and meta-heuristic search techniques, specifically evolutionary algorithms [5,10,28]. An evolutionary algorithm uses genetic operators such as cross-over and mutation to find optimal solution candidates in an efficient way. In this paper, we focus on mutation operators, which have been specified using graph transformation rules in these works.

Figure 1 depicts four mutation rules for the CRA problem, taken from the available MDEOptimiser solution [6]. The rules are specified as graph transformation rules [8] in the Henshin notation [1,29]: Rule elements are tagged as *delete, create, preserve* or *forbid*, which denotes them as being included in the LHS, the RHS, in both rule sides, or a NAC. Rule *assignFeature* assigns a randomly selected as-yet-unassigned feature to a class. Rule *createClass* creates a class and assigns an as-yet-unassigned feature to it. Rule *moveFeature* moves a feature between two classes. Rule *deleteEmptyClass* deletes a class to which no feature is assigned.

Solutions in an optimization problem such as the given one usually need to be consistent with regard to the constraints given by the problem domain. We consider three constraints for the CRA case:

(c_1) Every feature is contained in at most one class.
(c_2) Every class contains at least one feature.
(c_3) If a feature F_1 has a dependency to another feature F_2,
    and F_2 is contained in a different class than F_1,
    then F_1 must have a dependency to a feature F_3 in the same class.

Constraints c_1 and c_2 come from Fleck et al.'s formulation of the CRA problem [10]. Constraint c_3 can be considered a *helper constraint* (compare *helper objectives* [13]) that aims to enhance the efficiency of the search by formulating a constraint with a positive impact to the fitness function: Assigning dependent features to the same class is likely to improve coherence.

Given an arbitrary solution model (valid or invalid), mutations may introduce new violations. For example, applying *moveFeature* can leave behind an empty class, thus violating c_2. While constraint violations can potentially be removed using repair techniques [12,22,25], these can be computationally expensive and may involve strategies that lead to certain regions of the search space being preferred, threatening the efficiency of the search. Instead, it would be desirable to design mutation operators that impact consistency in a positive or at least neutral way. Each application of a mutation rule should contribute to some particular violations being removed, or at least ensure that the degree of consistency does not decrease. Currently, there exists no formal framework for identifying such rules. The established notions of constraint-preserving and constraint-guaranteeing rules [11] assume an already-valid model or a transformation that removes all violations at once; both are infeasible in our scenario.

3   Preliminaries

Our new contributions are based on typed graph transformation systems following the double-pushout approach [8]. We implicitly assume that all graphs, also the ones occurring in rules and constraints, are typed over a common type graph TG; that is, there is a class $Graph_{TG}$ of graphs typed over TG. A *nested graph constraint* [11] is a tree of injective graph morphisms.

Definition 1 ((Nested) graph conditions and constraints). *Given a graph P, a (nested) graph condition over P is defined recursively as follows:* **true** *is a graph condition over P and if* $a : P \hookrightarrow C$ *is an injective morphism and d is*

a graph condition over C, $\exists\,(a : P \hookrightarrow C, d)$ *is a graph condition over* P *again.*
If d_1 *and* d_2 *are graph conditions over* P, $\neg d_1$ *and* $d_1 \wedge d_2$ *are graph conditions*
over P. *A* (nested) graph constraint *is a condition over the empty graph* $\emptyset$.

A condition or constraint is called linear *if the symbol* $\wedge$ *does not occur, i.e., if*
it is a (possibly empty) chain of morphisms. The nesting level nl *of a condition* c
is recursively defined by setting $nl(\mathbf{true}) := 0$, $nl(\exists\,(a : P \hookrightarrow C, d)) := nl(d) + 1$,
$nl(\neg d) := nl(d)$, *and* $nl(d_1 \wedge d_2) := \max(nl(d_1), nl(d_2))$. *Given a graph condition* c
over P, *an injective morphism* $p : P \hookrightarrow G$ *satisfies* c, *written* $p \models c$, *if the following*
applies: Every morphism satisfies $\mathbf{true}$. *The morphism* p *satisfies a condition of*
the form $c = \exists\,(a : P \hookrightarrow C, d)$ *if there exists an injective morphism* $q : C \hookrightarrow G$
such that $p = q \circ a$ *and* q *satisfies* d. *For Boolean operators, satisfaction is defined*
as usual. A graph G *satisfies a graph constraint* c, *denoted as* $G \models c$, *if the empty*
morphism to G *does so. A graph constraint* c_1 *implies a graph constraint* c_2, *denoted*
as $c_1 \Rightarrow c_2$, *if* $G \models c_1 \Rightarrow G \models c_2$ *for all graphs* G. *The constraints are equivalent,*
denoted as $c_1 \equiv c_2$, *if* $c_1 \Rightarrow c_2$ *and* $c_2 \Rightarrow c_1$.

In the notation of graph constraints, we drop the domains of the involved morphisms and occurrences of $\mathbf{true}$ whenever they can unambiguously be inferred. For example, we write $\exists(C, \neg\exists C')$ instead of $\exists(\emptyset \hookrightarrow C, \neg\exists(a : C \hookrightarrow C', \mathbf{true}))$. Moreover, we introduce $\forall(C, d)$ as an abbreviation for the graph constraint $\neg\exists(C, \neg d)$. Further sentential connectives like $\vee$ or $\Rightarrow$ can be introduced as abbreviations as usual (which is irrelevant for linear constraints).

We define a normal form for graph conditions that requires that the occurring quantifiers alternate. For every linear condition there is an equivalent condition in this normal form [25, Fact 2].

Definition 2 (Alternating quantifier normal form (ANF)). *A linear condition* c *with* $nl(c) \geq 1$ *is in* alternating quantifier normal form *(ANF) when the occurring quantifiers alternate, i.e., if* c *is of the form* $Q(a_1, \bar{Q}(a_2, Q(a_3, \dots)\dots))$
with $Q \in \{\exists, \forall\}$ *and* $\bar{\exists} = \forall, \bar{\forall} = \exists$, *none of the occurring morphisms* a_i *is an isomorphism, and the only negation, if any, occurs at the innermost nesting level (i.e., the constraint is allowed to end with* $\mathbf{false}$). *If a constraint in ANF starts with* $\exists$, *it is called* existential, *otherwise it is called* universal.

Lemma 1 (Non-equivalence of constraints in ANF). *Let* $c_1 = \exists(C_1, d_1)$
and $c_2 = \forall(C_2, d_2)$ *be constraints in ANF. Then* $c_1 \not\equiv c_2$.

We have $c_1 \not\equiv c_2$ since $\emptyset \models c_2$ but $\emptyset \not\models c_1$. Lemma 1 implies that the first quantifier occurring in the ANF of a constraint separates linear constraints into two disjoint classes. This ensures that our definitions in Sect. 4 are meaningful.

Graph transformation is the rule-based modification of graphs. The following definition recalls graph transformation as a double-pushout.

Definition 3 (Rule and transformation). *A plain rule* r *is defined by* $p = (L \hookleftarrow K \hookrightarrow R)$ *with* $L, K,$ *and* R *being graphs connected by two graph inclusions. An* application condition *ac for* p *is a condition over* L. *A rule* $r = (p, ac)$ *consists of a plain rule* p *and an application condition* ac *over* L.

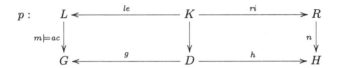

Fig. 2. Rule application

A transformation (step) $G \Rightarrow_{r,m} H$ *which applies rule* r *to a graph* G *consists of two pushouts as depicted in Fig. 2. Rule* r *is* applicable *at the injective morphism* $m : L \to G$ *called* match *if* $m \models ac$ *and there exists a graph* D *such that the left square is a pushout. Morphism* n *is called* co-match. *Morphisms* g *and* h *are called* transformation morphisms. *The* track morphism *[24] of a transformation step* $G \Rightarrow_{r,m} H$ *is the partial morphism* $tr : G \dashrightarrow H$ *defined by* $tr(x) = h(g^{-1}(x))$ *for* $x \in g(D)$ *and undefined otherwise.*

Obviously, transformations interact with the validity of graph constraints. Two well-studied notions are constraint-guaranteeing and -preserving transformations [11].

Definition 4 (c-guaranteeing and -preserving transformation). *Given a constraint* c, *a transformation* $G \Rightarrow_{r,m} H$ *is* c-guaranteeing *if* $H \models c$. *Such a transformation is* c-preserving *if* $G \models c \Rightarrow H \models c$. *A rule* r *is* c-guaranteeing *(c-preserving) if every transformation via* r *is.*

As we will present criteria for consistency sustainment and improvement based on conflicts and dependencies of rules, we recall these notions here as well. Intuitively, a transformation step *causes a conflict* on another one if it hinders this second one. A transformation step is *dependent* on another one if it is first enabled by that.

Definition 5 (Conflict). *Let a pair of transformations* $(t_1, t_2) : (G \Rightarrow_{m_1,r_1} H_1, G \Rightarrow_{m_2,r_2} H_2)$ *applying rules* $r_i = (L_i \hookleftarrow K_i \hookrightarrow R_i, ac_i)$, $i = 1, 2$ *be given such that* t_i *yields transformation morphisms* $G \xleftarrow{g_i} D_i \xrightarrow{h_i} H_i$. *Transformation pair* (t_1, t_2) *is* conflicting *(or* t_1 *causes a conflict on* t_2*) if there does not exist a morphism* $x : L_2 \to D_1$ *such that* $g_1 \circ x = m_2$ *and* $h_1 \circ x \models ac$. *Rule pair* (r_1, r_2) *is* conflicting *if there exists a conflicting transformation pair* $(G \Rightarrow_{m_1,r_1} H_1, G \Rightarrow_{m_2,r_2} H_2)$. *If* (r_1, r_2) *and* (r_2, r_1) *are both not conflicting, rule pair* (r_1, r_2) *is called* parallel independent.

Definition 6 (Dependency). *Let a sequence* $t_1; t_2 : G \Rightarrow_{m_1,r_1} H_1 \Rightarrow_{m_2,r_2} X$ *of transformations applying rules* $r_i = (L_i \hookleftarrow K_i \hookrightarrow R_i, ac_i)$, $i = 1, 2$ *be given such that* t_1 *yields transformation morphisms* $G \xleftarrow{g_1} D_1 \xrightarrow{h_1} H_1$. *Transformation* t_2 *is* dependent *on* t_1 *if there does not exist a morphism* $x : L_2 \to D_1$ *such that* $h_1 \circ x = m_2$ *and* $g_1 \circ x \models ac_2$. *Rule* r_2 *is* dependent *on rule* r_1 *if there exists a transformation sequence* $t_1; t_2 : G \Rightarrow_{m_1,r_1} H_1 \Rightarrow_{m_2,r_2} X$ *such that* t_2 *is dependent on* t_1. *If* r_1 *is not dependent on* r_2 *and* r_2 *is not dependent on* r_1, *rule pair* (r_1, r_2) *is called* sequentially independent.

A weak critical sequence *is a sequence* $t_1; t_2 : G \Rightarrow_{m_1, r_1} H_1 \Rightarrow_{m_2, r_2} X$ *of transformations such that* t_2 *depends on* t_1, n_1 *and* m_2 *are jointly surjective (where* n_1 *is the co-match of* t_1*), and* m_i *is* not *required to satisfy* ac_i *(*$i = 1, 2$*).*

As rule r_2 in a rule pair (r_1, r_2) will always be plain in this paper, a transformation step can cause a conflict on another one if and only if it deletes an element that the second transformation step matches. Similarly, a transformation step can depend on another one if and only if the first step creates an element that the second matches or deletes an edge that is adjacent to a node the second one deletes.

4 Consistency-Sustaining and Consistency-improving Rules and Transformations

In this section, we introduce our key new concepts. We do so in three stages, first introducing foundational definitions for partial consistency, followed by a generic definition of consistency sustainment and improvement. Finally, we give stronger definitions for which we will be able to provide a static analysis in Sect. 5.

4.1 Partial Consistency

To support the discussion and analysis of rules and transformations that improve graph consistency, but do not produce a fully consistent graph in one step, we introduce the notion of *partial consistency*. We base this notion on relating the number of *constraint violations* to the total number of *relevant occurrences* of a constraint. For the satisfaction of an existential constraint, a single valid occurrence is enough. In contrast, universal constraints require the satisfaction of some sub-constraint for every occurrence. Hence, the resulting notion is binary in the existential case, but graduated in the universal one.

> In the remainder of this paper, a *constraint* is always a linear constraint in ANF having a nesting level ≥ 1.[1] Moreover, all graphs are finite.

Definition 7 (Occurrences and violations). *Let* $c = Q(\emptyset \rightarrow C, d)$ *with* $Q \in \{\exists, \forall\}$ *be a constraint. An* occurrence *of* c *in a graph* G *is an injective morphism* $p : C \hookrightarrow G$, *and* $occ(G, c)$ *denotes the* number *of such occurrences.*

If c *is universal, its* number of *relevant occurrences in a graph* G, *denoted as* $ro(G, c)$, *is defined as* $ro(G, c) := occ(G, c)$ *and its* number of constraint violations, *denoted as* $ncv(G, c)$, *is the number of occurrences* p *for which* $p \not\models d$.

If c *is existential,* $ro(G, c) := 1$ *and* $ncv(G, c) := 0$ *if there exists an occurrence* $p : C \hookrightarrow G$ *such that* $p \models d$ *but* $ncv(G, c) := 1$ *otherwise.*

---

[1] Requiring nesting level ≥ 1 is no real restriction as constraints with nesting level 0 are Boolean combinations of `true` which means they are equivalent to `true` or `false`, anyhow. In contrast, restricting to linear constraints actually excludes some interesting cases. We believe that the extension of our definitions and results to also include the non-linear case will be doable. Restricting to the linear case first, however, makes the statements much more accessible and succinct.

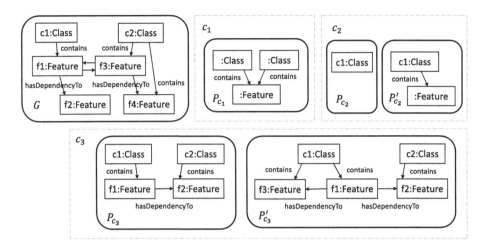

Fig. 3. Example constraints and graph.

Definition 8 (Partial consistency). *Given a graph G and a constraint c, G is consistent w.r.t. c if $G \models c$. The consistency index of G w.r.t. c is defined as*

$$ci(G, c) := 1 - \frac{ncv(G, c)}{ro(G, c)}$$

where we set $\frac{0}{0} := 0$. We say that G is partially consistent w.r.t. c if $ci(G, c) > 0$.

The next proposition makes precise that the consistency index runs between 0 and 1 and indicates the degree of consistency a graph G has w.r.t. a constraint c.

Fact 1 (Consistency index). *Given a graph G and a constraint c, then $0 \leq ci(G, c) \leq 1$ and $G \models c$ if and only if $ci(G, c) = 1$. Consistency implies partial consistency. Moreover, $ci(G, c) \in \{0, 1\}$ for an existential constraint.*

Example 1. Based on Fig. 3, we can express the three informal constraints from Sect. 2 as nested graph constraints. Constraint c_1 can be expressed as $\neg \exists P_{c_1}$, constraint c_2 becomes $\forall(P_{c_2}, \exists P'_{c_2})$, and constraint c_3 becomes $\forall(P_{c_3}, \exists P'_{c_3})$. Graph G (in the left top corner of Fig. 3) satisfies c_1 and c_2. It does not satisfy c_3, since we cannot find an occurrence of P'_{c_3} for the occurrence of P_{c_3} in G where $f1$ and $f2$ are mapped to $f1$ and $f3$, respectively. Graph G in Fig. 3 has the consistency index 0.5 with regard to c_3, since one violation exists, and two non-violating occurrences are required.

4.2   Consistency Sustainment and Improvement

In the remainder of this section, our goal is to introduce the notions of *consistency-sustaining* and *consistency-improving rule applications* which refine the established notions of preserving and guaranteeing applications [11].

Definition 9 (Consistency sustainment and improvement). *Given a graph constraint c and a rule r, a transformation* $t : G \Rightarrow_{r,m} H$ *is* consistency sustaining *w.r.t.* c *if* $ci(G, c) \leq ci(H, c)$. *It is* consistency improving *if it is consistency sustaining,* $ncv(G, c) > 0$, *and* $ncv(G, c) > ncv(H, c)$.

The rule r is consistency sustaining *if all of its applications are. It is* consistency improving *if all of its applications are consistency sustaining and there exists a graph* $G \in Graph_{TG}$ *with* $ncv(G, c) > 0$ *and a consistency-improving transformation* $G \Rightarrow_{r,m} H$. *A consistency improving rule is* strongly consistency improving *if all of its applications to graphs* G *with* $ncv(G, c) > 0$ *are consistency-improving transformations.*

In the above definition, we use the number of constraint violations (and not the consistency index) to define improvement to avoid an undesirable side-effect: Defining improvement via a growing consistency index would lead to consistency-improving transformations (w.r.t. a universal constraint) which do not repair existing violations but only create new valid occurrences of the constraint. Hence, there would exist infinitely long transformation sequences where every step increases the consistency index but validity is never restored. Consistency-improving transformations, and therefore *strongly* consistency improving rules, require that the number of constraint violations strictly decreases in each step. Therefore, using only such transformations and rules, we cannot construct infinite transformation sequences.

Any consistency-improving rule can be turned into a strongly consistency-improving rule if suitable pre-conditions can be added that restrict the applicability of the rule only to those cases where it can actually improve a constraint violation. This links the two forms of consistency-improving rules to their practical applications: in model repair [21,25] we want to use rules that will only make a change to a graph when there is a violation to be repaired—strongly consistency-improving rules. However, in evolutionary search [5], we want to allow rules to be able to make changes even when there is no need for repair, but to fix violations when they occur; consistency-improving rules are well-suited here as they can be applied even when no constraint violations need fixing.

4.3   Direct Consistency Sustainment and Improvement

While the above definitions are easy to state and understand, it turns out that they are inherently difficult to investigate. Comparing numbers of (relevant) occurrences and violations allows for very disparate behavior of consistency-sustaining (-improving) transformations: For example, a transformation is allowed to destroy as many valid occurrences as it repairs violations and is still considered to be consistency sustaining w.r.t. a universal constraint.

Next, we introduce further qualified notions of consistency sustainment and improvement. The idea behind this refinement is to *retain* the validity of occurrences of a universal constraint: valid occurrences that are preserved by a transformation are to remain valid. In this way, sustainment and improvement become more *direct* as it is no longer possible to compensate for introduced violations by introducing additional valid occurrences. The notions of (direct) sustainment and improvement are related to one another and also to the already known ones that preserve and guarantee constraints. In Sect. 5 we will show how these stricter definitions allow for static analysis techniques to identify consistency-sustaining and -improving rules.

The following definitions assume a transformation step to be given and relate occurrences of constraints in its start and result graph as depicted in Fig. 4. The existence of a morphism p_D such that the left triangle commutes (and p' might be defined as $h \circ p_D$) is equivalent to the tracking morphism $tr : G \dashrightarrow H$ being a total morphism when restricted to $p(C)$ which is equivalent to the transformation not destroying the occurrence p.

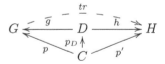

Fig. 4. Rule application with morphisms from a graph C, occurring in some constraint

Definition 10 (Direct consistency sustainment). *Given a graph constraint c, a transformation $t : G \Rightarrow_{m,r} H$ via rule r at match m with trace tr (Fig. 4) is directly consistency sustaining w.r.t. c if either c is existential and the transformation is c-preserving or $c = \forall(C, d)$ is universal and*

$$\forall p : C \hookrightarrow G\big((p \models d \wedge tr \circ p \text{ is total}) \Rightarrow tr \circ p \models d\big) \wedge$$
$$\forall p' : C \hookrightarrow H\big(\neg \exists p : C \hookrightarrow G\,(p' = tr \circ p) \Rightarrow p' \models d\big) \ .$$

A rule r is directly consistency sustaining w.r.t. c if all its applications are.

The first requirement in the definition checks that constraints that were already valid in G are still valid in H, unless their occurrence has been removed; that is, the transformation must not make existing valid occurrences invalid. Note, however, that we do not require that the constraint be satisfied by the same extension, just that there is still a way to satisfy the constraint at that occurrence. The second requirement in the definition checks that every "new" occurrence of the constraint in H satisfies the constraint; that is, the transformation must not introduce fresh violations.

Table 1. Properties of example rules.

Rule	Consistency sustaining			Consistency improving		
	c_1	c_2	c_3	c_1	c_2	c_3
assignFeature	+	+	-	-	+	-
createClass	+	+	-	-	-	-
moveFeature	(+)	-	-	-	-	-
deleteEmptyClass	+	+	+	-	+*	-

Legend: + denotes *directly*, (+) denotes *non-directly*, * denotes *strongly*

The following theorem relates the new notions of (direct) consistency sustainment to preservation and guarantee of constraints.

Theorem 2 (Sustainment relations). *Given a graph constraint c, every c-guaranteeing transformation is directly consistency-sustaining, every directly consistency-sustaining transformation is consistency sustaining, and every*

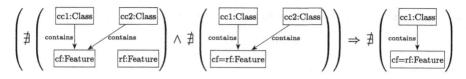

Fig. 5. Generated preserving application condition for *createClass* w.r.t. constraint c_1. The feature named rf is the one from the LHS of *createClass*.

consistency-sustaining transformation is c-preserving. The analogous implications hold on the rule level:

$$\text{constraint-preserving rule} \xleftarrow{\quad\quad [11] \quad\quad} \text{constraint-guaranteeing rule}$$

$$\Big\Uparrow Thm.\ 2 \qquad\qquad\qquad\qquad\qquad\qquad \Big\Downarrow Thm.\ 2$$

$$\text{consistency-sustaining rule} \xleftarrow[Thm.\ 2]{\quad\quad\quad\quad} \text{directly consistency-sustaining rule}$$

The following example illustrates these notions and shows that sustainment is different from constraint guaranteeing or preserving.

Example 2. Table 1 denotes for each rule from the running example if it is consistency sustaining w.r.t. each constraint. Rule *createClass* is directly consistency sustaining w.r.t. c_1 (no double assignments) and c_2 (no empty classes), since it cannot assign an already assigned feature or remove existing assignments. However, it is not consistency guaranteeing, since it cannot remove any violation either. Rule *moveFeature* is consistency sustaining w.r.t c_1, but not directly so, since it can introduce new violations, but only while at the same time removing another violation, leading to a neutral outcome. Starting with the plain version of rule *createClass* and computing a preserving application condition for constraint c_1 according to the construction provided by Habel and Pennemann [11] results in the application condition depicted in Fig. 5. By construction, equipping the plain version of *createClass* with that application condition results in a consistency-preserving rule. However, whenever applied to an invalid graph, the antecedent of this application condition evaluates to `false` and, hence, the whole application condition to `true`. In particular, the rule with this application condition might introduce further violations of c_1 and is, thus, not sustaining.

Similarly, the *direct* notion of consistency improvement preserves the validity of already valid occurrences in the case of universal constraints and degenerates to the known concept of constraint-guarantee in the existential case.

Definition 11 (Direct consistency improvement). *Given a graph constraint c, a transformation $t : G \Rightarrow_{m,r} H$ via rule r at match $m : L \hookrightarrow G$ with trace tr (Fig. 4) is* directly consistency improving *w.r.t. c if $G \nvDash c$, the transformation is directly consistency sustaining, and either c is existential and* *the transformation is c-guaranteeing or $c = \forall(C, d)$ is universal and*

$$\exists p : C \hookrightarrow G\big(p \not\models d \wedge p' := tr \circ p \text{ is total } \wedge p' \models d\big) \vee$$
$$\exists p : C \hookrightarrow G\big(p \not\models d \wedge p' := tr \circ p \text{ is not total}\big)$$

We lift the notion of directly consistency-improving transformations to the level of rules in the same way as in Definition 9. This leads to directly consistency-improving rules and a strong form of directly consistency-improving rules.

(Direct) consistency improvement is related to, but different from constraint guarantee and consistency sustainment as made explicit in the next theorem.

Theorem 3 (Improvement relations). *Given a graph constraint c, every directly consistency-improving transformation is a consistency-improving transformation and every consistency-improving transformation is consistency sustaining. Moreover, every c-guaranteeing transformation starting from a graph G that is inconsistent w.r.t. c is a directly consistency-improving transformation. The analogous implications hold on the rule level, provided that there exists a match for the respective rule r in a graph G with $G \not\models c$:*

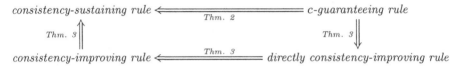

Example 3. Table 1 denotes for each rule of the running example if it is consistency improving w.r.t. each constraint. For example, the rule *deleteEmptyClass* is directly strongly consistency improving but not guaranteeing w.r.t. c_2 (no empty classes), since it always removes a violation (empty class), but generally not all violations in one step. Rule *assignFeature* is consistency improving w.r.t. c_2, but not directly so, since it can turn empty classes into non-empty ones, but does not do so in every possible application. Rule *createClass* is consistency sustaining but not improving w.r.t. c_2, as it cannot reduce the number of empty classes.

5   Static Analysis for Direct Consistency Sustainment and Improvement

In this section, we consider specific kinds of constraints and present a static analysis technique for direct consistency sustainment and improvement. We present criteria for rules to be *directly* consistency sustaining or *directly* consistency improving w.r.t. these kinds of constraint. The restriction to specific kinds of constraint greatly simplifies the presentation; at the end of the section we hint at how our results may generalize to arbitrary universal constraints.

The general idea behind our static analysis technique is to check for validity of a constraint by applying a trivial (non-modifying) rule that just checks for the existence of a graph occurring in the constraint. This allows us to present our analysis technique in the language of *conflicts and dependencies* which has been developed to characterise the possible interactions between rule applications [8,24]. As a bonus, since the efficient detection of such conflicts and dependencies has been the focus of recent theoretical and practical research [17,18], we obtain tool support for an automated analysis based on Henshin.

> In the remainder of this paper, we assume the following setting: Let $r = (L \hookleftarrow K \hookrightarrow R, ac)$ be a rule, c a graph constraint of the form $\neg \exists C = \forall (\emptyset \hookrightarrow C, \mathtt{false})$ and d a graph constraint of the form $\forall (C, \exists C') = \forall (\emptyset \hookrightarrow C, \exists a : C \hookrightarrow C')$. Given a graph G, there is the rule $check_G := G \xleftarrow{id_G} G \xrightarrow{id_G} G$ given.

For the statement of the following results, note that sequential independence of the (non-modifying) rule $check_C$ from r means that r cannot create a new match for C. Similarly, parallel independence of $check_{C'}$ from r means that r cannot destroy a match for C'. We first state criteria for direct consistency sustainment: If a rule cannot create a new occurrence of C, it is directly consistency sustaining w.r.t. a constraint of the form $\neg \exists C$. If, in addition, it cannot delete an occurrence of C', it is directly consistency sustaining w.r.t. a constraint of the form $\forall (C, \exists C')$.

Theorem 4 (Criteria for direct consistency sustainment). *Rule r is directly consistency sustaining w.r.t. constraint c if and only if $check_C$ is sequentially independent from r. If, in addition, $check_{C'}$ is parallel independent from r, then r is directly consistency sustaining w.r.t. constraint d.*

The above criterion is sufficient but not necessary for constraints of the form $\forall (C, \exists C')$. For example, it does not take into account the possibility of r creating a new valid occurrence of C. The next proposition strengthens the above theorem by partially remedying this.

Proposition 1. *If $check_{C'}$ is parallel independent from r and for every weak critical sequence $G \Rightarrow_{r,m} H \Rightarrow_{check_C, p''} H$ it holds that there is an injective morphism $q'' : C' \hookrightarrow H$ with $q'' \circ a = p''$, i.e., $p'' \models \exists C'$, then r is directly consistency sustaining w.r.t. constraint d.*

For consistency improvement we state criteria on rules as well: If a rule is directly consistency improving w.r.t. a constraint of the form $\forall (C, \exists C')$, it is either (1) able to destroy an occurrence of C (deleting a part of it) or (2) to bring about an occurrence of C' (creating a part of it). In case (2), we can even be more precise: The newly created elements do not stem from C but from the part of C' without C; this is what the formula in the next theorem expresses. For constraints of the form $\neg \exists C$, condition (1) is the only one that holds.

Table 2. Generalisation of the criteria from Theorems 4 and 5 to universal constraints up to nesting level 2. Here, ck_C is short for $check_C$, $r_1 <_D r_2$ denotes dependency of r_2 on r_1, $r_1 <_C r_2$ denotes r_2 causing a conflict for r_1, and crossed out versions denote the respective absence.

Type of constr.	Crit. for direct consist. sust.	Crit. for direct consist. impr.
$\forall(C, \texttt{false}) \equiv \neg\exists C$	$ck_C \not<_D r$	$ck_C <_C r$
$\forall(C_1, \exists C_2)$	$ck_{C_1} \not<_D r \wedge ck_{C_2} \not<_C r$	$ck_{C_1} <_C r \vee ck_{C_2} <_D r$
$\forall(C_1, \exists(C_2, \neg\exists C_3))$	$ck_{C_1} \not<_D r \wedge ck_{C_2} \not<_C r \wedge ck_{C_3} \not<_D r$	$ck_{C_1} <_C r \vee ck_{C_2} <_D r \vee ck_{C_3} <_C r$

Table 3. Applying the criteria from Table 2 to the example; ck_C is short for $check_C$.

Rule	Consis. sust. (suff. cr.)					Consis. impr. (necc. cr.)				
	seq. indep.			par. indep.		par. dep.			seq. dep.	
	$ck_{P_{c_1}}$	$ck_{P_{c_2}}$	$ck_{P_{c_3}}$	$ck_{P'_{c_2}}$	$ck_{P'_{c_3}}$	$ck_{P_{c_1}}$	$ck_{P_{c_2}}$	$ck_{P_{c_3}}$	$ck_{P'_{c_2}}$	$ck_{P'_{c_3}}$
assignFeature	-	+	-	+	+	-	-	-	+	+
createClass	-	-	-	+	+	-	-	-	+	+
moveFeature	-	+	-	-	-	+	-	+	+	+
deleteEmptyClass	+	+	+	+	+	-	+	-	-	-

Theorem 5 (Criteria for direct consistency improvement). *If rule r is directly consistency sustaining w.r.t. constraint c, then it is directly consistency improving w.r.t. c if and only if r causes a conflict for $check_C$. If r is directly consistency improving w.r.t. constraint d, then r causes a conflict for $check_C$ or $check_{C'}$ is sequentially dependent on r in such a way that*

$$n(R \setminus K) \cap p'(C') \subseteq p'(C' \setminus a(C))$$

where, in this dependency, n is the co-match of the first transformation applying r and p' is the match for $check_{C'}$.

The above criterion is not sufficient in case of constraint d. The existing conflicts or dependencies do not ensure that actually an *invalid* occurrence of C can be deleted or a new occurrence of C' can be created in such a way that an invalid occurrence of C is "repaired".

Looking closer to the criteria stated above, we can find some recurring patterns. Table 2 lists the kinds of universal constraints up to nesting level 2 and the corresponding criteria. While we have shown the criteria in the first two rows in Theorems 4 and 5, we conjecture the criteria in the last row of Table 2. To prove generalized theorems for nesting levels ≥ 2, however, is up to future work.

Example 4. We can use the criteria in Table 2 to semi-automatically reason about consistency sustainment and improvement in our example. To this end, we first apply automated conflict and dependency analysis (CDA, [18]) to the relevant pairs of mutation and check rules. Using the detected conflicts and dependencies, we infer parallel and sequential (in)dependence per definition, as shown in Table 3. For example, since no dependencies between *assignFeature* and *check*$_{P_{c_1}}$ exist, we conclude that these rules are sequentially independent.

Consistency Sustainment: Based on Table 3, we find that the sufficient criterion formulated in Theorem 4 is adequate to show direct consistency sustainment in four out of seven positive cases as per Table 1: rule *assignFeature* with constraint c_3 and rule *deleteEmptyClass* with constraints c_1, c_2 and c_3. Moreover, the stronger criterion in Proposition 1 allows to recognize the case of *createClass* with c_2. Discerning the remaining two positive cases (*assignFeature* with c_1; *createClass* with c_1) from the five negative ones requires further inspection.

Consistency Improvement: Based on Table 3, our necessary criterion allows to detect the two positive cases in Table 1: rules *deleteEmptyClass* and *assignFeature* with constraint c_2. The former is due to parallel dependence, the latter due to sequential dependence (where inspection of the CDA results reveals a critical sequence with a suitable co-match). The criterion is also fulfilled in six negative cases: *assignFeature* with c_3, *createClass* with c_2 and c_3, and *moveFeature* with c_1, c_2 and c_3. Four negative cases are correctly ruled out by the criterion.

6   Related Work

In this paper, we introduce a graduated version of a specific logic on graphs, namely of nested graph constraints. Moreover, we focus on the interaction of this graduation with graph transformations. Therefore, we leave a comparison with fuzzy or multi-valued logics (on graphs) to future work. Instead, we focus on works that also investigate the interaction between the validity of nested graph constraints and the application of transformation rules.

Given a graph transformation (sequence) $G \Rightarrow H$, the validity of graph H can be established with basically three strategies: (1) graph G is already valid and this validity is preserved, (2) graph G is not valid and there is a c-guaranteeing rule applied, and (3) graph G is made valid by a graph transformation (sequence) step-by-step.

Strategies (1) and (2) are supported by the incorporation of constraints in application conditions of rules as presented in [11] for nested graph constraints in general and implemented in Henshin [19]. As the applicability of rules enhanced in that way can be severely restricted, improved constructions have been considered of specific forms of constraints. For constraints of the form $\forall(C, \exists C')$, for example, a suitable rule scheme is constructed in [15]. In [2] refactoring rules are checked for the preservation of constraints of nesting level ≤ 2. In [19], two of the present authors also suggested certain simplifications of application conditions; the resulting ones are still constraint-preserving. In [20], we even showed that they result in the logically weakest application condition that is still directly consistency sustaining. However, the result is only shown for negative constraints of nesting level one. A very similar construction of negative application conditions from such negative constraints has very recently been suggested in [3].

Strategy (3) is followed in most of the rule-based graph repair or model repair approaches. In [22], the violation of mainly multiplicity constraints is considered.

In [12], Habel and Sandmann derive graph programs from graph constraints of nesting level ≤ 2. In [25], they extend their results to constraints in ANF which end with $\exists C$ or constraints of one of the forms $\exists(C, \neg\exists C')$ or $\neg\exists C$. They also investigate whether a given set of rules allows to repair such a given constraint. In [7] Dyck and Giese present an approach to automatically check whether a transformation sequence yields a graph that is valid w.r.t. specific constraints of nesting level ≤ 2.

Up to now, result graphs of transformations have been considered either valid or invalid w.r.t. to a graph constraint; intermediate consistency grades have not been made explicit. Thereby, c-preserving and c-guaranteeing transformations [11] focus on the full validity of the result graphs. Our newly developed notions of consistency-sustainment and improvement are located properly in between existing kinds of transformations (as proven in Theorems 2 and 3). These new forms of transformations make the gradual improvements in consistency explicit. While a detailed and systematic investigation (applying the static methods developed in this paper) is future work, a first check of the kinds of rules generated and used in [14] (model editing), [22] (model repair), and [5] (search-based model engineering) reveals that—in each case—at least some of them are indeed (directly) consistency-sustaining. We are therefore confident that the current paper formalizes properties of rules that are practically relevant in diverse application contexts. Work on partial graphs as in, e.g. [26], investigates the validity of constraints in families of graphs which is not our focus here and therefore, not further considered.

Stevens in [27] discusses similar challenges in the specific context of bidirectional transformations. Here, consistency is a property of a pair of models (loosely, graphs) rather than between a graph and constraint. In this sense, it may be argued that our formalisation generalises that of [27]. Several concepts are introduced that initially seem to make sense only in the specific context of bidirectional transformations (*e.g.,* the idea of $\overrightarrow{R}$ candidates), but may provide inspiration for a further extension of our framework with corresponding concepts.

7   Conclusions

In this paper, we have introduced a definition of graph consistency as a graduated property, which allows for graphs to be partially consistent w.r.t. a nested graph constraint, inducing a partial ordering between graphs based on the number of constraint violations they contain. Two new forms of transformation can be identified as consistency sustaining and consistency improving, respectively. They are properly located in between the existing notions of constraint-preserving and constraint-guaranteeing transformations. Lifting them to rules, we have presented criteria for determining whether a rule is consistency sustaining or improving w.r.t. a graph constraint. We have demonstrated how these criteria can be applied in the context of a case study from search-based model engineering.

While the propositions we present allow us to check a given rule against a graph constraint, their lifting to a set of constraints is the next step to go.

Furthermore, algorithms for constructing consistency-sustaining or -improving rules from a set of constraints are left for future work.

Acknowledgements. We thank the ICGT reviewers for their insightful and helpful comments. This work has been partially supported by DFG grants TA 294/17-1 and 413074939.

References

1. Arendt, T., Biermann, E., Jurack, S., Krause, C., Taentzer, G.: Henshin: advanced concepts and tools for in-place EMF model transformations. In: Petriu, D.C., Rouquette, N., Haugen, Ø. (eds.) MODELS 2010. LNCS, vol. 6394, pp. 121–135. Springer, Heidelberg (2010). https://doi.org/10.1007/978-3-642-16145-2_9
2. Becker, B., Lambers, L., Dyck, J., Birth, S., Giese, H.: Iterative development of consistency-preserving rule-based refactorings. In: Cabot, J., Visser, E. (eds.) ICMT 2011. LNCS, vol. 6707, pp. 123–137. Springer, Heidelberg (2011). https://doi.org/10.1007/978-3-642-21732-6_9
3. Behr, N., Saadat, M.G., Heckel, R.: Commutators for stochastic rewriting systems: theory and implementation in Z3 (2020). https://arxiv.org/abs/2003.11010
4. Bowman, M., Briand, L.C., Labiche, Y.: Solving the class responsibility assignment problem in object-oriented analysis with multi-objective genetic algorithms. IEEE Trans. Software Eng. **36**(6), 817–837 (2010)
5. Burdusel, A., Zschaler, S., John, S.: Automatic generation of atomic consistency preserving search operators for search-based model engineering. In: MODELS, pp. 106–116. IEEE (2019)
6. Burdusel, A., Zschaler, S., Strüber, D.: MDEOptimiser: a search based model engineering tool. In: MODELS, pp. 12–16 (2018)
7. Dyck, J., Giese, H.: k-inductive invariant checking for graph transformation systems. In: de Lara, J., Plump, D. (eds.) ICGT 2017. LNCS, vol. 10373, pp. 142–158. Springer, Cham (2017). https://doi.org/10.1007/978-3-319-61470-0_9
8. Ehrig, H., Ehrig, K., Prange, U., Taentzer, G.: Fundamentals of Algebraic Graph Transformation. MTCSAES. Springer, Heidelberg (2006). https://doi.org/10.1007/3-540-31188-2
9. Fleck, M., Troya, J., Wimmer, M.: Marrying search-based optimization and model transformation technology. In: NasBASE (2015)
10. Fleck, M., Troya Castilla, J., Wimmer, M.: The class responsibility assignment case. In: TTC (2016)
11. Habel, A., Pennemann, K.H.: Correctness of high-level transformation systems relative to nested conditions. Math. Struct. Comput. Sci. **19**, 245–296 (2009)
12. Habel, A., Sandmann, C.: Graph repair by graph programs. In: Mazzara, M., Ober, I., Salaün, G. (eds.) STAF 2018. LNCS, vol. 11176, pp. 431–446. Springer, Cham (2018). https://doi.org/10.1007/978-3-030-04771-9_31
13. Jensen, M.T.: Helper-objectives: using multi-objective evolutionary algorithms for single-objective optimisation. J. Math. Model. Algorithms **3**(4), 323–347 (2004)
14. Kehrer, T., Taentzer, G., Rindt, M., Kelter, U.: Automatically deriving the specification of model editing operations from meta-models. In: Van Van Gorp, P., Engels, G. (eds.) ICMT 2016. LNCS, vol. 9765, pp. 173–188. Springer, Cham (2016). https://doi.org/10.1007/978-3-319-42064-6_12

15. Kosiol, J., Fritsche, L., Nassar, N., Schürr, A., Taentzer, G.: Constructing constraint-preserving interaction schemes in adhesive categories. In: Fiadeiro, J.L. (ed.) WADT 2018. LNCS, vol. 11563, pp. 139–153. Springer, Cham (2019). https://doi.org/10.1007/978-3-030-23220-7_8

16. Kosiol, J., Strüber, D., Taentzer, G., Zschaler, S.: Graph consistency as a graduated property: consistency-sustaining and -improving graph transformations - Extended Version (2020). https://arxiv.org/abs/2005.04162

17. Lambers, L., Born, K., Kosiol, J., Strüber, D., Taentzer, G.: Granularity of conflicts and dependencies in graph transformation systems: a two-dimensional approach. J. Log. Algebr. Meth. Program. **103**, 105–129 (2019)

18. Lambers, L., Strüber, D., Taentzer, G., Born, K., Huebert, J.: Multi-granular conflict and dependency analysis in software engineering based on graph transformation. In: ICSE, pp. 716–727. ACM (2018)

19. Nassar, N., Kosiol, J., Arendt, T., Taentzer, G.: Constructing optimized validity-preserving application conditions for graph transformation rules. In: Guerra, E., Orejas, F. (eds.) ICGT 2019. LNCS, vol. 11629, pp. 177–194. Springer, Cham (2019). https://doi.org/10.1007/978-3-030-23611-3_11

20. Nassar, N., Kosiol, J., Arendt, T., Taentzer, G.: Constructing optimized validity-preserving application conditions for graph transformation rules. J. Log. Algebraic Meth. Program. (2020, to appear)

21. Nassar, N., Kosiol, J., Radke, H.: Rule-based repair of EMF models: formalization and correctness proof. In: GCM (2017)

22. Nassar, N., Radke, H., Arendt, T.: Rule-based repair of EMF models: an automated interactive approach. In: Guerra, E., van den Brand, M. (eds.) ICMT 2017. LNCS, vol. 10374, pp. 171–181. Springer, Cham (2017). https://doi.org/10.1007/978-3-319-61473-1_12

23. Nuseibeh, B., Easterbrook, S., Russo, A.: Making inconsistency respectable in software development. J. Syst. Softw. **58**(2), 171–180 (2001)

24. Plump, D.: Confluence of graph transformation revisited. In: Middeldorp, A., van Oostrom, V., van Raamsdonk, F., de Vrijer, R. (eds.) Processes, Terms and Cycles: Steps on the Road to Infinity. LNCS, vol. 3838, pp. 280–308. Springer, Heidelberg (2005). https://doi.org/10.1007/11601548_16

25. Sandmann, C., Habel, A.: Rule-based graph repair. CoRR abs/1912.09610 (2019). http://arxiv.org/abs/1912.09610

26. Semeráth, O., Varró, D.: Graph constraint evaluation over partial models by constraint rewriting. In: Guerra, E., van den Brand, M. (eds.) ICMT 2017. LNCS, vol. 10374, pp. 138–154. Springer, Cham (2017). https://doi.org/10.1007/978-3-319-61473-1_10

27. Stevens, P.: Bidirectionally tolerating inconsistency: partial transformations. In: Gnesi, S., Rensink, A. (eds.) FASE 2014. LNCS, vol. 8411, pp. 32–46. Springer, Heidelberg (2014). https://doi.org/10.1007/978-3-642-54804-8_3

28. Strüber, D.: Generating efficient mutation operators for search-based model-driven engineering. In: Guerra, E., van den Brand, M. (eds.) ICMT 2017. LNCS, vol. 10374, pp. 121–137. Springer, Cham (2017). https://doi.org/10.1007/978-3-319-61473-1_9

29. Strüber, D., et al.: Henshin: a usability-focused framework for EMF model transformation development. In: de Lara, J., Plump, D. (eds.) ICGT 2017. LNCS, vol. 10373, pp. 196–208. Springer, Cham (2017). https://doi.org/10.1007/978-3-319-61470-0_12

Formal Verification of Invariants for Attributed Graph Transformation Systems Based on Nested Attributed Graph Conditions

Sven Schneider$^{(\boxtimes)}$ ⓘ, Johannes Dyck ⓘ, and Holger Giese ⓘ

University of Potsdam, Hasso Plattner Institute, Potsdam, Germany
{sven.schneider,johannes.dyck,holger.giese}@hpi.de

Abstract. The behavior of various kinds of dynamic systems can be formalized using typed attributed graph transformation systems (GTSs). The states of these systems are then modelled using graphs and the evolution of the system from one state to another is described by a finite set of graph transformation rules. GTSs with small finite state spaces can be analyzed with ease but analysis is intractable/undecidable for GTSs inducing large/infinite state spaces due to the inherent expressiveness of GTSs. Hence, automatic analysis procedures do not terminate or return indefinite or incorrect results.

We propose an analysis procedure for establishing state-invariants for GTSs that are given by nested graph conditions (GCs). To this end, we formalize a symbolic analysis algorithm based on k-induction using Isabelle, apply it to GTSs and GCs over typed attributed graphs, develop support to single out some spurious counterexamples, and demonstrate the feasibility of the approach using our prototypical implementation.

Keywords: Formal static analysis · Symbolic state space abstraction · k-induction · Symbolic graphs · Isabelle

1 Introduction

The verification of formal models of complex dynamic systems w.r.t. to formal specifications is one of the grand challenges of model driven engineering. However, the expressiveness required to cover the multitude of complex actual and desired behaviors of such systems renders analysis often undecidable. Indeed, the formalism of graph transformation systems (GTSs) considered here is known to be Turing complete. Hence, fully-automatic procedures for establishing meaningful properties on the behavior of such systems are then guaranteed to be not terminating in general or to produce indefinite or even incorrect results.

Funded by the Deutsche Forschungsgemeinschaft (DFG, German Research Foundation) - 148420506.

F. Gadducci and T. Kehrer (Eds.): ICGT 2020, LNCS 12150, pp. 257–275, 2020.
https://doi.org/10.1007/978-3-030-51372-6_15

We subsequently focus on GTSs where an analysis using an explicit state space exploration using tools such as GROOVE [2] and HENSHIN [3] is not applicable due to infinite or intractably large sets of initial or reachable states.

We approach this problem by combining the symbolic static analysis techniques of k-induction and state abstractions to establish state invariants for dynamic systems with infinite state spaces modelled by GTSs. The idea of k-induction is to establish a state invariant by iteratively computing all shortest derivations from an initial state to a violating state. The use of state abstractions, which preserve and reflect the systems' behavior w.r.t. the invariant candidate, permits to handle GTSs with infinite sets of initial or violating states at the concrete level but finite (and sufficiently small) such sets at the abstract level.

As main contributions, we (a) formalize the principle of k-induction in the theorem prover Isabelle in the form of an analysis algorithm and (b) instantiate this analysis algorithm for the setting of (b1) invariant candidates formalized using the logic of nested graph conditions (GCs) and (b2) a suitable notion of typed attributed graph transformation. This instantiation based approach thereby also clearly separates aspects of k-induction from GTS related concepts.

To represent typed attributed graphs, we employ symbolic graphs [18–22], which are similar to E-GRAPHS [12]. These symbolic graphs also give rise to an instantiation of GCs that permits the specification of constraints on attributes throughout the GCs. We employ a graph transformation step relation on symbolic graphs that deviates from those formalized in [21,22] by being symmetric (allowing a backwards application used in the k-induction analysis algorithm) and by allowing for the removal of variables (not requiring that additional variables and their values must be guessed when computing backward steps).

As closest related work, approaches using k-induction have been used before without formal foundation in [4] and in [7–11] assuming $k = 1$, graphs without attributes, a single initial state, or a subclass of all GCs. Hence, we extend this line of research by formally treating the more general case of an arbitrary value of k, graphs with attributes, infinitely many initial states, and all GCs.

In [5,25,27,28], an abstraction of graphs results in shape graphs (which have limited expressiveness compared to GCs) where multiple nodes in the graph are represented by so called summary nodes in the shape graph and where multiplicity or even first-order logic constraints may further restrict this abstraction (see also [6]). Moreover, in [15], an abstraction of graphs is given in terms of consistent compasses (which can be encoded in GCs of depth one) containing a set of graphs of which one is matchable and a set of non-matchable graphs. Also, in [29], the tool ALLOY is used to establish state invariants for typed graphs.

Further related analysis approaches are as follows. The tool AUGUR2 [1] abstracts GTSs to Petri nets but imposes restrictions on graph transformation rules thereby limiting expressiveness. Lastly, static analysis of programs for GTSs w.r.t. pre/post conditions has been developed in [23] as well as [24].

In Sect. 2, we formalize the principle of k-induction in the form of an analysis algorithm. In Sect. 3, we discuss our running example, our notion of attributed graph transformation, and the logic of GCs. In Sect. 4, we instantiate the analysis algorithm for attributed graph transformation and apply our prototypical implementation of it to our running example demonstrating its feasibility. In Sect. 5, we provide a conclusion and a discussion of future work.

2   Invariant Verification Using k-Induction

We now introduce our formalization of the technique of k-induction for the verification of (state) invariants. For this purpose, we introduce labelled transition systems (LTS) as an abstract framework, which is instantiated later on for graph transformation. The results of this section have been formalized in the interactive theorem prover Isabelle and we therefore omit all proofs. An LTS consists of a set S of states, a set L of labels, a relation δ of labelled steps between states, and initial states identified via a state predicate Z.

Definition 1 (Labelled Transition System (LTS)). *If S and L are sets of states and labels, $\delta \subseteq S \times L \times S$, $Z : S \rightarrow \mathbf{B}$, and $\Gamma = (S, L, \delta, Z)$, then Γ is a labelled transition system, written $\Gamma \in \mathcal{S}^{\text{lts}}$.*

Moreover, a finite path $\pi \in \mathsf{paths}(\Gamma, n)$ of Γ of length n is a sequence of n states from S interleaved with labels from L where $s \cdot l \cdot s'$ in π implies $(s, l, s') \in \delta$. Also, π_S and π_L map indices to the states and labels of the path π.

In Sect. 4, we restrict the states of an LTS resulting in a sub-LTS as follows.

Definition 2 (Sub-LTS). *If $\Gamma = (S, L, \delta, Z) \in \mathcal{S}^{\text{lts}}$, $S' \subseteq S$, and $\Gamma' = (S', L, \delta \cap (S' \times L \times S'), Z \cap (S' \times \mathbf{B})) \in \mathcal{S}^{\text{lts}}$, then Γ' is a sub-LTS of Γ.*

A predicate I on the states of an LTS is an invariant for the LTS, if all states that are reachable from an initial state of the LTS satisfy I.

Definition 3 (Invariant). *If $\Gamma = (S, L, \delta, Z) \in \mathcal{S}^{\text{lts}}$, $I : S \rightarrow \mathbf{B}$, and $\forall n \in \mathbf{N}$. $\forall \pi \in \mathsf{paths}(\Gamma, n)$. $Z(\pi_\mathsf{S}(0)) \rightarrow I(\pi_\mathsf{S}(n))$, then Γ has invariant I, written invariant(Γ, I).*

Subsequently, we assume an invariant A (e.g. expressing earlier established invariants) for the LTS to improve applicability of the analysis approach as explained later on. For characterizing the k-induction algorithm below, we define shortest violations of a state predicate I as a finite path leading from an initial state to a state violating I visiting no further initial states and only passing through states satisfying I as well as A.

Definition 4 (Shortest Violation). *If $\Gamma = (S, L, \delta, Z) \in \mathcal{S}^{\text{lts}}$, $I : S \rightarrow \mathbf{B}$, $A : S \rightarrow \mathbf{B}$, $k \in \mathbf{N}$, $\pi \in \mathsf{paths}(\Gamma, k)$, $Z(\pi_\mathsf{S}(0))$, $\neg I(\pi_\mathsf{S}(k))$, $\forall 0 < j \leq k. \neg Z(\pi_\mathsf{S}(j))$, and $\forall j < k. I(\pi_\mathsf{S}(j)) \wedge A(\pi_\mathsf{S}(j))$, then π is a shortest violation of I by Γ of length k under A, written $\pi \in \mathsf{SVio}(\Gamma, A, I, k)$.*

The analysis algorithm $\mathcal{I}$ below checks for such shortest violations by (a) selecting all violating states s satisfying $\neg I(s)$ and by (b) computing up to k steps *backwards* ensuring that all k additional states s' visited on each of the paths obtained satisfy $A(s') \wedge I(s')$. Firstly, when a state s', which is visited in this process, satisfies $Z(s')$, a shortest violation is obtained. Secondly, when no such path of k steps exists, there cannot be a shortest violation of greater length. Note that this analysis process benefits from employing the assumed invariant A, which is used to rule out paths with states that are known to be unreachable from an initial state by not satisfying A.

The analysis algorithm $\mathcal{I}$ returns a value b with three different values where $b = \mathsf{i}$ represents a successful verification of the given state predicate I as an invariant (when no paths are left that may be extended to shortest violations), where $b = \mathsf{v}$ represents that at least one shortest violation was determined, and where $b = \mathsf{u}$ represents the situation that the analysis was unable to return one of the two former definite results for the provided value of k that is decremented in each recursive application of $\mathcal{I}^{\mathrm{inner}}$.

Definition 5 ($\mathcal{I}$). *If* $\Gamma = (S, L, \delta, Z) \in \mathcal{S}^{\mathrm{lts}}$, $A : S \rightharpoonup \mathbf{B}$, $I : S \rightharpoonup \mathbf{B}$, $k \in \mathbf{N}$, $i \in \mathbf{N}$, *paths* $\subseteq \mathsf{paths}(\Gamma, i)$, *then* $\mathcal{I}^{\mathrm{inner}}(\Gamma, A, I, k, i, \mathit{paths}) \subseteq \{(b, \mathit{violations}) \mid b \in \{\mathsf{i}, \mathsf{v}, \mathsf{u}\} \wedge \mathit{violations} \subseteq \mathsf{paths}(\Gamma, k+i)\}$ *as follows.*

$$\mathcal{I}^{\mathrm{inner}}(\Gamma, A, I, k, i, \mathit{paths}) = \begin{cases} \text{if } \mathit{paths} = \emptyset \text{ then } (\mathsf{i}, \emptyset) \\ \text{elseif } \mathsf{vio}(\mathit{paths}) \neq \emptyset \text{ then } (\mathsf{v}, \mathsf{vio}(\mathit{paths})) \\ \text{elseif } k = 0 \text{ then } (\mathsf{u}, \mathit{paths}) \\ \text{else } \mathcal{I}^{\mathrm{inner}}(\Gamma, A, I, k-1, i+1, \mathsf{ext}(\mathit{paths})) \end{cases}$$

where

$\mathsf{vio}(\mathit{paths}) = \{\pi \in \mathit{paths} \mid Z(\pi_{\mathsf{S}}(0))\}$

$\mathsf{ext}(\mathit{paths}) = \{s \cdot \ell \cdot \pi \mid \pi \in \mathit{paths} \wedge (s, \ell, \pi_{\mathsf{S}}(0)) \in \delta \wedge A(s) \wedge I(s)\}$

Moreover, if $k \in \mathbf{N}$ *and* $\mathit{paths}_0 = \{\pi \in \mathsf{paths}(\Gamma, 0) \mid \neg I(\pi_{\mathsf{S}}(0))\}$ *is the set of violating paths of length 0, then* $\mathcal{I}(\Gamma, A, I, k) = \mathcal{I}^{\mathrm{inner}}(\Gamma, A, I, k, 0, \mathit{paths}_0)$.

The following theorem states that the analysis algorithm $\mathcal{I}$ performs a sound state invariant analysis as just described above.

Theorem 1 (Soundness of $\mathcal{I}$). *If* $\Gamma = (S, L, \delta, Z) \in \mathcal{S}^{\mathrm{lts}}$, $A : S \rightharpoonup \mathbf{B}$, $I : S \rightharpoonup \mathbf{B}$, *invariant*$(\Gamma, A)$, $k \in \mathbf{N}$, *and* $\mathcal{I}(\Gamma, A, I, k) = (b, \mathit{paths})$, *then there is* $j \leq k$ *s.t.* *paths* $\subseteq \mathsf{paths}(\Gamma, j)$ *and one of the following items holds.*

- $b = \mathsf{u}$, $j = k$, $\mathit{paths} \neq \emptyset$, *and* $\bigcup\{\mathsf{SVio}(\Gamma, A, I, i) \mid i \leq k\} = \emptyset$.
- $b = \mathsf{i}$, *invariant*(Γ, I) *and* $\mathit{paths} = \emptyset$.
- $b = \mathsf{v}$, $\neg$*invariant*(Γ, I), $\mathit{paths} = \mathsf{SVio}(\Gamma, A, I, j) \neq \emptyset$.

The analysis algorithm $\mathcal{I}$ is implementable when the set of paths considered is finite throughout its computation. This is guaranteed when the LTS has violations for at most finitely many states (finite initial set of paths handed to $\mathcal{I}^{\mathrm{inner}}$) and when every state has at most finitely many predecessors (each path can only be extended backwards to finitely many paths in $\mathcal{I}^{\mathrm{inner}}$).

Definition 6 (Finitely Backwards Branching LTS). *If* $\Gamma = (S, L, \delta, Z) \in \mathcal{S}^{\mathrm{lts}}$, $I : S \rightharpoonup \mathbf{B}$, *finite*$(\{s \in S \mid \neg I(s)\})$, *and* $\forall s' \in S.\,\mathsf{finite}(\{s \mid (s, \ell, s') \in \delta\})$, *then* Γ *is finitely backwards branching for* I.

The *concrete* instantiation of LTSs for GTSs in Sect. 4 is not finitely backwards branching in general because invariant candidates I may be violated by infinitely many states. Hence, we apply in Sect. 4 an abstraction leading to an *abstract* instantiation of LTSs for GTSs where the corresponding invariant candidate I' is violated by finitely many states. We then establish a connection between both instantiations in terms of an LTS abstraction relation (LTSAR), which permits to analyze the abstract instantiation using $\mathcal{I}$ instead of the concrete instantiation.

Intuitively, the paths considered using $\mathcal{I}$ for the concrete LTS are symbolically represented by the finite set of paths considered using $\mathcal{I}$ for the abstract LTS. Formally, an LTSAR consists of two subrelations R_S relating states and R_L relating labels of the underlying concrete and abstract LTSs. Note that we state suitable requirements on the relations R_S and R_L of an LTSAR in the following theorem and define only the type of an LTSAR here.

Definition 7 (LTS Abstraction Relation (LTSAR)). *If* $\Gamma = (S, L, \delta, Z) \in \mathcal{S}^{\mathrm{lts}}$, $\Gamma' = (S', L', \delta', Z') \in \mathcal{S}^{\mathrm{lts}}$, $R_S \subseteq S \times S'$, $R_L \subseteq L \times L'$, *then* (R_S, R_L) *is an LTS Abstraction Relation from* Γ *to* Γ', *written* $\Gamma \leq_{R_S, R_L} \Gamma'$.

For invariant candidates I and I' for Γ and Γ', the following theorem states six requirements on an LTSAR (R_S, R_L), which guarantee that (a) a violation of I' in Γ' implies the existence of a violation of I in Γ and (b) the absence of violations of I' in Γ' implies the absence of violations of I in Γ.

Theorem 2 (Preservation/Reflection of Invariants using LTS Abstraction Relations). *If* $\Gamma = (S, L, \delta, Z) \in \mathcal{S}^{\mathrm{lts}}$, $\Gamma' = (S', L', \delta', Z') \in \mathcal{S}^{\mathrm{lts}}$, $A : S \rightarrow \mathbf{B}$, invariant$(\Gamma, A)$, $I : S \rightarrow \mathbf{B}$, $I' : S' \rightarrow \mathbf{B}$, *and* $\Gamma \leq_{R_S, R_L} \Gamma'$, *then both of the following items hold.*

- *Part1: R1, R2, R3, R4, R5, and not* invariant(Γ', I') *imply not* invariant(Γ, I).
- *Part2: R1, R2, R3, R4, R6, and* invariant(Γ', I') *imply* invariant(Γ, I).

The requirements R1–R6 used in these items are as follows.

- R1: $\forall (s, s') \in R_S.\ I(s) \leftrightarrow I'(s')$ (R_S is compatible with invariant satisfaction)
- R2: $\forall (s, s') \in R_S.\ Z(s) \leftrightarrow Z'(s')$ (R_S is compatible with initial states)
- R3: $\forall s' \in S'.\ \exists s \in S.\ (s, s') \in R_S$
  (R_S relates a concrete state $s \in S$ to each abstract state $s' \in S'$)
- R4: $\forall s \in S.\ \left(\exists k \in \mathbf{N}.\ \exists \pi \in \mathsf{SVio}(\Gamma, A, I, k).\ \pi_S(k) = s \right) \rightarrow \left(\exists s' \in S'.\ (s, s') \in R_S \right)$
  (R_S relates an abstract state $s' \in S'$ to each concrete state $s \in S$ for which a shortest violation of I exists)
- R5: $\forall (s, s') \in R_S.\ \forall (s', l', \bar{s}') \in \delta'.\ \exists (s, l, \bar{s}) \in \delta.\ (l, l') \in R_L \wedge (\bar{s}, \bar{s}') \in R_S$
  (every forward step of the abstract LTS Γ' can be mimicked (forwards) by the concrete LTS Γ for two related source states (s, s') to allow for the concretization of a violating path)
- R6: $\forall (\bar{s}, \bar{s}') \in R_S.\ \neg Z(\bar{s}) \rightarrow \forall (s, l, \bar{s}) \in \delta.\ \exists (s', l', \bar{s}') \in \delta'.\ (l, l') \in R_L \wedge (s, s') \in R_S$
  (every backward step of the concrete LTS Γ (except for those leading to initial states) can be mimicked (backwards) by the abstract LTS Γ' for two related target states $(\bar{s}, \bar{s}')$ to allow for the abstraction of a violating path)

3   Modelling and Specifying Graph Transformation

As a running example, we consider a single shuttle travelling on a network of tracks (see Fig. 1a for the type graph used) where subsequent tracks are connected using *next* edges. The graph attribution stores the velocity v and acceleration a of the shuttle and, moreover, the constants for minimal, maximal, and safe velocities as well as the constant track length s in a *System* node. The rules refer to the attributes to describe the velocity v' of a shuttle after travelling over a track based on its current velocity v, acceleration a, and the constant track length s using the standard equation $v'^2 = v^2 + 2as$. The velocity of the shuttle should be below the safe velocity on tracks with flag *signal*, the velocity of the shuttle should be constant on tracks with flag *const*, and the flag *warning* on a track indicates that a track with flag *signal* is to be expected ahead. Analysis should establish the fact that the shuttle never violates *signal* and *const* flags as an invariant, which is formalized in Fig. 1c using graph conditions explained below. Note that tracks with flag *const* between tracks with flag *warning* and tracks with flag *signal* may prevent timely deceleration. We employ an assumed invariant to (a) specify the constant attribute values of the system node, (b) to rule out track networks with dead ends and loops, and (c) to ensure warnings n tracks ahead of signals for a parameter $n \in \mathbf{N}$ in all considered track networks.

We now recall attribute conditions (ACs) used by symbolic graphs and then revisit GTSs and GCs over symbolic graphs for describing actual and desired behavior in terms of a concrete LTS and state predicates from before.

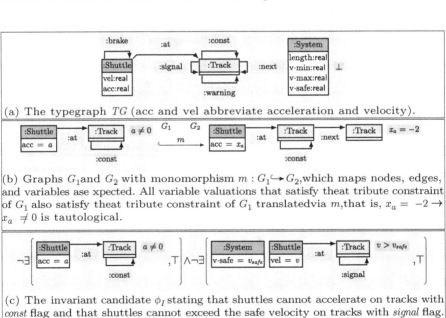

(a) The typegraph TG (acc and vel abbreviate acceleration and velocity).

(b) Graphs G_1 and G_2 with monomorphism $m : G_1 \hookrightarrow G_2$, which maps nodes, edges, and variables ase xpected. All variable valuations that satisfy thea ttribute constraint of G_1 also satisfy thea ttribute constraint of G_1 translatedvia m, that is, $x_a = -2 \rightarrow x_a \neq 0$ is tautological.

(c) The invariant candidate ϕ_I stating that shuttles cannot accelerate on tracks with *const* flag and that shuttles cannot exceed the safe velocity on tracks with *signal* flag.

Fig. 1. Type graph and invariant candidate for the shuttle scenario.

The attribute logic AL contains ACs $\gamma \in \mathcal{S}_X^{\mathrm{AC}}$ of first-order logic (FOL) ranging over a set X of variables. The satisfaction of γ by a valuation $\alpha : X \rightarrow \mathcal{V}$ is denoted by $\alpha \models_{\mathrm{AC}} \gamma$. The SMT solver Z3 [17] supports ACs constructed using a restricted set of operators for the sorts bool, int, real, and string. When Z3 is unable to determine an answer to the satisfiability problem (note that AL satisfaction is undecidable), which does not occur for the examples considered here, we would notify the user in our prototypical implementation.

Symbolic graphs (called graphs subsequently) [18] are an adaptation of E-GRAPHS [12]. A finite graph G (such as those depicted in Fig. 1b) contains nodes, edges, variables $G.\mathrm{X}$, and an AC $G.\mathrm{ac}$ ranging over $G.\mathrm{X}$. Moreover, nodes and edges are equipped with node and edge attributes, which are connected to variables for which values are specified in the AC $G.\mathrm{ac}$. A morphism $m : G_1 \rightarrow G_2$ from graph G_1 to G_2 (see e.g. Fig. 1b) maps nodes, edges, variables, node attributes, and edge attributes of G_1 to those of G_2. The mappings of m must be compatible with the source and target functions of G_1 and G_2 as usual and $G_2.\mathrm{ac}$ must imply the translation $m(G_1.\mathrm{ac})$ of $G_1.\mathrm{ac}$ for all variable valuations to ensure that m characterizes a restriction of attributes (cf. Fig. 1b where this implication is discussed). Moreover, the class of all finite graphs typed (as usual using a typing morphism) over a given type graph TG is given by $\mathcal{S}_{\mathrm{fin},TG}^{\mathrm{graphs}}$ or simply $\mathcal{S}_{\mathrm{fin}}^{\mathrm{graphs}}$ when TG is known. In the remainder, we only employ monomorphisms, written $m : G_1 \hookrightarrow G_2$, with only injective mappings. The unique monomorphism from the empty graph $\emptyset$ to a graph G is denoted $\mathrm{i}(G) : \emptyset \hookrightarrow G$. Finally, the special monomorphism $\mathrm{a}(G) : G' \hookrightarrow G$ describes that G' is obtained from G by setting the AC $G.\mathrm{ac}$ to true (i.e., G' equals G except that $G'.\mathrm{ac} = \top$).

The graph logic GL [14, 26] supports the specification of the (non)existence of certain subgraphs in a given host graph G. Besides propositional operators for (finite) conjunction and negation, GL features the *exists* operator $\exists$, which specifies for a given match $m : H \hookrightarrow G$ of a (context) graph H into the host graph G that m can be extended to a match $m' : H' \hookrightarrow G$ by using a monomorphism $f : H \hookrightarrow H'$ that explains how H is extended to the (context) graph H'.

The graph G_2 from Fig. 1b does not satisfy ϕ_I because the initial monomorphism $\mathrm{i}(G_2) : \emptyset \hookrightarrow G_2$ can be extended to m from Fig. 1b, which is forbidden by the left part $\exists(\mathrm{i}(G_1), \top)$ of ϕ_I.

Definition 8 (Graph Logic (GL)). *If $H \in \mathcal{S}_{\mathrm{fin}}^{\mathrm{graphs}}$ is a finite graph, $m : H \hookrightarrow G$ is a monomorphism, then ϕ' is a graph condition over H, written $\phi' \in \mathcal{S}_H^{\mathrm{GC}}$, which is satisfied by m, written $m \models \phi'$, if an item applies.*

- *$\phi' = \wedge S$, $S \subseteq_{\mathrm{fin}} \mathcal{S}_H^{\mathrm{GC}}$, and (for satisfaction) $\forall \phi \in S. \, m \models \phi$.*
- *$\phi' = \neg \phi$, $\phi \in \mathcal{S}_H^{\mathrm{GC}}$, and (for satisfaction) $m \not\models \phi$.*
- *$\phi' = \exists(f : H \hookrightarrow H', \phi)$, $\phi \in \mathcal{S}_{H'}^{\mathrm{GC}}$ is a GC over the extended graph H', and (for satisfaction) there is $m' : H' \hookrightarrow G$ s.t. $m = m' \circ f$ and $m' \models \phi$.*

Moreover, we define the following notions.

- *Derived operators: (true)* $\top$, *(false)* $\bot$, *(disjunction)* $\vee S$, *and (for all)* $\forall(f, \phi)$.
- *Graph Satisfaction: If* $\phi \in \mathcal{S}_\emptyset^{GC}$ *is a GC over the empty graph satisfied by the initial morphism* $i(G)$ *(i.e.,* $i(G) \models \phi$) *then* ϕ *is satisfied by* G, *written* $G \models \phi'$.
- *Satisfying morphisms: If* $\phi \in \mathcal{S}_H^{GC}$ *is a GC, then* $[\![\phi]\!] = \{m : H \hookrightarrow G \mid m \models \phi\}$.

Moreover, we define that two GCs ϕ_1 and ϕ_2 are consistent, when ϕ_1 only describes elements also described by ϕ_2 or none of them.

Definition 9 (Consistent GCs). *If* $\{\phi_1, \phi_2\} \subseteq \mathcal{S}_\emptyset^{GC}$ *and* $[\![\phi_1]\!] \cap [\![\phi_2]\!] \neq \emptyset$ *implies* $[\![\phi_1]\!] \subseteq [\![\phi_2]\!]$, *then* ϕ_1 *is consistent with* ϕ_2, *written* $\mathsf{cons}(\phi_1, \phi_2)$.

To check satisfiability of a GC and consistency of two GCs, we employ the automated reasoning technique for GL in the form of the algorithm $\mathcal{A}$ for which tool support is available in AUTOGRAPH as introduced in [26]. The algorithm $\mathcal{A}$ takes a GC ϕ as input, is known to terminate for unsatisfiable GCs (i.e., it is refutationally complete), and incrementally generates the set of minimal graphs satisfying ϕ (this set is empty for unsatisfiable GCs). As for the case of AL and Z3, we carefully handle cases where $\mathcal{A}$ does not terminate and also generates no minimal graph as discussed later on.

Fact 1 (Algorithm $\mathcal{A}$). *If* $\phi \in \mathcal{S}_\emptyset^{GC}$ *is a GC over the empty graph and* $\mathcal{A}$ *terminates for* ϕ, *it returns the finite set of all minimal graphs satisfying* ϕ.

The standard operation shift from [13] is also applicable to symbolic graphs [26]. It defines an adaptation of a GC ϕ with context graph H for a monomorphism $m : H \hookrightarrow H'$ resulting in an equivalent GC with context graph H' in the sense of the following fact (by considering how additional elements of H' may be used in a satisfaction proof for the given GC ϕ).

Fact 2 (Operation shift). *If* $m_1 : H \hookrightarrow H'$, $m_2 : H' \hookrightarrow H''$, *and* $\phi \in \mathcal{S}_H^{GC}$, *then* $m_2 \circ m_1 \models \phi$ *iff* $m_2 \models \mathsf{shift}(m_1, \phi)$.

Graph transformation steps are defined using rules specifying structural and attribute transformations. A rule ρ contains for the structural part (as in the DPO approach) two monomorphisms $\rho.\mathrm{del} : K \hookrightarrow L$ and $\rho.\mathrm{add} : K \hookrightarrow R$ where K, $L - \rho.\mathrm{del}(K)$, and $R - \rho.\mathrm{add}(K)$ contain the preserved/deleted/added elements. For the attribute part, L, K, and R have the trivial ACs $\top$ and a rule ρ contains an AC $\rho.\mathrm{ac}$ instead, which is defined over the disjoint union V (i.e., the coproduct, written II where $\rho.\mathrm{lX}$ and $\rho.\mathrm{rX}$ map variables to the disjoint union V) of the variables of L and R. Intuitively, variables originating from L are used as unprimed variables and variables originating from R are used as primed variables. Finally, a rule contains left and right hand side application conditions $\rho.\mathrm{lC}$ and $\rho.\mathrm{rC}$ defined over the graphs L and R and checked during the transformation as in the DPO approach. See Fig. 3 for two simple rules[1] and, for our running example, Fig. 2 for two of the total nine rules (see [8, Section C.1.6, p. 336] for a full description of the assumed invariants and rules of the considered GTS).

Definition 10 (Graph Transformation Rules). *If $\rho.\mathrm{del} : K \hookrightarrow L$, $\rho.\mathrm{add} : K \hookrightarrow R$ are to monomorphisms, $\Pi(\rho.\mathrm{lX} : L.X \hookrightarrow V, \rho.\mathrm{rX} : R.X \hookrightarrow V)$ is a coproduct, $\rho.\mathrm{ac} \in \mathcal{S}_V^{\mathrm{AC}}$, $\rho.\mathrm{lC} \in \mathcal{S}_L^{\mathrm{GC}}$, $\rho.\mathrm{rC} \in \mathcal{S}_R^{\mathrm{GC}}$, and $L.\mathrm{ac} = K.\mathrm{ac} = R.\mathrm{ac} = \top$, then $\rho = (\rho.\mathrm{del}, \rho.\mathrm{add}, \rho.\mathrm{lX}, \rho.\mathrm{rX}, \rho.\mathrm{ac}, \rho.\mathrm{lC}, \rho.\mathrm{rC})$ is a rule, written $\rho \in \mathcal{S}^{\mathrm{rules}}$. Moreover, we define the following abbreviations.*

- *$\rho.\mathrm{lG} = L$ and $\rho.\mathrm{rG} = R$ are the left and right hand side graphs of the rule ρ.*
- *$\mathcal{S}_{\mathrm{fin}}^{\mathrm{rules}}$ is the set of all rules where L, K, and R are finite.*

Graph transformations systems then contain a finite set of finite rules (used for graph transformation steps) and initial states described by a GC.

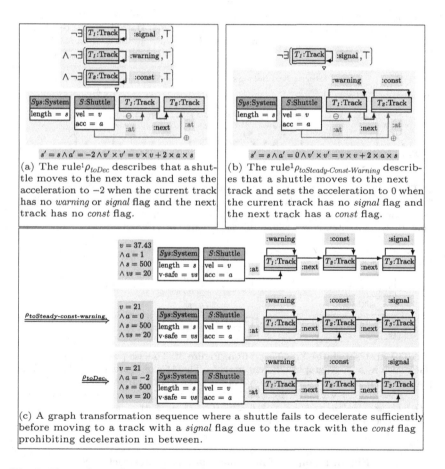

(a) The rule[1] ρ_{toDec} describes that a shuttle moves to the nex track and sets the acceleration to -2 when the current track has no *warning* or *signal* flag and the next track has no *const* flag.

(b) The rule[1] $\rho_{toSteady\text{-}Const\text{-}Warning}$ describes that a shuttle moves to the next track and sets the acceleration to 0 when the current track has no *signal* flag and the next track has a *const* flag.

(c) A graph transformation sequence where a shuttle fails to decelerate sufficiently before moving to a track with a *signal* flag due to the track with the *const* flag prohibiting deceleration in between.

Fig. 2. Two rules and a graph transformation sequence for our shuttle scenario.

---

[1] Here, L, K, and R are given in a single graph and preserved/deleted/added elements are colored black/red/green and deleted/added elements are marked with $\ominus/\oplus$.

$$\neg\exists\left(\,e{:}E_2\ \boxed{\;c{:}C\;},\top\right)\triangleright\quad L_1\ \boxed{\begin{array}{c}a{:}A\\ id=x\end{array}}\ \overset{\ell_1}{\hookleftarrow}\ K_1\ \boxed{\begin{array}{c}\bar a{:}A\\ id=\bar x\end{array}}\ \overset{r_1}{\hookrightarrow}\ R_1\ \boxed{\begin{array}{c}a'{:}A\\ e'{:}E_1\end{array}\ b'{:}B}\ \begin{array}{l}\text{ite}(5<x,\\ x'=0,\\ 2<x\wedge x'=x+4)\end{array}$$

(a)The rule ρ_1 .It is not applicable to graphs that contain a C node with E_2 loop. It adds a B node, adds loop on a, and changes the value of the id attribute (given by variable x) of node a using an AC that uses the *if-then-else* operation ite.

$$\neg\exists\left(\,e{:}E_4\ \boxed{\;a{:}A\;},\top\right)\triangleright\quad L_2\ \boxed{\begin{array}{c}a{:}A\\ id=x\end{array}}\ \overset{\ell_2}{\hookleftarrow}\ K_2\ \boxed{\begin{array}{c}\bar a{:}A\\ id=\bar x\end{array}}\ \overset{r_2}{\hookrightarrow}\ R_2\ \boxed{\begin{array}{c}a'{:}A\\ id=x'\end{array}\ \overset{e'{:}E_3}{\longrightarrow}\ \boxed{c'{:}C}}\ \begin{array}{l}6<x\\ \wedge x'=x+1\end{array}$$

(b) The rule ρ_2. It is not applicable to graphs where the matched A node has an E_4 loop or an id attribute of at most 6. It adds a C node c', adds an edge from the matched A node to c',and increases the id attribute (given by variable x) of node a.

$\phi_Z = \exists\left(\,e{:}E_2\ \boxed{\;c{:}C\;},\top\right)$	$\phi_A = \neg\exists\left(\boxed{\begin{array}{c}a{:}A\\ id=x\end{array}}\ x=0\,,\top\right)$	$\phi_I = \neg\exists\left(\boxed{b{:}B}\,,\top\right)$
(c)The initial states ϕ_Z have a C node with E_2 loop.	(d)The assumed invariant ϕ_A states that no A node may have an id of 0.	(e)The invariant candidate ϕ_I states reachable graph contain no B node.

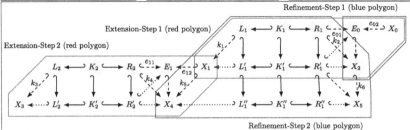

(f)The analysis using $\mathcal{I}$ starts with the path $\pi_0 = X_0$ of length 0 where $X0=(\emptyset,\neg\phi I)$. Using Ext, a path $\pi_1 = X_1\cdot(k_1, \rho_1, k_2)\cdot X2$ of length 1 is constructed by extending π_0. Using Ext, a path $\pi_2 = X_3 \cdot (k_3, \rho_2, k_4) \cdot X_4 \cdot (k_5 \circ k_1, \rho_1, k_6 \circ k_2) \cdot X_5$ of length 2 is constructed by extending $\pi_1$where the second step of π_2 is obtained by refinement of π_1 via Ref.

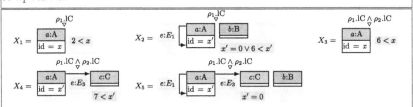

(g) The abstract states $X_i = (G_i, \phi_i)$ from Figure 3f. To ease presentation, we use GCs such as $\rho_1.$IC on graphs different from L_1. The ACs of G_i are obtained according to the step relation considering the AC of the given source/target graph. (G_1) $G_1.\mathrm{ac}\equiv\exists x'\,.\,\rho_1.\mathrm{ac}\equiv 5<x\vee 2<x\equiv 2<x.$ (G_2) $G_2.\mathrm{ac}\equiv\exists\,x.\rho_1.\mathrm{ac}\wedge 2<x\equiv x'=0\vee 6<x'.$ (G_3) $G_3.\mathrm{ac}\equiv\exists x'.\ \rho_2.\mathrm{ac}\wedge 2<x'\equiv 6<x.$ (G_4) $G_4.\mathrm{ac}\equiv\exists x.\ \rho_2.\mathrm{ac}\wedge 6<x\equiv 7<x'.$ (G_5) $G_5.\mathrm{ac}\equiv\exists x.\ \rho_1.\mathrm{ac}\wedge 7<x\equiv x'=0.$

Fig. 3. Example of invariant analysis for abstract LTS.

Definition 11 (Graph Transformation System (GTS)). *If $P \subseteq_{fin} \mathcal{S}_{fin}^{rules}$ and $\phi_Z \in \mathcal{S}_\emptyset^{GC}$, then (P, ϕ_Z) is a graph transformation system.*

Deviating from [22], we now introduce a notion of graph transformation steps in which structural and attribute transformations are decoupled. The defined step relation is symmetric and supports the removal as well as addition of variables, which is also relevant when attribute values are to be modified.

Definition 12 (Steps). *There is a step $G_1 \stackrel{\sigma}{\Longrightarrow} G_2$ with label σ, whenever*

- *there is a rule $\rho \in \mathcal{S}_{fin}^{rules}$ with $\rho.\mathrm{lG} = L$ and $\rho.\mathrm{rG} = R$ as depicted below,*
- *the graph L can be matched to G_1 using $m_1 : L \hookrightarrow G_1$ that satisfies the left-hand side application condition $\rho.\mathrm{lC}$,*
- *the graph $\bar{G}_1$ is obtained from G_1 by setting the AC of G_1 to $\top$ inducing the morphisms c_1 and $\mathsf{a}(G_1)$ compatible with m_1,*
- *the graphs D and $\bar{G}_2$ are constructed according to the double pushout approach as pushout complement and pushout from left to right,*
- *the graph G_2 is obtained from $\bar{G}_2$ by setting the AC of G_2 according to the AC $\rho.\mathrm{ac}$ of the rule inducing morphisms m_2 and $\mathsf{a}(G_2)$ compatible with c_2, and*
- *the morphism m_2 satisfies the right-hand side application condition $\rho.\mathrm{rC}$.*

For this construction, $\sigma = (\sigma.\mathrm{rule}, \sigma.\mathrm{drule}, \sigma.\mathrm{match}, \sigma.\mathrm{comatch}) = (\rho, \bar{\rho}, m_1, m_2)$ is the used label where $\bar{\rho}$ is the derived rule (cf. [13]) with $\bar{\rho}.\mathrm{del} = b_1$, $\bar{\rho}.\mathrm{add} = b_2$, $\bar{\rho}.\mathrm{lC} = \mathsf{shift}(c_1, \rho.\mathrm{lC})$, $\bar{\rho}.\mathrm{rC} = \mathsf{shift}(c_2, \rho.\mathrm{lC})$, and where the AC $\bar{\rho}.\mathrm{ac}$ is adapted from $\rho.\mathrm{ac}$ according to the renamings of c_1 and c_2.

For our running example, see Fig. 2c for a graph transformation sequence applying the two rules from Fig. 2a and Fig. 2b. Note that the last graph of this sequence violates the invariant candidate from Fig. 1c as the shuttle exceeds the permitted velocity on a track with *signal* flag.

The steps defined by this construction immediately induce a concrete LTS (see Definition 1) for a given GTS where the initial states are given by all graphs satisfying the GC characterizing initial graphs of the GTS.

Definition 13 (Concrete LTS of Graph Transformation). *If (P, ϕ_Z) is a GTS then $\mathsf{cLTS}((P, \phi_Z)) = \Gamma = (S, L, \delta, Z)$ is the concrete LTS of (P, ϕ_Z) with*

- $S = \mathcal{S}_{fin}^{graphs}$ *is the set of all finite graphs,*
- $L = \mathcal{S}^{steps}$ *is the set of all step labels,*
- $\delta = \{(G, \sigma, H) \mid G \stackrel{\sigma}{\Longrightarrow} H\}$ *is given by graph transformation steps of (P, ϕ_Z),*
- $Z(\bar{G}) = \bar{G} \models \phi_Z$ *uses the GC satisfaction relation,*

Moreover, a state G of Γ (i.e., a finite graph) satisfies a state predicate (cf. the last item above) given by a GC ϕ defined over the empty graph $\emptyset$ iff $G \models \phi$.

Finally, the operations left and the reverse operation right introduced in [13] can be adapted to symbolic graphs. The operation left inductively propagates a GC ϕ over the right hand side graph $\rho.\mathrm{rG} = R$ (such as the application condition $\rho.\mathrm{lC}$) of a rule ρ to the left hand side graph $\rho.\mathrm{lG} = L$ of ρ by applying the renaming of graph elements according to $\rho.\mathrm{del}$ and $\rho.\mathrm{add}$ to the graphs in the GC ϕ. The two operations ensure the following compatibility with steps (cf. [13]).

Fact 3 (Operations left and right). *If $\rho \in \mathcal{S}_{\mathrm{fin}}^{\mathrm{rules}}$ is a finite rule with the left and right hand side graphs L and R, $\phi_L \in \mathcal{S}_L^{\mathrm{GC}}$ and $\phi_R \in \mathcal{S}_R^{\mathrm{GC}}$ are GCs over L and R, and $G \xrightarrow{(\rho,\bar{\rho},m,\bar{m})} H$ is a graph transformation step, then $\bar{m} \models \phi_R$ iff $m \models \mathsf{left}(\rho, \phi_R)$ and $m \models \phi_L$ iff $\bar{m} \models \mathsf{right}(\rho, \phi_L)$.*

4   Invariant Analysis for Graph Transformation Systems

Based on the preliminaries from the previous section on graph transformation and graph specification using GCs, we now apply our theory on k-induction from Sect. 2. Note that the instantiation presented here is specific to the step relation for graph transformation presented in the previous section due to the decoupling of transformation of structure and ACs. For this instantiation, we construct an LTS that is finitely backwards branching (see Definition 6) and that is related to the concrete LTS Γ from the previous section via a suitable LTSAR (see Definition 7) to permit an application of Theorem 2 for enabling the analysis of the GTS using $\mathcal{I}$ according to Theorem 1. For this purpose, we assume a fixed GTS (P, ϕ_Z), the induced LTS $\mathsf{cLTS}((P, \phi_Z)) = \Gamma$ (see Definition 13), an assumed invariant $\phi_A \in \mathcal{S}_\emptyset^{\mathrm{GC}}$, and an invariant candidate $\phi_I \in \mathcal{S}_\emptyset^{\mathrm{GC}}$.

For demonstration purposes, we consider the GTS $(\{\rho_1, \rho_2\}, \phi_Z)$ with assumed invariant ϕ_A and invariant candidate ϕ_I from Fig. 3.

As an initial candidate for the LTS to be constructed, we define the LTS Γ' in which each state $(\bar{G}, \bar{\phi})$ is given by a GC $\bar{\phi}$ and the graph $\bar{G}$ over which $\bar{\phi}$ is defined for improved readability. The LTS Γ' induces an LTSAR in which the relation R_S contains pairs $(G, (\bar{G}, \bar{\phi}))$ for which some monomorphism $m : \bar{G} \hookrightarrow G$ with $m \models \bar{\phi}$ exists. The steps of Γ' adapt states (G_1, ϕ_1) to states (G_2, ϕ_2) using a rule ρ of the GTS (P, ϕ_Z) for matches $k_1 : \rho.\mathrm{lG} \hookrightarrow G_1$ and $k_2 : \rho.\mathrm{rG} \hookrightarrow G_2$ at the abstract level by considering all concrete steps of graphs H_1 and H_2 that are related to G_1 and G_2 via R_S (by means of instantiation morphisms m_1 and m_2). That is, the same rule ρ can be applied to each graph covered by (G_1, ϕ_1) and, vice versa, (G_2, ϕ_2) covers only the graphs reachable using such steps.

Definition 14 (Abstract LTS of GC Transformation). *If (P, ϕ_Z) is a GTS then $\mathsf{aLTS}((P, \phi_Z)) = \Gamma' = (S', L', \delta', Z')$ is the abstract LTS of (P, ϕ_Z) with*

- $S' = \{(\bar{G}, \bar{\phi}) \mid \bar{G} \in \mathcal{S}_{\mathrm{fin}}^{\mathrm{graphs}} \wedge \bar{\phi} \in \mathcal{S}_{\bar{G}}^{\mathrm{GC}}\}$,
- $L' = \{(k_1 : \rho.\mathrm{lG} \hookrightarrow G_1, \rho, k_2 : \rho.\mathrm{rG} \hookrightarrow G_2) \mid \rho \in P, \{G_1, G_2\} \subseteq \mathcal{S}_{\mathrm{fin}}^{\mathrm{graphs}}\}$,

- $((G_1, \phi_1), (k_1, \rho, k_2), (G_2, \phi_2)) \in \delta'$ *iff* $\rho \in P$, $k_1 : L \hookrightarrow G_1$, $k_2 : R \hookrightarrow G_2$,
  - $\forall m_1 \in [\![\phi_1]\!]. \exists m_2 \in [\![\phi_2]\!]. P(m_1 \circ k_1, m_2 \circ k_2, \rho)$ *and*
  - $\forall m_2 \in [\![\phi_2]\!]. \exists m_1 \in [\![\phi_1]\!]. P(m_1 \circ k_1, m_2 \circ k_2, \rho)$ *using the abbreviation P:*
  - $P(n_1, n_2, \rho) = (\exists \sigma. H_1 \overset{\sigma}{\Longrightarrow} H_2 \wedge \sigma.\text{rule} = \rho \wedge \sigma.\text{match} = n_1 \wedge \sigma.\text{comatch} = n_2)$,

- $Z'((\bar{G}, \bar{\phi})) = [\![\exists(\text{i}(\bar{G}), \bar{\phi}) \wedge \phi_Z]\!] \neq \emptyset$.

Moreover, a state $(\bar{G}, \bar{\phi})$ of Γ' satisfies a state predicate (cf. the last item above) given by a GC ϕ defined over the empty graph $\emptyset$ iff $[\![\exists(\text{i}(\bar{G}), \bar{\phi}) \wedge \phi]\!] \neq \emptyset$.[2]

We state that each sub-LTS Γ'' of Γ' induces a certain LTSAR for the LTS Γ.

Lemma 1 (LTSAR for GTS). *If (P, ϕ_Z) is a GTS, $\Gamma = \text{cLTS}((P, \phi_Z))$, Γ'' is a sub-LTS of $\Gamma' = \text{aLTS}((P, \phi_Z))$, $R_S = \{(G, (\bar{G}, \bar{\phi})) \mid \exists m : \bar{G} \hookrightarrow G. m \models \bar{\phi}\}$, and $R_L = \{(\sigma, (k_1, \rho, k_2)) \mid \sigma.\text{rule} = \rho \in P\}$, then $\Gamma \leq_{R_S, R_L} \Gamma''$ by Definition 7.* □

Selecting the entire LTS $\Gamma'' = \Gamma'$ results in an LTSAR, which does not satisfy the requirements of Theorem 2 in general. Instead, we obtain a suitable sub-LTS Γ'' of Γ' in an on-the-fly manner during an application of $\mathcal{I}$ (see Definition 5): Γ'' then describes precisely the paths maintained by $\mathcal{I}^{\text{inner}}$ in its parameter *paths* at any point in the computation. Hence, the initial candidate is the sub-LTS Γ''_0 that contains the single state $(\emptyset, \neg \phi_I)$ violating ϕ_I. Note that Γ''_0 induces an LTSAR satisfying the requirements R1–R5 already. See Fig. 3f where node X_0 represents this initial state inducing the path π_0 of length 0.

Inside an application of $\mathcal{I}^{\text{inner}}(\Gamma', \phi_A, \phi_I, k, i, paths)$ (see Definition 5), we extend paths in *paths* w.r.t. Γ' and thereby adapt Γ''_i to Γ''_{i+1} such that the LTSAR for Γ''_{i+1} (see Lemma 1) also satisfies the requirements R1–R5 of Theorem 2. The satisfaction of requirement R6 for the backwards simulation may require that further path extensions are computed in subsequent iterations of $\mathcal{I}^{\text{inner}}$. In Fig. 3f, the path π_0 is extended to paths π_1 and π_2 where the last nodes X_2 and X_5 are then incrementally more specific than X_0 (w.r.t. the monomorphisms that satisfy their GCs).

When the application of $\mathcal{I}$ terminates with a definite result $b \in \{\text{i}, \text{v}\}$, the obtained sub-LTS Γ''_i constructed up to this point induces an LTSAR, which meets the relevant requirements listed in Theorem 2. In particular (see also Theorem 3 later on), (a) for the result (v, *paths*) meaning that the invariant candidate ϕ_I is violated by Γ', we can apply *Part1* of Theorem 2 because R1–R5 are satisfied and (b) for the result (i, $\emptyset$) meaning that ϕ_I is established as an invariant for Γ', we can apply *Part2* of Theorem 2 because there are no further backward

---

[2] Definition 16 resolves cases where $\exists(\text{i}(\bar{G}), \bar{\phi})$ and ϕ are not consistent (Definition 8).

steps that require consideration since all paths constructed so far were discarded for not having any further relevant step implying also R6 as required.

In the remainder, we discuss the backwards construction of paths of Γ' for a GTS using the operation Ext. This extension operation (see Definition 16) entails a refining operation Ref (see Definition 15) used to adapt paths in line with the operation ext($paths$) used in $\mathcal{I}$ to ensure that the requirements R1–R5 are satisfied by the corresponding sub-LTS Γ''_{i+1} constructed so far.

Extending a path π of Γ' starting in a state $(\bar{G}, \bar{\phi})$ adding a backwards step for a rule ρ may result in a refinement due to (a) additional graph elements when the comatch of the step does not only match elements of $\bar{G}$, (b) additional restrictions originating from the application conditions of ρ, and (c) fewer variable valuations satisfying the AC of the start graph of the path.

For example, in Fig. 3f, the path $\pi_1 = X_1 \cdot (k_1, \rho_1, k_2) \cdot X_2$ (in the second line) is refined to $X_4 \cdot (k_5 \circ k_1, \rho_1, k_6 \circ k_2) \cdot X_5$ according to the monomorphism $k_5 : G_1 \hookrightarrow G_4$ for the application of ρ_2 in *Extension-Step 2*. Considering the elements X_1 and X_4 given in more detail in Fig. 3g, we see that X_4 is much more specific than X_2 due to the additional GC originating from ρ_2, the inclusion of node c and edge e, and a more restrictive AC.

The following operation Ref refines the path π starting in $(\bar{G}, \bar{\phi}_G)$ to a path π' starting in $(\bar{X}, \bar{\phi}_X)$ for a monomorphism $m : \bar{G} \hookrightarrow \bar{X}$ and a GC $\bar{\phi}_X$ defined on $\bar{X}$, which describe the effect of the backwards step on π. It does so by adapting the monomorphisms contained in the labels of the steps in π, performs a step leading to a graph $\bar{Y}$ to propagate attribute restrictions given by the AC of $\bar{X}$, and propagates the additional GC $\bar{\phi}_X$ to the resulting graph $\bar{Y}$.

Definition 15 (Refinement of Abstract Paths). *If $\Gamma' = (S', L', \delta', Z') \in \mathcal{S}^{\text{lts}}$, $\pi \in \text{paths}(\Gamma', n)$, $m : \bar{G} \hookrightarrow \bar{X}$, $\bar{\phi}_X \in \mathcal{S}_{\bar{X}}^{\text{GC}}$, $\pi' \in \text{paths}(\Gamma', n)$, then π' is the refinement of π via m and $\bar{\phi}_X$, written $\pi' = \text{Ref}(\pi, m, \bar{\phi}_X)$, if an item applies.*

- $n = 0$, $\pi = (\bar{G}, \bar{\phi}_G)$, and $\pi' = (\bar{X}, \bar{\phi}_X \wedge \text{shift}(m, \bar{\phi}_G))$.
- $n > 0$, $\pi = (\bar{G}, \bar{\phi}_G) \cdot (k_1, \rho, k_2) \cdot \tilde{\pi}$, $\pi_{\mathsf{S}}(1) = (\bar{H}, \bar{\phi}_H)$, $\bar{X} \xrightarrow{(\rho, \bar{\rho}, m \circ k_1, \bar{m} \circ k_2)} \bar{Y}$, $\bar{\phi}_Y = \text{shift}(a(\bar{Y}), \text{right}(\bar{\rho}, \exists(a(\bar{X}), \bar{\phi}_X)))$, and $\pi' = (\bar{X}, \bar{\phi}_X \wedge \text{shift}(m, \bar{\phi}_G)) \cdot (m \circ k_1, \rho, \bar{m} \circ k_2) \cdot \text{Ref}(\tilde{\pi}, \bar{m}, \bar{\phi}_Y)$.

Concrete violating paths of Γ (such as in Fig. 2c for our running example) can be constructed from symbolic violating paths of Γ' starting in $(\bar{G}, \bar{\phi}_G)$ by (a) running the algorithm $\mathcal{A}$ from Fact 1 to obtain some monomorphism $m : \bar{G} \hookrightarrow \bar{X}$ satisfying $\bar{\phi}_G$, (b) employing Z3 to determine a variable valuation satisfying the AC of $\bar{X}$ resulting in some monomorphism $m' : \bar{G} \hookrightarrow \bar{Y}$, and (c) applying the operation Ref for m' and $\bar{\phi}_Y = \top$. Besides such concrete violating paths, we return all *symbolic* violating paths to the user for which $\mathcal{A}$ or Z3 fail to determine definite results (which does not occur in the examples considered here).

We now introduce the operation Ext for extending a path of Γ' by adding a further backwards step. To ensure that we construct all paths, we follow the definition of E-concurrent rules from [13] to generate all minimal overlaps for each successive rule application and to adjust GCs to the application conditions of the rules. Moreover, in item (8), we employ the operation Ref to adapt the given path of Γ' to the additional step. Finally, in item (9), item (10), and item (11), we further split and filter the constructed paths to ensure that the state predicate satisfaction is compatible with R_S (see Theorem 3).

Definition 16 (Extension of Abstract Paths). *If (P, ϕ_Z) is a GTS, $\Gamma' =$ aLTS$((P, \phi_Z))$, $\pi \in$ paths(Γ', n), then Ext(π) computes the possibly empty set of all path extensions $\pi' \in$ paths$(\Gamma', n+1)$ of π using the following procedure.*

(1) $(\bar{G}, \bar{\phi}_G)$ is the first state of π.

(2) $\rho \in P$ is some rule of the GTS with $\rho.\text{lG} = L$ and $\rho.\text{rG} = R$.

(3) $(e_1 : R \hookrightarrow E, e_2 : \bar{G} \hookrightarrow E) \in \mathcal{E}'$ is a minimal overlapping of R and $\bar{G}$ (cf. [13]).[3]

(4) $E \xrightarrow{(\text{rev}(\rho), \bar{\rho}, e_1, m)} \bar{X}$ is a step of the GTS where ρ is reversed using rev and applied forwards to E using match e_1 to obtain the required backwards step.

(5) $\bar{\phi}_X = \text{shift}(\text{a}(\bar{X}), \bar{\rho}.\text{lC} \wedge \text{left}(\bar{\rho}, \bar{\rho}.\text{rC} \wedge \exists(\text{a}(E), \text{shift}(e_2, \bar{\phi}_G))))$ is the GC for $\bar{X}$ obtained using GC propagation as in [13].

(6) $\bar{X} \xrightarrow{(\rho, \bar{\rho}, m, \bar{m} \circ e_1)} \bar{Y}$ is a step of the GTS using the rule ρ possibly further restricting the AC from E to $\bar{Y}$.

(7) $\bar{\phi}_Y = \text{shift}(\text{a}(\bar{Y}), \bar{\rho}.\text{rC} \wedge \text{right}(\bar{\rho}, \bar{\rho}.\text{lC})) \wedge \text{shift}(\bar{m} \circ e_2, \bar{\phi}_G)$ is the GC for $\bar{Y}$ obtained using GC propagation as in [13].

(8) $\bar{\pi}_0 = (\bar{X}, \bar{\phi}_X) \cdot (m, \rho, n = \bar{m} \circ e_1) \cdot \text{Ref}(\pi, \bar{m} \circ e_2, \bar{\phi}_Y)$ is obtained by prepending the new step to the path refinement of π according to $\bar{m} \circ e_2$ and $\bar{\phi}_Y$.

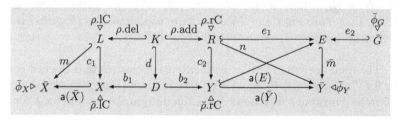

(9) (Disambiguation of Abstraction for ϕ_I) If $\exists(\text{i}(\bar{X}), \bar{\phi}_X)$ is consistent with ϕ_I (see Definition 8), which can be checked using $\mathcal{A}$, we know that $(\bar{X}, \bar{\phi}_X)$ either only covers graphs satisfying ϕ_I or no such graphs. In this case, $\tilde{\pi}_1$ is $\text{Ref}(\tilde{\pi}, \text{id}(\bar{X}), \phi_I)$ or $\text{Ref}(\tilde{\pi}, \text{id}(\bar{X}), \neg\phi_I)$ (where $\text{id}(\bar{X})$ is the identity morphism on $\bar{X}$) and $\tilde{\pi}_1 = \tilde{\pi}$ otherwise.

(10) (Disambiguation of Abstraction for ϕ_Z) Analogous to item (9) for the GC ϕ_Z representing the initial state of the GTS at hand obtaining $\tilde{\pi}_2$ from $\tilde{\pi}_1$.

---

[3] $\mathcal{E}'$ denotes the set of pairs of monomorphisms that are jointly epimorphic, that is, two monomorphisms that map to each graph element of their common target graph.

Table 1. Results of invariant analysis for the abstract LTS for shuttle scenario.

| | | | Outcome (b, *paths*) of analysis algorithm $\mathcal{I}$ | |
Lookahead n	Path length k	Duration	Element b	Size of element *paths*
2	2	1 s	u	6
3	3	2 s	u	12
4	4	12 s	u	8
5	5	63 s	i	0

(11) *(Nonemptyness of Abstraction) If $\mathcal{A}$ and Z3 determine that $\tilde{\pi}_2$ represents at least one concrete violation (as discussed subsequent to Definition 15) compatible with ϕ_A, then π' is equal to $\tilde{\pi}_2$ (otherwise $\tilde{\pi}_2$ results in no path extension).*

Figure 3f depicts two applications of Ext both requiring applications of Ref (the first refinement regarding the empty path π_0 is trivial and the second has been discussed above). The first extension uses ρ_1 from Fig. 3a, constructs the overlapping E_0 where the two B nodes are identified (not explicitly depicted), applies the reversal of rule ρ_1 using the match e_{01} to obtain X_1, and then applies ρ_1 to obtain the AC refinement $X_2 = (G_2, \phi_2)$ of E_0 depicted in Fig. 3g. Note that $X_2 = (G_2, \phi_2)$ still violates the invariant candidate ϕ_I (for all monomorphisms $m : G_2 \hookrightarrow H$). The further extension using ρ_2 then results in path π_2 ending in $X_5 = (G_5, \phi_5)$, which does not need to be considered further as X_5 violates the assumed invariant ϕ_A (for all monomorphisms $m : G_5 \hookrightarrow H$).

Finally, $\mathcal{I}$ from Definition 5 can be used to check a GTS against an invariant candidate ϕ_I by applying $\mathcal{I}$ using the described instantiation.

Theorem 3 (Instantiation of k-Induction for GTSs). *If (P, ϕ_Z) is a GTS, $\phi_A \in \mathcal{S}_\emptyset^{GC}$ is an assumed invariant, $\phi_I \in \mathcal{S}_\emptyset^{GC}$ is an invariant candidate, $k \in \mathbf{N}$, and the application of the algorithm $\mathcal{I}$ using the described instantiation Γ' for Γ, Ext (from Definition 16) for ext, and $\{(\emptyset, \neg\phi_I)\}$ for paths$_0$ terminates with $(b, paths)$, then Theorem 1 and Theorem 2 are applicable and $(b, paths)$ is a sound judgement on whether ϕ_I is an invariant for (P, ϕ_Z).*

Proof. The used operation Ext for path extension ensures that the last computed sub-LTS Γ'' of Γ' results in an LTSAR (see Lemma 1) meeting the requirements R1–R5 from Theorem 2 as follows (by induction on the parameter k for R4).

- Requirements *R1* and *R2* (preservation of invariant and initial state): Ensured by item (9) and item (10) in Definition 16.
- Requirement *R3* (R_S is right total): Ensured by item (11) in Definition 16.
- Requirement *R4* (R_S is left total on violating states): R4 means that each state G that violates ϕ_I in Γ via some shortest violation is covered by some state $(\bar{G}, \bar{\phi}_G)$ of Γ''. R4 is obviously satisfied by the initial LTS candidate that has the only state $(\emptyset, \neg\phi_I)$. Moreover, every extension (entailing the described refinement) of the set of paths in each iteration preserves this property because the refinement only excludes paths that are known to be only covering paths not representing shortest violations.

- Requirement *R5* (forward steps of Γ' are simulated by Γ): Ensured by applying the path refinement operation Ref in the operation Ref.

Lastly, the requirement R6 is satisfied for all states that are not at the beginning of a path in Γ'' since Ext considers all possible backward steps. □

For our running example, we applied our prototypical implementation of the analysis algorithm $\mathcal{I}$. For $k = 2$, we obtained the indefinite result (u, *paths*) where the sequence from Fig. 2c is a concretization of a path in *paths* that could not be ruled out. As stated in Table 1, a path length of $k = 5$ (i.e., 5-induction) was required to establish that ϕ_I is an invariant. While the time required for invariant analysis increases exponentially with longer values of k due to the exponentially increasing number of paths of that length, we believe that the analysis times required for the running example already demonstrate feasibility albeit a potential for further optimizations of our prototypical implementation. Also note that the number of path extensions in each step grows exponentially with the size of the rules.

5   Conclusion and Future Work

We formalized the static analysis approach of k-induction using Isabelle for the abstract setting of LTSs establishing sufficient conditions for the preservation/reflection of invariants by means of an abstraction relation. We then applied this analysis approach to typed attributed GTSs by abstracting graphs by nested graph conditions (GCs) and by applying k-induction on these GCs. Our results extend the state of the art by permitting *attributes* as well as *nested* GCs for the specification of initial states, assumed invariants, and invariant candidates.

In the future, we want to develop support for probabilistic/timed GTSs such as [16]. Moreover, we strive to develop further abstractions to improve support for GTSs with multiple active components such as shuttles. Finally, heuristics guiding the computation of paths in the analysis procedure using parameterizations may improve performance by e.g. prioritizing path extension over checking for violations of attribute constraints of assumed invariants.

References

1. Augur 2 (2008). http://www.ti.inf.uni-due.de/en/research/tools/augur2
2. Graphs for Object-Oriented Verification (GROOVE) (2011). http://groove.cs.utwente.nl
3. EMF Henshin (2013). http://www.eclipse.org/modeling/emft/henshin
4. Becker, B., Giese, H.: On safe service-oriented real-time coordination for autonomous vehicles. In: 11th IEEE International Symposium on Object-Oriented Real-Time Distributed Computing (ISORC 2008), 5–7 May 2008, Orlando, Florida, USA, pp. 203–210. IEEE Computer Society (2008). https://doi.org/10.1109/ISORC.2008.13

5. Boneva, I., Rensink, A., Kurbán, M.E., Bauer, J.: Graph abstraction and abstract graph transformation. Technical report LNCS4549/TR-CTIT-07-50, July 2007
6. Corradini, A., Heindel, T., König, B., Nolte, D., Rensink, A.: Rewriting abstract structures: materialization explained categorically. In: Bojańczyk, M., Simpson, A. (eds.) FoSSaCS 2019. LNCS, vol. 11425, pp. 169–188. Springer, Cham (2019). https://doi.org/10.1007/978-3-030-17127-8_10
7. Dyck, J.: Increasing expressive power of graph rules and conditions and automatic verification with inductive invariants. Master's thesis, University of Potsdam, Hasso Plattner Institute, Potsdam, Germany (2012)
8. Dyck, J.: Verification of graph transformation systems with k-inductive invariants. Ph.D. thesis, University of Potsdam, Hasso Plattner Institute, Potsdam, Germany (2020). https://doi.org/10.25932/publishup-44274
9. Dyck, J., Giese, H.: Inductive invariant checking with partial negative application conditions. In: Parisi-Presicce, F., Westfechtel, B. (eds.) ICGT 2015. LNCS, vol. 9151, pp. 237–253. Springer, Cham (2015). https://doi.org/10.1007/978-3-319-21145-9_15
10. Dyck, J., Giese, H.: k-inductive invariant checking for graph transformation systems. In: de Lara, J., Plump, D. (eds.) ICGT 2017. LNCS, vol. 10373, pp. 142–158. Springer, Cham (2017). https://doi.org/10.1007/978-3-319-61470-0_9
11. Dyck, J., Giese, H.: k-inductive invariant checking for graph transformation systems. Technical report 119, Hasso Plattner Institute at the University of Potsdam, Potsdam, Germany (2017)
12. Ehrig, H., Ehrig, K., Prange, U., Taentzer, G.: Fundamentals of Algebraic Graph Transformation. Springer, Berlin (2006). https://doi.org/10.1007/3-540-31188-2
13. Ehrig, H., Golas, U., Habel, A., Lambers, L., Orejas, F.: $\mathcal{M}$-adhesive transformation systems with nested application conditions. part 1: parallelism, concurrency and amalgamation. Math. Struct. Comput. Sci. **24**(4) (2014). https://doi.org/10.1017/S0960129512000357
14. Habel, A., Pennemann, K.: Correctness of high-level transformation systems relative to nested conditions. Math. Struct. Comput. Sci. **19**(2), 245–296 (2009). https://doi.org/10.1017/S0960129508007202
15. Kulcsár, G.: A compass to controlled graph rewriting. Ph.D. thesis, Technische Universität Darmstadt, January 2019. http://tuprints.ulb.tu-darmstadt.de/9304/
16. Maximova, M., Giese, H., Krause, C.: Probabilistic timed graph transformation systems. J. Log. Algebr. Meth. Program. **101**, 110–131 (2018). https://doi.org/10.1016/j.jlamp.2018.09.003
17. Microsoft Corporation: Z3. https://github.com/Z3Prover/z3
18. Orejas, F.: Attributed graph constraints. In: Ehrig, H., Heckel, R., Rozenberg, G., Taentzer, G. (eds.) ICGT 2008. LNCS, vol. 5214, pp. 274–288. Springer, Heidelberg (2008). https://doi.org/10.1007/978-3-540-87405-8_19
19. Orejas, F.: Symbolic graphs for attributed graph constraints. J. Symb. Comput. **46**(3), 294–315 (2011). https://doi.org/10.1016/j.jsc.2010.09.009
20. Orejas, F., Lambers, L.: Delaying constraint solving in symbolic graph transformation. In: Ehrig, H., Rensink, A., Rozenberg, G., Schürr, A. (eds.) ICGT 2010. LNCS, vol. 6372, pp. 43–58. Springer, Heidelberg (2010). https://doi.org/10.1007/978-3-642-15928-2_4
21. Orejas, F., Lambers, L.: Symbolic attributed graphs for attributed graph transformation. ECEASST **30** (2010). http://journal.ub.tu-berlin.de/index.php/eceasst/article/view/405
22. Orejas, F., Lambers, L.: Lazy graph transformation. Fundam. Inform. **118**(1–2), 65–96 (2012). https://doi.org/10.3233/FI-2012-706

23. Pennemann, K.: Development of correct graph transformation systems. Ph.D. thesis, University of Oldenburg, Germany (2009). http://oops.uni-oldenburg.de/884/. URN http://nbn-resolving.de/urn:nbn:de:gbv:715-oops-9483

24. Poskitt, C.M., Plump, D.: Verifying monadic second-order properties of graph programs. In: Giese, H., König, B. (eds.) ICGT 2014. LNCS, vol. 8571, pp. 33–48. Springer, Cham (2014). https://doi.org/10.1007/978-3-319-09108-2_3

25. Rensink, A.: Canonical graph shapes. In: Schmidt, D. (ed.) ESOP 2004. LNCS, vol. 2986, pp. 401–415. Springer, Heidelberg (2004). https://doi.org/10.1007/978-3-540-24725-8_28

26. Schneider, S., Lambers, L., Orejas, F.: Automated reasoning for attributed graph properties. STTT **20**(6), 705–737 (2018). https://doi.org/10.1007/s10009-018-0496-3

27. Steenken, D.: Verification of infinite-state graph transformation systems via abstraction. Ph.D. thesis, University of Paderborn (2015). https://nbn-resolving.de/urn:nbn:de:hbz:466:2--15768

28. Steenken, D., Wehrheim, H., Wonisch, D.: Sound and complete abstract graph transformation. In: Simao, A., Morgan, C. (eds.) SBMF 2011. LNCS, vol. 7021, pp. 92–107. Springer, Heidelberg (2011). https://doi.org/10.1007/978-3-642-25032-3_7

29. Wang, X., Büttner, F., Lamo, Y.: Verification of graph-based model transformations using alloy. ECEASST **67** (2014). https://doi.org/10.14279/tuj.eceasst.67.943

Optimistic and Pessimistic On-the-fly Analysis for Metric Temporal Graph Logic

Sven Schneider$^{(\boxtimes)}$, Lucas Sakizloglou , Maria Maximova ,
and Holger Giese

University of Potsdam, Hasso Plattner Institute, Potsdam, Germany
{sven.schneider,lucas.sakizloglou,maria.maximova,holger.giese}@hpi.de

Abstract. The nonpropositional *Metric Temporal Graph Logic* (MTGL)
specifies the behavior of timed dynamic systems given by *timed graph
sequences* (TGSs), which contain typed attributed graphs representing
system states and the elapsed time between states. MTGL satisfaction can
be analyzed for *finite* TGSs by translating its satisfaction problem to the
satisfaction problem of nested graph conditions using a folding operation
(aggregating a TGS into a *graph with history*) and a reduction operation
(translating an MTGL condition into a nested graph condition).

In this paper, we introduce an analysis procedure for MTGL to allow
for an *on-the-fly* analysis of *finite/infinite* TGSs. To this end, we introduce a further (optimistic) reduction of MTGL conditions, which leads
to violations during the on-the-fly analysis only when non-satisfaction is
guaranteed in the future whereas the former (pessimistic) reduction leads
to violations when satisfaction is not guaranteed in the future. We motivate the relevance of our analysis procedure, which uses both reduction
operations, by means of a running example. Finally, we discuss prototypical support in the tool AUTOGRAPH.

Keywords: Graph logic with binding · Nonpropositional metric
temporal logic · Runtime monitoring · Three-valued logic

1 Introduction

The challenges for developing embedded real-time systems with a high degree
of parallelism, data dependencies, and timing constraints that must adhere to a
given specification are manifold. The formal verification of such systems given
by formal models is often intractable and, moreover, such formal models cannot
be obtained for systems with unpredictable behaviors such as human-in-the-loop
systems. Model-based testing and runtime monitoring are two standard model-
driven approaches supporting the engineering of such systems.

Funded by the Deutsche Forschungsgemeinschaft (DFG, German Research Foundation)
- 241885098, 148420506, 158230677.

F. Gadducci and T. Kehrer (Eds.): ICGT 2020, LNCS 12150, pp. 276–294, 2020.
https://doi.org/10.1007/978-3-030-51372-6_16

In this paper, we check the conformance of a *timed graph sequence* (TGS), given by a sequence of states where time elapses between the states, against a formal specification. TGSs can be (a) generated using timed graph transformation [3,7,17] or can be (b) returned incrementally by a monitor. While offline analysis procedures suffice for finite TGSs, infinite TGSs require an on-the-fly conformance analysis to determine violations as early and as precise as possible.

As a running example, we consider an operating system as an advanced embedded real-time system in which tasks are executed by handlers computing results. For such an operating system, we require the following property **P**.

P: Whenever a task T is created in a system S, S must have a handler H applicable to T (based on a common id). Moreover, within 10 timeunits, H must produce a result R with value *ok* and, during the computation of R, no other handler H' applicable to T (based on a common id) may exist in S.

For specifying desired system behavior such as the property **P** from above, we employ an extension of the *Metric Temporal Graph Logic* (MTGL) [8]. This logic permits to concisely express (a) state properties to specify single graphs in a TGS and (b) sequence properties relating graphs at different timepoints in a TGS by their attributes and inner structure. For state properties, MTGL subsumes the graph logic GL of nested graph conditions [9], which is as expressive as first-order logic on graphs. For sequence properties, MTGL has metric temporal operators that refer to matches of graph patterns in graphs in a TGS as first-class citizens. Due to these operators, MTGL is more expressive compared to metric temporal logics such as MTL [12] only relying on atomic propositions since MTGL allows to keep track of an unbounded number of elements. For example, for the property **P**, we must separately track tasks T and T' for which *corresponding* results R and R' must be created before the *corresponding* deadline expires.

The main contributions of this paper are as follows. Firstly, we integrate the *metric-exists-new* operator as a first-class citizen into MTGL, which was not explicitly done in [8]. It matches graph patterns as early as possible to fix a timepoint from which a deadline can be started as in our running example when a new task is matched. Secondly, we formally integrate attribute quantification into MTGL and GL. Thirdly, we develop an on-the-fly checking procedure, which takes a formal MTGL specification and considers a TGS incrementally returning a lower (optimistic) and an upper (pessimistic) bound of the set of *true violations*, which determine a ground truth of violations that would be obtained ideally. The optimistic lower bound does not contain all true violations to handle TGSs that are continued in a way leading to satisfaction in the future whereas the pessimistic upper bound contains additional false violations to handle TGSs that are continued in a way not leading to satisfaction in the future. Returning the optimistic bound in addition to the pessimistic bound as computed in [8] results essentially in a three valued logic[1] where an intervention (e.g. by a user) may depend on whether a pessimistic violation is also an optimistic violation.

---

[1] At each timepoint during the on-the-fly analysis, we return either no violation, only a pessimistic violation, or a pessimistic and an optimistic violation.

In Sect. 2, we recall symbolic graphs and the logic GL of nested graph conditions. In Sect. 3, we extend MTGL by integrating the operator *metric-exists-new*. We present our on-the-fly analysis procedure and discuss its prototypical tool support by AUTOGRAPH in Sect. 4. Finally, we discuss related work in Sect. 5 and conclude the paper with a summary and remarks on future work in Sect. 6.

2   Symbolic Graphs and Graph Logic

We now recall typed attributed graphs and nested graph conditions used for representing system states and properties on these states, respectively.

We use *symbolic graphs* (see e.g. [18–20,24]), called graphs subsequently, to encode typed attributed graphs. Symbolic graphs are an adaptation of E-GRAPHS [5] where nodes and edges of a graph G are connected to (sorted) variables X_G instead of data nodes representing actual values. To specify the possible values of variables, graphs are equipped with an *attribute constraint (AC)* Θ_G over the variables X_G (e.g. $x = 5$, $\exists y.\, x \le y$, and $z =$ "aabb").

An AC θ is constructed using variables from a set X and the usual operators for the sorts bool, int, real, and string, which range over the set $\mathcal{V}$ of all values. Satisfaction of an AC θ by a valuation $\alpha : X \to \mathcal{V}$ is denoted by $\alpha \models_{\mathrm{AC}} \theta$.[2] If an AC θ is satisfiable or tautological, we write $\mathrm{sat}_\exists(\theta)$ or $\mathrm{sat}_\forall(\theta)$, respectively.

In the following, we consider graphs that are typed over a type graph TG using a typing morphism $type : G \to TG$. Type graphs restrict attributed graphs to an admitted subclass Graphs(TG). The empty graph is denoted by $\emptyset$. For our running example, we employ the type graph TG from Fig. 1a. Examples of graphs that are typed over TG are given in Fig. 3b.

Morphisms $f : G_1 \to G_2$ between graphs G_1 and G_2 are defined as usual (see e.g. [24] for a formal definition) and consist of mappings between the components of G_1 and G_2. In the remainder of this paper, we only use morphisms $f : G_1 \hookrightarrow G_2$ for which all mappings are injective.

Moreover, we distinguish between two kinds of morphisms $f : G_1 \hookrightarrow G_2$. Firstly, *restrictive (mono)morphisms* $f : G_1 \hookrightarrow_r G_2$ must ensure that the AC of G_2 is more restrictive compared to the AC of G_1. This means that each valuation that satisfies the AC of G_2 also satisfies the f-translated AC of G_1 (i.e., $\mathrm{sat}_\forall(\Theta_{G_2} \to f_X(\Theta_{G_1}))$ where f_X is the mapping contained in f between the variables of G_1 and G_2). Secondly, *consistent morphisms* $f : G_1 \hookrightarrow_c G_2$ must ensure that the AC of G_2 is compatible with the AC of G_1. This means that there is at least one valuation that satisfies the AC of G_2 as well as the f-translated AC of G_1 (i.e., $\mathrm{sat}_\exists(\Theta_{G_2} \wedge f_X(\Theta_{G_1}))$). See Fig. 1b for examples of restrictive and consistent morphisms. The initial (mono)morphism $i_G : \emptyset \to G$ for graph G is restrictive and, when the AC of G is satisfiable, also consistent.

---

[2] The solver Z3 [16] has support for checking satisfiability of ACs but is known to return indefinite results because satisfiability is undecidable for ACs of unrestricted form. While Z3 always succeeds for our running example, we handle this special case in our prototypical implementation by providing warnings to the user.

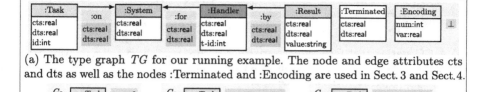

(a) The type graph TG for our running example. The node and edge attributes cts and dts as well as the nodes :Terminated and :Encoding are used in Sect. 3 and Sect. 4.

(b) The restrictive monomorphism $m_1 : G_1 \hookrightarrow_r G_2$ refines the requirements on x_0 and y_0 (i.e., $(x_1 = 4 \wedge y_1 \geq 6) \to (x_1 \geq 2 \wedge y_1 \geq 4)$ is tautological). The consistent morphism $m_2 : G_1 \hookrightarrow_c G_3$ checks for a variable valuation satisfying the constraints of both graphs (e.g. $\{x_2 \mapsto 2, y_2 \mapsto 6\}$ satisfies $(y_2 = 6) \wedge (x_2 \geq 2 \wedge y_2 \geq 4)$).

(c) A GC stating that for every task there is a handler that pertains to the task based on the same attribute values for t-id and id.

Fig. 1. The type graph TG of our running example, two morphisms, and a GC

The *graph logic GL* over nested graph conditions (GCs) can be used to specify graphs. GL features propositional connectives and the operator $\exists$ (called *exists*) to extend given matches of graph patterns (given by graphs) in a graph (called *host graph*) with a satisfiable AC. Technically, the *exists* operator describes the extension of a graph pattern H using a restrictive monomorphism $f : H \hookrightarrow_r H'$. See Fig. 1c for an example of a GC. For improved readability, we visualize restrictive monomorphisms $f : H \hookrightarrow_r H'$ occurring in GCs by omitting graph elements from H not connected to graph elements in $H' \setminus f(H)$.

Definition 1 (Nested Graph Conditions (GCs)). *If H is a graph and $f : H \hookrightarrow_r H'$ is a restrictive monomorphism, then ϕ_H is a nested graph condition (GC) over H, written $\phi_H \in \Phi_H^{GC}$, as follows.*

$$\phi_H ::= \top \mid \neg \phi_H \mid \phi_H \wedge \phi_H \mid \exists(f, \phi_{H'})$$

We also make use of the operators false ($\bot$), disjunction ($\vee$), *and* universal quantification ($\forall$), *which can be derived from the operators above.*

The semantics of GL is given by the satisfaction relation below, which deviates from [24] by using (a) variable valuations $\alpha : X_H \to \mathcal{V}$ and (b) consistent instead of restrictive matches $m : H \hookrightarrow_c G$. Most notably, the GC $\phi = \exists(f : H \hookrightarrow_r H', \phi')$ is satisfied by a match m and a valuation α when they can be extended to a match $m' : H' \hookrightarrow_c G$ and a valuation $\alpha' : X_{H'} \to \mathcal{V}$ that are consistent with f. To ensure that the valuation α' is used consistently when evaluating ACs occurring in graphs in ϕ, we require that the consistency condition is satisfied

$$\forall \begin{bmatrix} \boxed{\begin{array}{c} x{:}Ax{:}\text{int} \\ 0 \le x \le 2 \end{array}}, \exists \begin{bmatrix} \boxed{\begin{array}{cc} x{:}Ax{:}\text{int} & y{:}Bx{:}\text{int} \\ y = x \wedge 0 \le x \le 2 \end{array}}, \top \end{bmatrix} \end{bmatrix}$$

$$\boxed{\begin{array}{cc} \bar{x}{:}A\bar{x}{:}\text{int} & y_0{:}B\bar{x}{:}\text{int} \\ y_1{:}B\bar{x}{:}\text{int} & y_2{:}B\bar{x}{:}\text{int} \\ y_0 = 0 \wedge y_1 = 1 \wedge y_2 = 2 \end{array}}$$

(a) A GC stating that for each int variable x of type Ax where $x \in \{0, 1, 2\}$ there is an int variable y of type Bx with the value of x.

(b) A graph G satisfying the GC from Fig. 2a.

$\theta = (y_0 = 0 \wedge y_1 = 1 \wedge y_2 = 2) \rightarrow (\forall \bar{x}{:}\text{int}.\ 0 \le \bar{x} \le 2 \rightarrow \vee\{\bar{y} = \bar{x} \mid \bar{y} \in \{y_0, y_1, y_2\}\})$

(c) The satisfiability of θ shows that G from Fig. 2b satisfies the GC from Fig. 2a.

Fig. 2. Satisfaction of GCs by graphs

by m' (i.e., $\text{sat}_\exists(m'_X(\Theta_{H'}) \wedge \Theta_G)$) using a valuation compatible with α'. Finally, as in [9, 24], a graph G satisfies a GC defined over $\emptyset$ when the initial morphism $i_G : \emptyset \hookrightarrow_c G$ and the empty valuation $\alpha : \emptyset \rightarrow \mathcal{V}$ satisfy the GC.

Definition 2 (Satisfaction of GCs). *If $\phi \in \Phi_H^{\text{GC}}$ is a GC, $m : H \hookrightarrow_c G$ is a consistent morphism, and $\alpha : X_H \rightarrow \mathcal{V}$ is a valuation, then m satisfies ϕ via α, written $(m, \alpha) \models \phi$, if an item applies.*

- *$\phi = \top$.*
- *$\phi = \neg\phi'$ and $(m, \alpha) \not\models \phi'$.*
- *$\phi = \phi_1 \wedge \phi_2$, $(m, \alpha) \models \phi_1$, and $(m, \alpha) \models \phi_2$.*
- *$\phi = \exists(f : H \hookrightarrow_r H', \phi')$ and there are $m' : H' \hookrightarrow_c G$ and $\alpha' : X_{H'} \rightarrow \mathcal{V}$ s.t. $m' \circ f = m$, $\alpha' \circ f_X = \alpha$, $(m', \alpha') \models \phi'$, and $\text{sat}_\exists(\alpha'(\Theta_{H'} \wedge m'^{-1}_X(\Theta_G)))$.*[3]

If $\phi \in \Phi_\emptyset^{\text{GC}}$, $i_G : \emptyset \hookrightarrow_c G$, $\alpha : \emptyset \rightarrow \mathcal{V}$, and $(i_G, \alpha) \models \phi$, then $G \models \phi$.

This novel adaptation of the satisfaction relation (informally handled in [8]) allows to express quantification over attribute values as required for our on-the-fly analysis procedure in Sect. 4. For example, we can state that "for each $x \in \mathbf{Z}$ satisfying $0 \le x \le 2$ there is a variable $y \in \mathbf{Z}$ with the value of x" using the GC ϕ in Fig. 2a, which is satisfied by the graph G from Fig. 2b containing a matchable copy of x. In our implementation in the tool AUTOGRAPH, we operationalize the satisfaction check by constructing the AC from Fig. 2c incorporating all possible matches for the variable y for which satisfiability then implies that G satisfies ϕ.

3   Metric Temporal Graph Logic

The *Metric Temporal Graph Logic (MTGL)* over metric temporal graph conditions (MTGCs) [8] extends GL with two metric temporal operators. In particular, (a) the *until* operator U is well-known from MTL [12] and (b) the formal integration of the *metric-exists-new* operator $\exists^N$ allows to match graph patterns at a future timepoint (restricted by an interval) when the match is first available.

---

[3] The partial function $m'^{-1}_X : X_G \rightharpoonup X_{H'}$ obtained as the reversal of the injective function $m'_X : X_{H'} \rightarrow X_G$ does not replace variables in Θ_G that are not mapped to by m'_X. For simplicity, we assume that G and H' have disjoint sets of variables.

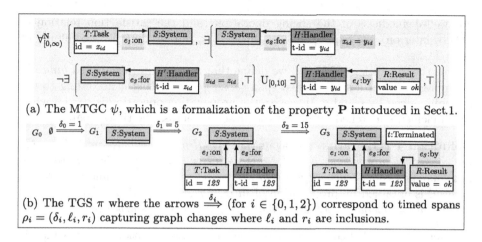

(a) The MTGC ψ, which is a formalization of the property **P** introduced in Sect. 1.

(b) The TGS π where the arrows $\overset{\delta_i}{\Longrightarrow}$ (for $i \in \{0,1,2\}$) correspond to timed spans $\rho_i = (\delta_i, \ell_i, r_i)$ capturing graph changes where ℓ_i and r_i are inclusions.

Fig. 3. The MTGC ψ and the TGS π from our running example

Definition 3 (Metric Temporal Graph Conditions (MTGCs)).

If H is a graph, $f : H \hookrightarrow_r H'$ is a restrictive monomorphism, and I is an interval over $\mathbf{R}_0^+$, then ϕ_H is a metric temporal graph condition (MTGC) over H, written $\phi_H \in \Phi_H^{\mathrm{MTGC}}$, as follows.

$$\phi_H ::= \top \mid \neg\phi_H \mid \phi_H \wedge \phi_H \mid \exists(f, \phi_{H'}) \mid \exists_I^N(f, \phi_{H'}) \mid \phi_H \mathbf{U}_I \phi_H$$

The derived operator metric-forall-new $(\forall^N)$ is the dual operator to $\exists^N$. Also, operators such as eventually *and* globally *can be derived using the operator* U.

The integration of graph pattern matching of GCs (for state properties) and the two metric temporal operators (for sequence properties) allows for the formalization of properties where a match into a graph (established using the operators $\exists$ or $\exists^N$) is preserved/extended over multiple timed steps of the system (using the operator U) and where the duration of these timed steps can be specified using intervals. For our running example, see Fig. 3a for a formalization of the property **P** introduced in Sect. 1 in the form of an MTGC ψ.

The semantics of MTGL is defined over *Timed Graph Sequences* (TGSs) [8] (e.g. π in Fig. 3b), which describe a single (possibly infinite) evolution of a system in terms of its visited states and the durations between these states. A TGS π starting in a graph G (written $\pi \in \Pi_G$ or $\pi \in \Pi_G^{fin}$ when π is finite) is a sequence of timed spans $(\delta, \ell : D \hookrightarrow_r G_1, r : D \hookrightarrow_r G_2)$ where δ is the relative time at which the successor state G_2 is reached from G_1 and where ℓ and r describe the deletion and addition of graph elements as usual. We also assume the reversal operation $\mathrm{rev}(\pi)$, the concatenation operation $\pi_1; \pi_2$, the prefix relation $\pi_1 \sqsubseteq \pi_2$, the length operation $\mathrm{length}(\pi) \in \mathbf{N} \cup \{\infty\}$ (which counts the timed spans), the duration operation $\mathrm{dur}(\pi) \in \mathbf{R}_0^+ \cup \{\infty\}$ (which sums the durations δ of all timed spans), the projection operation π_t to the graph at timepoint t, and the slicing operation $\pi_{[t_1,t_2]}$ delivering the TGS between timepoints t_1 and t_2.

For well-definedness of the slicing operation (and the satisfaction relation of MTGL later on), we require that $\mathsf{length}(\pi) = \infty$ implies $\mathsf{dur}(\pi) = \infty$ ruling out Zeno TGSs.

For the semantics of MTGL, we define that a match m (also called *binding*) is propagated over a single timed span (δ, ℓ, r) by adapting it according to the renaming given by ℓ and r. However, the propagation operation is partial when graph elements matched by m are not preserved across the timed span.

Definition 4 (Operation propagateMatch). *If $m : H \hookrightarrow_c G_1$, $m' : H \hookrightarrow_c G_2$ are consistent morphisms, $\rho = (\delta, \ell : D \hookrightarrow_r G_1, r : D \hookrightarrow_r G_2)$ is a timed span, and there is a consistent morphism $\bar{m} : H \hookrightarrow_c D$ s.t. $\ell \circ \bar{m} = m$ and $r \circ \bar{m} = m'$, then m' is obtained by propagation of m over ρ, written $m' \in \mathsf{PM}(m, \rho)$. Also, we extend the operation propagateMatch to finite TGSs as expected.*

The semantics of MTGL is given by the satisfaction relation below, which is defined as for GL for the operators *conjunction*, *negation*, and *exists* and uses a current observation timepoint t (which does not exceed the duration of the TGS) in addition to a consistent match $m : H \hookrightarrow_c G$ and a valuation $\alpha : X_H \to \mathcal{V}$. Note that the target of the match m is adapted (via propagation over timed spans from TGS π) and the current timepoint t is modified (according to the provided intervals) only in the cases of the *metric-exists-new* and *until* operators. For these two metric temporal operators, we provide further informal explanations below. Finally, a TGS satisfies an MTGC defined over $\emptyset$ when the TGS, the initial observation timepoint 0, the initial morphism $i_{\pi_0} : \emptyset \hookrightarrow_c \pi_0$ representing an empty binding, and the empty valuation $\alpha : \emptyset \to \mathcal{V}$ satisfy the MTGC.

Definition 5 (Satisfaction of MTGCs by TGSs). *If $\phi \in \Phi_H^{\mathrm{MTGC}}$ is an MTGC, $\pi \in \Pi_G$ is a TGS, $0 \leq t \leq \mathsf{dur}(\pi)$ is a timepoint before the end of π, $m : H \hookrightarrow_c \pi_t$ is a consistent morphism into the graph at timepoint t, and $\alpha : X_H \to \mathcal{V}$ is a valuation, then $(\pi, t, m, \alpha) \models_{TGS} \phi$, if an item applies.*

- $\phi = \top$.
- $\phi = \neg\phi'$ *and* $(\pi, t, m, \alpha) \not\models_{TGS} \phi'$.
- $\phi = \phi_1 \wedge \phi_2$, $(\pi, t, m, \alpha) \models_{TGS} \phi_1$, *and* $(\pi, t, m, \alpha) \models_{TGS} \phi_2$.
- $\phi = \exists(f : H \hookrightarrow_r H', \phi')$ *and there are* $m' : H' \hookrightarrow_c \pi_t$ *and* $\alpha' : X_{H'} \to \mathcal{V}$ *s.t.* $m' \circ f = m$, $\alpha' \circ f_X = \alpha$, $(\pi, t, m', \alpha') \models_{TGS} \phi'$, *and* $\mathsf{sat}_\exists(\alpha'(\Theta_{H'} \wedge m_X'^{-1}(\Theta_{\pi_t})))$.
- $\phi = \exists_I^N(f : H \hookrightarrow_r H', \phi')$ *and there are* $t' \in t + I$, $m' \in \mathsf{PM}(m, \pi_{[t,t']})$, *and* $m'' : H' \hookrightarrow_c \pi_{t'}$ *s.t.*
  - *there is* $\alpha' : X_{H'} \to \mathcal{V}$ *s.t.* $m'' \circ f = m'$, $\alpha' \circ f_X = \alpha$, $(\pi, t', m'', \alpha') \models_{TGS} \phi'$, $\mathsf{sat}_\exists(\alpha'(\Theta_{H'} \wedge m_X''^{-1}(\Theta_{\pi_{t'}})))$, *and*
  - *for each* $t'' \in [0, t')$ *there is no* $m''' \in \mathsf{PM}(m'', \mathsf{rev}(\pi_{[t'',t']}))$.
- $\phi = \phi_1' \mathbin{\mathsf{U}}_I \phi_2'$ *and there is* $t' \in t + I$ *s.t.*
  - *there is* $m' \in \mathsf{PM}(m, \pi_{[t,t']})$ *s.t.* $(\pi, t', m', \alpha) \models_{TGS} \phi_2'$ *and*
  - *for each* $t'' \in [t, t')$ *there is* $m'' \in \mathsf{PM}(m, \pi_{[t,t'']})$ *s.t.* $(\pi, t'', m'', \alpha) \models_{TGS} \phi_1'$.

If $\phi \in \Phi_\emptyset^{\mathrm{MTGC}}$, $i_{\pi_0} : \emptyset \hookrightarrow_c \pi_0$, $\alpha : \emptyset \to \mathcal{V}$, *and* $(\pi, 0, i_{\pi_0}, \alpha) \models_{TGS} \phi$, *then* $\pi \models_{TGS} \phi$.

For the *metric-exists-new* operator, we state that there is some timepoint t' in the future that is compatible with the given interval I where the propagated match m' can be extended to a match m'' such that (first item) this extension is compatible with f as required for the case of *exists* and (second item) the extended match m'' *cannot* be propagated to any timepoint in the strict past. For the *until* operator, we state (first item) that ϕ'_2 is eventually satisfied for some timepoint t' in the future that is compatible with the given interval I and (second item) that ϕ'_1 is permanently satisfied for all timepoints between the current observation timepoint t and the timepoint t' except for t'.

For our running example, see Fig. 3b for a TGS π, which does not satisfy the MTGC ψ from Fig. 3a. In this TGS, we find a fresh match of a task on a system at timepoint $t = 6$ (after step 2). Moreover, we find immediately a (unique) handler for this task (based on the common id *123*) such that there is no second handler for that task (with common id) until we find at timepoint $t = 21$ a result with the successful attribute value *ok* obtained by the handler. However, this result is not obtained within the specified interval of at most 10 timeunits, which corresponds to the global time interval $6 + [0, 10]$ in this case.

4   On-the-fly Analysis for MTGL

We now present an on-the-fly analysis procedure for checking MTGCs against *finite and infinite* TGSs as our main contribution. For this aim, we build upon [8] where an operation $\mathcal{R}$educe for translating an MTGC ψ into a GC ψ' and an operation $\mathcal{F}$old for translating a finite TGS π into a so-called *graph with history (GH)* G were presented. These two operations ensured that $\pi \models_{\text{TGS}} \psi$ iff $G \models \psi'$, which allows for an efficient check of MTGL satisfaction for *finite* TGSs by checking GL satisfaction for finite GHs instead. Note that the problem of checking MTGL satisfaction (as for other metric temporal logics) becomes particularly difficult when instances of *until* operators are nested.

When considering a TGS π in an on-the-fly scenario where timed spans are added one-by-one, we cannot simply apply the procedure from [8] to all prefixes of π because the MTGL satisfaction relation is inherently pessimistic not returning the desirable results. For example, the MTGC ψ from Fig. 3a would be violated by the TGS π from Fig. 3b not only at timepoint 21 when the violation of the deadline is detected (as discussed before) but also at timepoint 6 since the prefix of length 2 of π does not contain any node of type Result. Note that we would indeed expect a violation at timepoint 6 already when the prefix of length 2 of π would be the entire TGS to be considered. As a ground truth of violations that would ideally be returned by our procedure, we define *true violations* later on in Definition 12 where also all subsequent behavior given by the timed spans in the TGS, which is not available to the on-the-fly procedure, is also taken into account.

For our on-the-fly analysis procedure (see Fig. 4 for an overview), we employ extensions/adaptations of the operations $\mathcal{R}$educe and $\mathcal{F}$old from [8]. As inputs, we consider an MTGC ψ and a finite/infinite TGS π that is incrementally considered. If the given TGS π is finite, it may include a node of type Terminated in

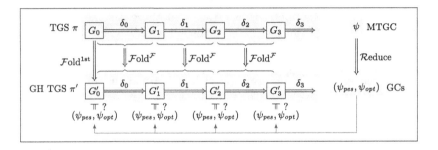

Fig. 4. Overview of the on-the-fly analysis procedure

its last graph to denote that it is not going to be continued (see Fig. 3b). Firstly, we employ a modification of the operation $\mathcal{R}$educe (see Definition 9) to obtain for ψ (via a parameter $mode \in \{\mathcal{P}es, \mathcal{O}pt\}$), in addition to the pessimistic GC ψ_{pes} as in [8], a second optimistic GC ψ_{opt}. Note that we apply the operation $\mathcal{R}$educe offline (once for each mode) before considering the timed spans of the TGS for increased efficiency. Secondly, we split the operation $\mathcal{F}$old into two operations $\mathcal{F}$old^{1st} and $\mathcal{F}$old$^{\mathcal{F}}$ to allow for an incremental rather than a batch folding of a TGS. That is, $\mathcal{F}$old^{1st} produces a first GH from the start graph of the TGS and $\mathcal{F}$old$^{\mathcal{F}}$ modifies a given GH G'_i into a GH G'_{i+1} for each timed span from the TGS as soon as that timed span is available. The sequence of GHs constructed in this way results in a *GH TGS* π' that corresponds to the prefix of the TGS available so far. Thirdly, we check for the conformance of each G'_i with ψ_{pes} and ψ_{opt} as soon as G'_i is available by separately applying the GL satisfaction relation to G'_i and ψ_{pes} as well as to G'_i and ψ_{opt}. Each determined non-satisfaction is a violation for which we add the global time of its occurrence (which is given by the sum of all δ_i so far) to the resulting sets of pessimistic and optimistic violations.

The operations $\mathcal{F}$old^{1st} and $\mathcal{F}$old$^{\mathcal{F}}$ ensure that each GH resulting from folding a TGS contains for each node/edge occurring in the TGS the timepoints of its creation and (if it was deleted) its deletion using additional cts and dts attributes. For our running example, the type graph TG from Fig. 1a contains these cts and dts attributes already and the GH eventually obtained for the entire finite terminated TGS π from Fig. 3b is given in Fig. 5c.

The operation $\mathcal{R}$educe returns GCs that encode the satisfaction checks for MTGL operators according to Definition 5 using ACs. These ACs make use of (a) the cts and dts attributes as added to the GH to control the matching of elements, (b) additional variables for quantifying over observation timepoints as in Definition 5, and (c) a variable x_{tv} storing the current global time (i.e., the duration of the considered TGS prefix π'). As for the variable $x:Ax:$int in Fig. 2, the additional quantified variables and the variable x_{tv} for the current global time are required to be contained in the GHs. The operation $\mathcal{R}$educe returns for this purpose, besides the GC, also a graph G_{tv} containing these additional variables including the distinguished global time variable x_{tv}. For our running example, the graph G_{tv} is given in Fig. 5b where all variables are associated with :Encoding nodes to decrease the number of matches via the num attributes that need to be considered when we check whether a GH subsuming G_{tv} satisfies the obtained GC. Note that the construction of :Encoding nodes is omitted later in Definition 9 to ease presentation. While G_{tv} is a subgraph of each constructed GH, we add the AC $x_{tv} = \mathsf{dur}(\pi')$ to the current GH assigning the current global time to the variable x_{tv} just before checking GL satisfaction whenever a new GH has been constructed during the on-the-fly analysis.

We now define the operations used in the on-the-fly analysis procedure. The predicate Gtv identifies a unique global time variable x_{tv} in a GH G (subsuming the graph G_{tv} obtained from folding) and the operation Gta adds the AC that assigns the current global time to the x_{tv} into the graph G.

Definition 6 (Predicate Gtv and Operation Gta). *If G is a GH with a variable x of sort real, then G is a graph with global time variable x, written $\mathsf{Gtv}(G, x)$. If, additionally, $t \in \mathbf{R}_0^+$ is a global timepoint and G' is the graph obtained from G by adding the AC $x = t$, then G' is the time-assigned version of G for the timepoint t, written $\mathsf{Gta}(G, x, t) = G'$.*

For incrementally folding a TGS starting in a graph H into a GH, we use in the first step the following operation $\mathcal{F}\mathrm{old}^{1\mathrm{st}}$, which joins the graphs G_{tv} and H as well as adds cts attributes to all nodes and edges originating from H.

Definition 7 (Operation $\mathcal{F}\mathrm{old}^{1\mathrm{st}}$). *If G' is the componentwise disjoint union of the graphs G_{tv} and H where the attributes $\mathrm{cts}(\alpha) = 0$ and $\mathrm{dts}(\alpha) = -1$ are added to each node and edge α originating from H, then $\mathcal{F}\mathrm{old}^{1\mathrm{st}}(G_{tv}, H) = G'$.*

The operation $\mathcal{F}\mathrm{old}^{\mathcal{F}}$ adapts a GH G reached at global timepoint t to a GH G' by incorporating the changes described by a timed span $\rho = (\delta, \ell, r)$. Firstly, dts attributes of nodes and edges deleted by ℓ are updated to the new global timepoint $t + \delta$. Secondly, cts attributes with new global timepoint $t + \delta$ and dts attributes with default value -1 are added to all nodes and edges created by r.

Definition 8 (Operation $\mathcal{F}\mathrm{old}^{\mathcal{F}}$). *If G is a graph, $t \in \mathbf{R}_0^+$ is the current global timepoint, $\rho = (\delta, \ell : D \hookrightarrow_r H_1, r : D \hookrightarrow_r H_2)$ is a timed span, and G' is a graph constructed from G by (a) changing the attribute $\mathrm{dts}(\alpha)$ to $t + \delta$ for each node or edge $\alpha \in H_1 - \ell(D)$, (b) renaming each node and edge $\alpha \in \ell(D)$ according to ℓ, (c) renaming each node and edge $\alpha \in r(D)$ according to r, (d) adding each node and edge $\alpha \in H_2 - r(D)$, and (e) adding the attributes $\mathrm{cts}(\alpha) = t + \delta$ and $\mathrm{dts}(\alpha) = -1$ to each node and edge $\alpha \in H_2 - r(D)$, then $\mathcal{F}\mathrm{old}^{\mathcal{F}}(G, t, \rho) = G'$.*

The operations $\mathcal{F}\mathrm{old}^{1\mathrm{st}}$ and $\mathcal{F}\mathrm{old}^{\mathcal{F}}$ preserve the predicate Gtv implying that the variable x_{tv} for the current global time can still be identified and used after the folding steps during the on-the-fly analysis. Moreover, the operation $\mathcal{F}\mathrm{old}^{\mathcal{F}}$ induces a timed span between the GHs G and G' leading to a GH TGS as discussed above. For our running example, the result of applying $\mathcal{F}\mathrm{old}^{1\mathrm{st}}$ and

S. Schneider et al.

then incrementally $\mathcal{F}\text{old}^{\mathcal{F}}$ to the TGS π from Fig. 3b leads to the graph G_3' in Fig. 5c where the elements of the graph G_{tv} from Fig. 5b are omitted for brevity. Since no elements are deleted in π, all elements in G_3' have dts attributes of value -1.

We apply the operation $\mathcal{R}\text{educe}$, which is equipped with a *mode* parameter, to separately construct the GCs ψ_{pes} and ψ_{opt} for a given MTGC ψ. The following definition of $\mathcal{R}\text{educe}$ extends the definition in [8] by additionally covering the optimistic reduction for *mode* $= \mathcal{O}\text{pt}$ and explicitly integrating the formal reduction for the MTGL operator *metric-exists-new* introduced in Definition 3.

Definition 9 (Operation $\mathcal{R}\text{educe}$). *If $\psi \in \Phi_{\emptyset}^{\text{MTGC}}$ is an MTGC and mode $\in \{\mathcal{P}\text{es}, \mathcal{O}\text{pt}\}$ is the chosen mode for reduction, then $\mathcal{R}\text{educe}(\text{mode}, \psi) = (G_{tv}, x_{tv}, \exists(i_{G_0}, \mathcal{R}\text{educe}_{\text{rec}}(\psi, x_0)))$ where the graph G_0 contains the variable x_{tv} and an additional variable x_0 for the initial observation timepoint 0. The employed recursive operation $\mathcal{R}\text{educe}_{\text{rec}}$ is homomorphic for* true, conjunction, *and* negation *and adds* cts *and* dts *attributes to all nodes and edges in the resulting GC. For the remaining operators, $\mathcal{R}\text{educe}_{\text{rec}}$ is defined as follows where the formal parameter x_t corresponds to the variable t in Definition 5 capturing the timepoint at which the MTGC provided to $\mathcal{R}\text{educe}_{\text{rec}}$ is checked for satisfaction.*

- $\mathcal{R}\text{educe}_{\text{rec}}(\exists(f : H \hookrightarrow_r H', \phi'), x_t) = \exists(f', \mathcal{R}\text{educe}_{\text{rec}}(\phi', x_t))$ *where f' additionally requires* alive$(x_t, H')^4$ *ensuring that H' is matchable at timepoint x_t.*
- $\mathcal{R}\text{educe}_{\text{rec}}(\exists_I^N(f : H \hookrightarrow_r H', \phi'), x_t) = \exists(f', \mathcal{R}\text{educe}_{\text{rec}}(\phi', x_t'))$ *where f' additionally requires a variable x_t' satisfying $x_t' \in (x_t + I) \land$ alive$(x_t', H') \land$ earliest$(x_t', H')^5$ ensuring that H' is matchable at timepoint x_t' but not earlier.*
- $\mathcal{R}\text{educe}_{\text{rec}}(\phi_1' \cup_I \phi_2', x_t) = \psi_{pes} \lor \psi_{ext}$ *with*
  - $\psi_{pes} = \exists(f_0, \mathcal{R}\text{educe}_{\text{rec}}(\phi_2', x_t')) \land \forall(f_1, \mathcal{R}\text{educe}_{\text{rec}}(\phi_1', x_t''))$ *where f_0 additionally requires a variable x_t' satisfying $x_t' \in [0, x_{tv}] \cap (x_t + I)$ ensuring that x_t' is a future timepoint where ϕ_2' is satisfied and f_1 additionally requires a variable x_t'' satisfying $x_t'' \in [x_t, x_t')$ for checking that ϕ_1' is satisfied until timepoint x_t'.*
  - $\psi_{ext} = \bot$ *for mode $= \mathcal{P}\text{es}$ disabling the optimistic check.*
  - $\psi_{ext} = \neg\exists(f_0, \top) \land \exists(f_1, \top) \land \forall(f_2, \mathcal{R}\text{educe}_{\text{rec}}(\phi_1', x_t''))$ *for mode $= \mathcal{O}\text{pt}$ where f_0 additionally requires a :Terminated node disabling the optimistic check when the TGS corresponding to the GH against which the resulting GC is checked is known to have ended, f_1 additionally requires a variable x_t' satisfying $x_t' \in (x_{tv}, \infty) \cap (x_t + I)$ ensuring that there is still a timepoint in the strict future at which ϕ_2' could be satisfied, and f_2 additionally requires a variable x_t'' satisfying $x_t'' \in [x_t, x_{tv}]$ for checking that ϕ_1' was satisfied at least until the current global time x_{tv}.*

The returned graph G_{tv} contains all additional variables used in the reduction.^{6}

---

^{4} alive(x, H) is an AC based on cts and dts attributes stating that all nodes and edges in H are created and not yet deleted at timepoint x.

^{5} earliest(x, H) is an AC stating that the highest cts attribute value in H is x.

^{6} Note that the predicate $\text{Gtv}(G_{tv}, x_{tv})$ is satisfied by construction.

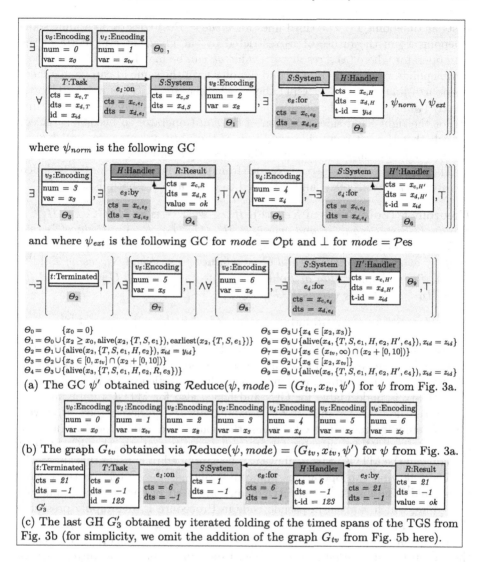

$\Theta_0 = \{x_0 = 0\}$
$\Theta_1 = \Theta_0 \cup \{x_2 \ge x_0, \text{alive}(x_2, \{T, S, e_1\}), \text{earliest}(x_2, \{T, S, e_1\})\}$
$\Theta_2 = \Theta_1 \cup \{\text{alive}(x_2, \{T, S, e_1, H, e_2\}), x_{id} = y_{id}\}$
$\Theta_3 = \Theta_2 \cup \{x_3 \in [0, x_{tv}] \cap (x_2 + [0, 10])\}$
$\Theta_4 = \Theta_3 \cup \{\text{alive}(x_3, \{T, S, e_1, H, e_2, R, e_3\})\}$

$\Theta_5 = \Theta_3 \cup \{x_4 \in [x_2, x_3)\}$
$\Theta_6 = \Theta_5 \cup \{\text{alive}(x_4, \{T, S, e_1, H, e_2, H', e_4\}), x_{id} = z_{id}\}$
$\Theta_7 = \Theta_2 \cup \{x_5 \in (x_{tv}, \infty) \cap (x_2 + [0, 10])\}$
$\Theta_8 = \Theta_2 \cup \{x_6 \in [x_2, x_{tv}]\}$
$\Theta_9 = \Theta_8 \cup \{\text{alive}(x_6, \{T, S, e_1, H, e_2, H', e_4\}), x_{id} = z_{id}\}$

(a) The GC ψ' obtained using $\mathcal{R}educe(\psi, mode) = (G_{tv}, x_{tv}, \psi')$ for ψ from Fig. 3a.

(b) The graph G_{tv} obtained via $\mathcal{R}educe(\psi, mode) = (G_{tv}, x_{tv}, \psi')$ for ψ from Fig. 3a.

(c) The last GH G'_3 obtained by iterated folding of the timed spans of the TGS from Fig. 3b (for simplicity, we omit the addition of the graph G_{tv} from Fig. 5b here).

Fig. 5. The results of reducing the MTGC ψ from Fig. 3a and the result of incrementally folding the entire TGS π from Fig. 3b

For our running example, the result of the reduction of the MTGC ψ from Fig. 3a is given in Fig. 5a for both reduction modes. The first line requires that the initial observation timepoint x_0 is 0 and that the variable x_{tv} is matched into the GH thereby binding it to the current global time. The second line (a) attempts to match T, e_1, and S for some observation timepoint x_2 in the future of x_0 such that the matched elements are alive at timepoint x_2 and not all of them are alive at any earlier timepoint (encoding the *metric-forall-new* operator) and (b) then checks whether a suitable handler with the same id exists at timepoint x_2. The third line (for *mode* = $\mathcal{P}$es) tries to determine some timepoint x_3 in the future of x_2 restricted by the given interval $[0, 10]$ of the *until* operator where (a) a result with value *ok* can be matched and where (b) at all timepoints $x_4 \in [x_2, x_3)$ no second handler with the same id can be matched. The fourth line (for *mode* = $\mathcal{O}$pt) checks whether the TGS has not terminated already (which disables the optimistic checking), whether there is still some time left for the right-hand side condition of the *until* operator to become satisfied, and whether the left-hand side condition of the *until* operator was satisfied at all timepoints x_6 since x_2 and up-to the current global time.

The theorem on soundness of $\mathcal{R}$educe from [8] is now adapted for the included operator *metric-exists-new* and extended to cover also the optimistic mode.

Theorem 1 (Soundness of $\mathcal{R}$educe). *If $\psi \in \Phi_{\emptyset}^{\mathrm{MTGC}}$ is an MTGC, $\pi \in \Pi_H^{fin}$ is a finite TGS starting in H and ending in H', $(G_{tv}, x_{tv}, \psi_{pes}) = \mathcal{R}educe(\mathcal{P}es, \psi)$ is the result of pessimistic reduction, $(G_{tv}, x_{tv}, \psi_{opt}) = \mathcal{R}educe(\mathcal{O}pt, \psi)$ is the result of optimistic reduction, and G is obtained from π and G_{tv} using $\mathcal{F}old^{1st}$ and the iterated application of $\mathcal{F}old^{\mathcal{F}}$, then (a) $\pi \models_{TGS} \psi$ iff $G \models \psi_{pes}$ and (b) $\exists \pi' \in \Pi_{H'}. \pi; \pi' \models_{TGS} \psi$ only if $G \models \psi_{opt}$.*

Proof Idea. Straightforward inductions on ψ and π along the lines of [8].   □

The reverse direction of the item (b) in the theorem above does not hold for the MTGC $\phi = \top \, U_{[2,2]} \perp$ even though non-satisfaction in the future is guaranteed. Moreover, no other reduction can achieve the equivalence for item (b) since satisfiability is undecidable for GCs and hence also for MTGCs implying that guaranteed non-satisfiability as for ϕ cannot always be determined. However, we may simplify generated GCs using the sound and refutationally complete procedure from [24] for checking GL satisfiability. For example, the GC obtained by reducing the MTGC ϕ from above can be simplified to $\perp$ using this approach.

The following two operations $\mathcal{A}nalyze^{1st}$ and $\mathcal{A}nalyze^{\mathcal{F}}$ rely on the operations $\mathcal{F}old^{1st}$ and $\mathcal{F}old^{\mathcal{F}}$, respectively, and are used in our on-the-fly analysis procedure, which is given in pseudo code in Procedure 1. To simplify presentation, Procedure 1 describes the on-the-fly analysis using only one of both modes requiring that two instances of Procedure 1 are executed concurrently to cover the optimistic and the pessimistic mode. Procedure 1 maintains during its execution a triple (G, t, V) consisting of the current GH G, the current global time t, and the set of computed violations V. The operation $\mathcal{A}nalyze^{1st}$ constructs the first triple (G, t, V) by applying $\mathcal{F}old^{1st}$ and by inserting the initial timepoint 0 into the set of violations when the resulting GH G does not satisfy the GC ψ'.

Definition 10 (Operation $\mathcal{A}nalyze^{1st}$). *If (G_{tv}, x_{tv}, ψ') was returned by an application of the operation $\mathcal{R}educe$, $\pi \in \Pi_H^{fin}$ is a finite TGS starting in graph H, G is the GH obtained using $\mathcal{F}old^{1st}(G_{tv}, H)$, and if $\mathsf{Gta}(G, x_{tv}, 0) \not\models \psi'$ then $V = \{0\}$ else $V = \emptyset$, then $\mathcal{A}nalyze^{1st}((G_{tv}, x_{tv}, \psi'), \pi) = (0, G, V)$.*

The operation $\mathcal{A}nalyze^{\mathcal{F}}$ modifies a triple (G, t, V) to a triple (G', t', V') according to a timed span ρ by modifying the GH G to a GH G', by increasing the

---

Procedure 1. On-the-fly Analysis Procedure

Input: *mode* {A mode parameter from $\{\mathcal{P}es, \mathcal{O}pt\}$}
Input: ψ {An MTGC from $\Phi_\emptyset^{\text{MTGC}}$}
Input: H {A start graph}
Input: *stream* {A stream of timed spans starting with H}
  1: $(G_{tv}, x_{tv}, \psi') \leftarrow \mathcal{R}educe(mode, \psi)$
  2: $t \leftarrow 0$ {Current global time}
  3: $G \leftarrow \mathcal{F}\text{old}^{\text{1st}}(G_{tv}, H)$ {Current GH}   ⎤
  4: **loop**   ⎥ $\mathcal{A}nalyze^{\text{1st}}$
  5:     **if** $\text{Gta}(G, x_{tv}, t) \not\models \psi'$ **then**   ⎥
  6:         **print** violation at timepoint t {Output violation} ⎥
  7:     **end if**   ⎦
  8:     **if** *stream*.$\text{hasNext}()$ **then**
  9:         $\rho \leftarrow$ *stream*.$\text{next}()$ {Get next timed span ρ}
 10:         $G \leftarrow \mathcal{F}\text{old}^{\mathcal{F}}(G, t, \rho)$ {Modify GH according to ρ}   ⎤ $\mathcal{A}nalyze^{\mathcal{F}}$
 11:         $t \leftarrow t + \rho.\delta$ {Modify current global time according to ρ}
 12:     **else**
 13:         **return** {Reached end of the TGS}
 14:     **end if**
 15: **end loop**

---

global time t to t', and by inserting the global time t' into the set of violations V when G' does not satisfy the GC ψ'.

Definition 11 (Operation $\mathcal{A}nalyze^{\mathcal{F}}$). *If* (G_{tv}, x_{tv}, ψ') *was returned by an application of the operation* $\mathcal{R}educe$, $\rho = (\delta, \ell, r)$ *is a timed span,* $t \in \mathbf{R}_0^+$ *is the global time before* ρ, G *is a GH with time-storing variable* x_{tv} *satisfying* $\text{Gtv}(G, x_{tv})$, $V \subseteq \mathbf{R}_0^+$ *is a set of violations computed until the timepoint* t, $t' = t + \delta$ *is the global time after* ρ, $G' = \mathcal{F}\text{old}^{\mathcal{F}}(G, t, \rho)$ *is the modification of the GH* G *according to* ρ, *and if* $\text{Gta}(G', x_{tv}, t') \not\models \psi'$ *then* $V' = V \cup \{t'\}$ *else* $V' = V$ *implements the conditional addition of a violation* t' *to the set of violations* V, *then the triple* (t, G, V) *is modified to the triple* (t', G', V') *according to the timed span* ρ, *written* $(t, G, V) \, \mathcal{A}nalyze^{\mathcal{F}}_{x_{tv}, \psi', \rho} \, (t', G', V')$.

Moreover, we extend the operation $\mathcal{A}nalyze^{\mathcal{F}}$ *to finite TGSs using its iterated application starting with the triple obtained using* $\mathcal{A}nalyze^{\text{1st}}((G_{tv}, x_{tv}, \psi'), \pi)$.

For our running example, Procedure 1 returns the sets $\{6, 21\}$ and $\{21\}$ of violations for the pessimistic and the optimistic mode, respectively. Note that the given MTGC is violated at each timepoint in $(16, \infty)$, i.e., the violation is detected with a delay of about 5 timeunits. However, we believe that assuming a sufficiently high timed span rate (checking for violations permanently in a hot loop) mitigates this problem of delayed detection. Also, we assume a user-provided policy for deciding whether timepoints are recorded as violations when employed solvers such as Z3 are unable to decide the satisfaction problem in time.

Subsequently, we formally evaluate Procedure 1 by comparing the violations printed by it with a ground truth of violations that would be returned ideally. This ground truth is defined by the operation trueViolations delivering the set of true violations of an MTGC ψ in a TGS π for a maximum checking index n. This set of true violations contains the durations of all prefixes π' of the TGS π where the MTGC ψ is not satisfied subsequently by any continuation $\tilde{\pi}$ that is a prefix of π. Observe that a *precise* detection of violations thereby requires the knowledge of future steps, which is unavailable in the context of our proposed on-the-fly analysis approach.

Definition 12 (Operation trueViolations). *If $\psi \in \Phi_\emptyset^{\mathrm{MTGC}}$ is an MTGC, $\pi \in \Pi_H$ is a TGS, $n \in \mathbf{N}$ is a maximum checking index, and $\mathsf{length}(\pi) \geq n$, then* $\mathsf{trueViolations}(\psi, \pi, n) = \{ t \mid \exists \pi'. (\pi' \sqsubseteq \pi) \wedge (\mathsf{length}(\pi') \leq n) \wedge (t = \mathsf{dur}(\pi')) \wedge (\forall \tilde{\pi}. (\pi' \sqsubseteq \tilde{\pi} \sqsubseteq \pi) \rightarrow (\tilde{\pi} \not\models_{TGS} \psi)) \}$.

To enable a comparison of Procedure 1 with the set of true violations, we now introduce the operation check, which gathers the optimistic and pessimistic violations printed by Procedure 1 up to a given index n.

Definition 13 (Operation check). *If $(G_{tv}, x_{tv}, \psi') = \mathcal{R}educe(mode, \psi)$ is the reduction obtained using the operation $\mathcal{R}educe$, $\pi \in \Pi_H$ is a TGS of length at least $n \in \mathbf{N}$ starting in graph H, $\bar{\pi}$ is the prefix of length n of π, $(t, G, V) = \mathcal{A}nalyze^{1st}((G_{tv}, x_{tv}, \psi'), \pi)$ is the first triple for the on-the-fly analysis, and $(t, G, V)\ \mathcal{A}nalyze_{x_{tv}, \psi', \bar{\pi}}^{\mathcal{F}}\ (t', G', V')$ computes the violations V' using the on-the-fly analysis on the prefix $\bar{\pi}$, then $\mathsf{check}(mode, \psi, \pi, n) = V'$.*

Based on the presented definitions, we introduce our main result stating that the optimistic and the pessimistic on-the-fly analysis carried out using Procedure 1 (and formalized using the operation check) determines under- and over-approximations of the set of true violations, respectively. That is, optimistic violations *must* be true violations and pessimistic violations *may* be true violations.

Theorem 2 (Approximate Detection of True Violations using Optimistic and Pessimistic On-the-fly Analysis). *If $\psi \in \Phi_\emptyset^{\mathrm{MTGC}}$ is an MTGC, $\pi \in \Pi_H$ is a TGS, $n \in \mathbf{N}$ is a maximum checking index, and $\mathsf{length}(\pi) \geq n$, then*

- $\mathsf{check}(\mathcal{O}\mathrm{pt}, \psi, \pi, n) \subseteq \mathsf{trueViolations}(\psi, \pi, n)$ *and*
- $\mathsf{trueViolations}(\psi, \pi, n) \subseteq \mathsf{check}(\mathcal{P}\mathrm{es}, \psi, \pi, n)$.

Proof Idea. The first item holds using Theorem 1 since every violation of the GC ψ' obtained using reduction for $mode = \mathcal{O}\mathrm{pt}$ ensures that the future evolution used in the operation trueViolations is also guaranteed to violate the MTGC ψ as the violation cannot be undone in any continuation. The second item holds using Theorem 1 because (via contraposition) when the GC ψ' obtained using reduction for $mode = \mathcal{P}\mathrm{es}$ is not violated, this means that the considered TGS prefix π' satisfies the MTGC ψ and therefore at least the empty continuation $\pi' = \tilde{\pi}$ satisfies ψ as well implying that the current global timepoint is no true violation.   □

For our running example from Fig. 3 where the TGS π has length 3, we observe that $\mathsf{check}(\mathcal{O}\mathrm{pt}, \psi, \pi, 3) = \{21\} = \mathsf{trueViolations}(\psi, \pi, 3) \subseteq \{6, 21\} = \mathsf{check}(\mathcal{P}\mathrm{es}, \psi, \pi, 3)$. This means that the optimistic on-the-fly analysis detects only one true violation at global timepoint 21 but the pessimistic on-the-fly analysis returns additionally a second violation at global timepoint 6 indicating that the MTGC ψ may not be satisfied by all continuations, which indeed is the case for the TGS π. When changing δ_2 to 1 in Fig. 3b, the pessimistic on-the-fly analysis still generates the violation at global timepoint 6 but 6 is no true violation since the result R is then generated in time.

Based on Theorem 2, we obtain, as mentioned in Sect. 1, a three-valued interpretation of MTGL when using Procedure 1 (formalized using the operation check) to generate optimistic and pessimistic violations. In this interpretation, the absence of a pessimistic violation t guarantees satisfaction, each optimistic violation t guarantees nonsatisfaction, and a pessimistic violation t that is no optimistic violation is an indifferent judgement on satisfaction.

Definition 14 (Three-valued Interpretation of MTGL). *If $\psi \in \Phi_\emptyset^{MTGC}$ is an MTGC, $\pi \in \Pi_H$ is a TGS of length at least $n \in \mathbf{N}$, and $\bar{\pi}$ is the prefix of length n of π, then $[\![\pi, \psi]\!]_n = true$ if $\mathsf{dur}(\bar{\pi}) \notin \mathsf{check}(\mathcal{P}es, \psi, \pi, n)$, $[\![\pi, \psi]\!]_n = false$ if $\mathsf{dur}(\bar{\pi}) \in \mathsf{check}(\mathcal{O}pt, \psi, \pi, n)$, and $[\![\pi, \psi]\!]_n = indifferent$ otherwise.*

From a practical point of view, Procedure 1 solves the satisfaction problem for three kinds of TGSs. Firstly, non-terminating systems can be analyzed throughout their entire runtime in an on-the-fly manner. Secondly, finite slices of TGSs generated by long-running systems can be analyzed in an offline manner producing pessimistic but no optimistic violations for cases where the ongoing evolution of the system may be admissible. Lastly, terminating systems where a Terminated node is added by the last timed span can be analyzed where pessimistic and optimistic violations coincide after the last step.

However, for formal specifications given by complex MTGCs, we intend to equip violations given by timepoints with human-readable explanations. For this purpose, we may use partial MTGC satisfaction trees following GC satisfaction trees from [25]. We expect that this would also permit an analysis of the causal dependencies among violations and their *origins*. For our running example, the violation at timepoint 21 has the pessimistic violation at timepoint 6 as an origin because the problematic task T connected to system S was freshly matched at timepoint 6 leading to the violated deadline at timepoint 21.

We implemented Procedure 1 in the tool AUTOGRAPH [24], which supports GL and MTGL and relies on the constraint solver Z3 [16] for checking the satisfiability of ACs. For a high level of confidence, we applied extensive testing of the implemented functionality for diverse and deeply nested MTGCs also covering our running example. Overall, the implementation is promising and demonstrates the feasibility of our approach.

5 Related Work

Verification approaches for graph transformation systems are incomplete due to their expressiveness. For example, logics such as CTL and PTCTL can be applied to entire state spaces [6,15] but have limited support for sequence properties relying on atomic propositions. Similarly, invariant verification [4,23] as an example of static analysis considers graph sequences but only state invariants.

On-the-fly analysis of dynamic systems is used (before deployment) in testing and (after deployment) in specification-based monitoring using specifications given by temporal logics, automata with quantification, and rule-based systems [1]. These approaches are difficult to compare due to highly domain-specific requirements regarding expressiveness, efficiency, and usability. Logic-based approaches e.g. [11,14] often lack support for key features of MTGL such as data elements, bindings, or metric bounds in temporal operators. A notable exception is the *Metric First-Order Temporal Logic* (MFOTL) [2], supported by the tool MonPoly, that represents a system state as a set of relations, supports the binding of elements and uses a point based rather than an interval based semantics as MTGL. Note that the encoding of MTGCs in MFOTL is highly technical and error-prone (similarly, the logic of nested graph conditions [9] is advantageous compared to FOL on graphs in graph centered scenarios). Finally, MonPoly imposes syntactic limitations on MFOTL conditions to ensure that provided conditions are satisfied/violated by a finite future.

Logics such as MTGL and MFOTL can be directly applied in the context of runtime monitoring [1,13]. A roadmap towards such an application using an extended subset of MTGL is presented in an informal way in [21,22].

6   Conclusion and Future Work

We introduced an on-the-fly analysis procedure for the satisfaction of MTGCs by infinite TGSs (generated by non-terminating systems) and finite TGSs (generated by terminating systems or representing prefixes of infinite TGSs). The analysis procedure results in a three-valued interpretation of MTGL where unavoidable non-satisfaction is detected via optimistic violations and where potential non-satisfaction in the future is detected via pessimistic violations as soon as possible. The two sets of violations approximate the ground truth given by the set of true violations, which can only be determined by offline analysis for finite terminated TGSs. The on-the-fly analysis procedure including both sets of violations is supported by our extension of the tool AutoGraph.

In the future, we will (a) integrate the *since* operator into MTGL and the proposed analysis procedure, (b) improve applicability of our approach using more detailed violations, (c) improve the optimistic reduction by simplifying the reduced MTGC using the constraint solver approach from [24], (d) employ incremental GC checking to improve the on-the-fly checking performance, and (e) compare our approach to other tools such as MonPoly w.r.t. efficiency.

References

1. Bartocci, E., Deshmukh, J., Donzé, A., Fainekos, G., Maler, O., Ničković, D., Sankaranarayanan, S.: Specification-based monitoring of cyber-physical systems: a survey on theory, tools and applications. In: Bartocci, E., Falcone, Y. (eds.) Lectures on Runtime Verification. LNCS, vol. 10457, pp. 135–175. Springer, Cham (2018). https://doi.org/10.1007/978-3-319-75632-5_5

2. Basin, D.A., Klaedtke, F., Müller, S., Zalinescu, E.: Monitoring metric first-order temporal properties. J. ACM **62**(2), 15:1–15:45 (2015). https://doi.org/10.1145/2699444

3. Becker, B., Giese, H.: On safe service-oriented real-time coordination for autonomous vehicles. In: ISORC 2008, pp. 203–210. IEEE Computer Society (2008). https://doi.org/10.1109/ISORC.2008.13

4. Dyck, J., Giese, H.: k-inductive invariant checking for graph transformation systems. In: de Lara, J., Plump, D. (eds.) ICGT 2017. LNCS, vol. 10373, pp. 142–158. Springer, Cham (2017). https://doi.org/10.1007/978-3-319-61470-0_9

5. Ehrig, H., Ehrig, K., Prange, U., Taentzer, G.: Fundamentals of Algebraic Graph Transformation. Springer, Heidelberg (2006). https://doi.org/10.1007/3-540-31188-2

6. Ghamarian, A.H., de Mol, M., Rensink, A., Zambon, E., Zimakova, M.: Modelling and analysis using GROOVE. STTT **14**(1), 15–40 (2012). https://doi.org/10.1007/s10009-011-0186-x

7. Giese, H.: Modeling and verification of cooperative self-adaptive mechatronic systems. In: Kordon, F., Sztipanovits, J. (eds.) Monterey Workshop 2005. LNCS, vol. 4322, pp. 258–280. Springer, Heidelberg (2007). https://doi.org/10.1007/978-3-540-71156-8_14

8. Giese, H., Maximova, M., Sakizloglou, L., Schneider, S.: Metric temporal graph logic over typed attributed graphs. In: Hähnle and van der Aalst [10], pp. 282–298 (2019). https://doi.org/10.1007/978-3-030-16722-6_16

9. Habel, A., Pennemann, K.: Correctness of high-level transformation systems relative to nested conditions. Math. Struct. Comput. Sci. **19**(2), 245–296 (2009). https://doi.org/10.1017/S0960129508007202

10. Hähnle, R., van der Aalst, W. (eds.): FASE 2019. LNCS, vol. 11424. Springer, Cham (2019). https://doi.org/10.1007/978-3-030-16722-6

11. Havelund, K., Peled, D.: Runtime verification: from propositional to first-order temporal logic. In: Colombo, C., Leucker, M. (eds.) RV 2018. LNCS, vol. 11237, pp. 90–112. Springer, Cham (2018). https://doi.org/10.1007/978-3-030-03769-7_7

12. Koymans, R.: Specifying real-time properties with metric temporal logic. Real-Time Syst. **2**(4), 255–299 (1990). https://doi.org/10.1007/BF01995674

13. Leucker, M., Schallhart, C.: A brief account of runtime verification. J. Log. Algebr. Program. **78**(5), 293–303 (2009). https://doi.org/10.1016/j.jlap.2008.08.004

14. Maler, O., Nickovic, D.: Monitoring temporal properties of continuous signals. In: Lakhnech, Y., Yovine, S. (eds.) FORMATS/FTRTFT 2004. LNCS, vol. 3253, pp. 152–166. Springer, Heidelberg (2004). https://doi.org/10.1007/978-3-540-30206-3_12

15. Maximova, M., Giese, H., Krause, C.: Probabilistic timed graph transformation systems. J. Log. Algebr. Meth. Program. **101**, 110–131 (2018). https://doi.org/10.1016/j.jlamp.2018.09.003

16. Microsoft Corporation: Z3. https://github.com/Z3Prover/z3

17. Neumann, S.: Modellierung und Verifikation zeitbehafteter Graphtransformationssysteme mittels Groove. Master's thesis, University of Paderborn (2007)

18. Orejas, F.: Symbolic graphs for attributed graph constraints. J. Symb. Comput. **46**(3), 294–315 (2011). https://doi.org/10.1016/j.jsc.2010.09.009

19. Orejas, F., Lambers, L.: Symbolic attributed graphs for attributed graph transformation. ECEASST 30 (2010). http://journal.ub.tu-berlin.de/index.php/eceasst/article/view/405

20. Orejas, F., Lambers, L.: Lazy graph transformation. Fundam. Inf. **118**(1–2), 65–96 (2012). https://doi.org/10.3233/FI-2012-706

21. Sakizloglou, L., Ghahremani, S., Brand, T., Barkowsky, M., Giese, H.: Towards highly scalable runtime models with history. In: SEAMS 2020. IEEE Computer Society (2020). (to appear)
22. Sakizloglou, L., Ghahremani, S., Brand, T., Barkowsky, M., Giese, H.: Towards highly scalable runtime models with history. Technical Report (2020). arxiv: 2004.03727
23. Schneider, S., Dyck, J., Giese, H.: Formal verification of invariants for attributed graph transformation systems based on nested attributed graph conditions. In: Gadducci, F., Kehrer, T. (eds.) ICGT 2020, LNCS. Springer, Heidelberg (2020). (to appear)
24. Schneider, S., Lambers, L., Orejas, F.: Automated reasoning for attributed graph properties. STTT **20**(6), 705–737 (2018). https://doi.org/10.1007/s10009-018-0496-3
25. Schneider, S., Lambers, L., Orejas, F.: A logic-based incremental approach to graph repair. In: Hähnle and van der Aalst [10], pp. 151–167 (2019). https://doi.org/10.1007/978-3-030-16722-6_9

Tool Presentations

A Flexible and Easy-to-Use Library for the Rapid Development of Graph Tools in Java

H. J. Sander Bruggink[1], Barbara König[2], Marleen Matjeka[2], Dennis Nolte[2], and Lara Stoltenow[2(✉)]

[1] GEBIT Solutions, Düsseldorf, Germany
[2] Abteilung für Informatik und Angewandte Kognitionswissenschaft, Universität Duisburg-Essen, Duisburg, Germany
{barbara_koenig,lara.stoltenow}@uni-due.de

Abstract. We present a programming library for the rapid development of graph tools, with applications in graph transformation and related fields. Features include working with graphs, graph morphisms, basic categorical constructions such as computing pushouts and pushout complements or enumerating all morphisms with certain properties, but also applications such as executing graph transformation steps. Additionally, we offer graphical user interface widgets for visualization and manipulation of graphs, morphisms and categorical diagrams.

Our objective is to allow users to quickly develop graph tools for both simple and complex problems, to allow easy embedding into existing software, and to have comprehensible code especially for the main algorithms. Existing tools that demonstrate the versatility and ease of use of the library include: DPOdactic (a didactic tool for teaching double-pushout graph transformation), DrAGoM (a tool to handle multiply annotated type graphs for abstract graph rewriting), and Grez (termination analysis of graph transformation systems).

Keywords: Graph transformation · Rapid development · Graph tools

1 Introduction

The graph transformation community has always been strong in the development of tools, for support of generic graph rewriting, for supporting software development and for verification and analysis (see for instance [2,6,8,15,16] for a non-exhaustive enumeration).

However, to our knowledge there is no publicly available, easily accessible and flexible library that provides a backbone and toolbox for the rapid development of graph tools, including both the support of various constructions and visualization of graphs. We have developed such a library and are still in the process of extending it. Since we believe that there may be a wider interest, we will here present it as a community service and describe existing tools that are already

© Springer Nature Switzerland AG 2020
F. Gadducci and T. Kehrer (Eds.): ICGT 2020, LNCS 12150, pp. 297–306, 2020.
https://doi.org/10.1007/978-3-030-51372-6_17

based on such a library. It does not contain groundbreaking new functionality, but in our opinion we present a nice, comprehensive package.

Our design principles while developing this library are as follows:

- It is designed to have a low learning curve, which we have tested by its successful use in several student projects. It has been integrated both into tools that have been implemented from scratch, and the continued development of existing tools.
- Simple tasks should be easy to implement quickly. We will illustrate this with two suggestive examples:
  - If you need to compute the composition of two given graph cospans, it should – after becoming acquainted with the library – take you as much time to do it by hand on paper, as it does to just write a small prototype program.
  - Assume you develop a nice theory about commuting hexagons of graph morphisms and you want beautiful renderings of them. Using generic utility functions, you can convert the abstract representation into a displayable graph using no more than 50 well-formatted lines of code.
- We aim for readability and favor clear and understandable code over raw speed. It should e.g. be possible to learn how to compute pushouts of graph morphisms just by reading the code.
- We provide automatic visualization of entities such as graph morphisms, commuting squares, etc., which are typically not supported by general purpose graph display libraries.
- We aim for easy integration of the library into your own application.

The library can be downloaded from https://www.uni-due.de/theoinf/research/tools_javagraph.php.

Related Work. We have evaluated some of the tools that are commonly used in the context of graph transformation and offer similar functionality, in particular, PROGRES [16], GRAJ [6], ENFORCE [2], AGG [15], and a tool for graph transformation by computational category theory [13].

PROGRES is a suite of tools that focuses on the specific application of graph grammars and graph rewriting systems. It is, however, not designed as a generic library. Together with the fact that no source code is available (but only non-portable binaries), it is probably hard to embed it into other applications.

GRAJ is a tool for the execution of graph programs. It features a modular design that facilitates embedding into custom tools. ENFORCE builds on GRAJ to prove correctness not only for graph programs, but also other weak adhesive HLR categories. Notably, it also supports graph conditions and constraints. Both tools are, however, currently unmaintained and not publicly available for download. Thus we could not evaluate its suitability for e.g. prototyping purposes.

AGG also focuses on graph grammars and graph transformation. It too has support for graph conditions, and has an extensible architecture. However, it appears to be designed as a standalone tool. While it is a very powerful tool and the components that do the actual computations feature useful algorithms, the programming interface does not appear to be specifically designed for use as a library.

In [13], a library for carrying out graph transformation in an abstract categorical setting is proposed. In this way, it is similar to the CatLib component of our work. The full code was not available for evaluation, but their focus is less on graphs and more on the categorical side, which makes it potentially more difficult to work with the library. In addition the library does not offer a ready-to-use visualization component.

Our justification for the development of a new library is not just to avoid these particular problems, but to also focus on additional aspects (ease of use, prioritize clear and understandable code over efficient implementations, make it easy to embed into your own tools) as detailed above. In this regard, it is similar in spirit to the SiTra library [1] which focuses not on practical applications, but to "aid a programmer in learning the concept of writing transformation rules".

Outline of the Paper. The article is structured as follows: In Sect. 2, we describe the architecture and the features of our library. In Sect. 3, we give a detailed overview of existing tools that are using the library. We conclude in Sect. 4 with an outline of future work.

2 Components and Features

2.1 Components Overview

In this section, we give a detailed overview of the components that together make up the library and the features that are available. The components can be used together or independently of each other as needed.

The *Java-Graph* component provides the computational foundations. It provides abstract representations of graphs, graph morphisms, graph conditions and related objects; categorical constructions such as pushout complements; enumeration of morphisms with certain properties; graph transformation; loading and saving of objects to files in a plaintext format that is easy to read and write. We give a more detailed description of this component in Sect. 2.2.

Java-Graph by itself provides no graphical user interface and can therefore be used for batch processing tasks, or as part of tools that already build on different frameworks. Graphical output is provided by a separate component.

The *VisiGraph* component is responsible for displaying graphs to the user, and provides a similar feature set as other graph display libraries. It automatically layouts graphs that can then be shown to the user. Currently, it provides display and editor widgets for Swing-based graphical user interfaces (however, it does not have a strong dependency on Swing and can be quickly ported to work with other GUI toolkits). It is also possible to export the graph to image files.

As a companion component, *VisiGraphJS* is a reimplementation in Javascript and can be used to provide the same type of visualizations in web applications.

It is also possible to do the layouting process in Java using VisiGraph and then only display the result in a web browser.

The *VxToolbox* component serves as a bridge between Java-Graph and Visi-Graph. It is responsible for creating useful visualizations not just for ordinary graphs, but for the various objects that are supported by Java-Graph. As an example, the visualization of a pushout square should put the four graphs at the four corners of an appropriately-sized square, and to make the output more easily graspable, common elements (e.g. nodes that are in both the domain and codomain of some morphism) should be positioned in a consistent way. VxToolbox provides not only visualization routines for the objects supported by Java-Graph, but also basic building blocks to make it easy to generate visualizations of custom objects (as a rule of thumb, visualizing e.g. commuting hexagons should require no more than 50 lines of code).

Finally, *CatLib* is a generalization of Java-Graph to arbitrary categories. CatLib can be used independently of, or together with, Java-Graph. Prototype tools can thus be implemented in a generic way, doing computations on arbitrary categories, where Java-Graph is used to showcase the generically implemented tool for a specific example category. Currently, CatLib implements the categories **Set** and, using Java-Graph, the category of finite (hyper)graphs **Graph**$_{\text{fin}}$.

2.2   Detailed Description of the Java-Graph Component

At the core, we have the de.uni_due.inf.ti.graph package (prefix abbreviated hereinafter as ...ti.graph), with classes for the basic entities. Graphs are represented with the Graph class, containing collections of Nodes and (hyper-)Edges with Labels. We provide the usual methods for construction and manipulation of graphs such as graph.addEdge(new Label("A"),n0,n1,n2) to add a ternary hyperedge or graph.getNodes() to obtain a (read-only) List of the nodes in a Graph. Although edges are generally hyperedges, we provide additional methods as simplifications for the common case of directed edges (e.g. edge.getTarget() as an alternative to edge.getNodes().get(1)).

Using the ...ti.graph.io package, all supported objects (graphs, conditions etc.) can be read and written in a custom text-based file format named SGF. The textual representation of SGF resembles the way a graph would be written on paper. The SGF code graph { n0 --A-> n1 --A-> n2 --A-> n0; }; describes a graph with three nodes (n0 to n2) that are connected by directed *a*-labeled edges in a circle. Objects can be loaded from files or from strings. In our example below, we use the latter, in conjunction with the Java 13 Text Blocks feature, to obtain very concise prototype code.

Graph morphisms map elements of one graph to compatible elements of another one, where the map can be either total or partial. A Morphism has a Map-like interface (mor.get(node0), mor.getPreimage(edgeA) and the like) with additional functionality; for instance, mor.put(domEdge, codomEdge) maps not only the edge, but also creates mappings for all nodes that are incident to the given edge (unless this mapping would conflict with the node mapping of the graph, in which case an exception is thrown). Morphisms can be created easily by explicitly giving the node and edge mappings, either using the put method in Java, or using => in SGF (see the example at the end of this section).

Graph conditions can be used to specify additional properties of graphs such as the existence or absence of certain elements. They come in two flavours: the *nested conditions* (roughly, first-order formulas on graphs) as introduced in [9] for weak adhesive HLR categories; and *cospan conditions*, which use a slightly different tree-based structure, as introduced in [3] for adhesive categories. As an example for the former, the condition $\exists(m_1, \text{true}) \lor \exists(m_2, \text{true})$, where m_1, m_2 are morphisms describing the elements that should exist at some point, can be written in SGF as follows: c = or [exists(m1,true), exists(m2,true)];

We provide various fundamental categorical constructions (...ti.graph.ext). Given a span, a cospan, or a pair of composable morphisms, it is possible to compute the pushout, pullback, or pushout complement, respectively. It is possible to enumerate all morphisms between two graphs with certain properties (examples include enumeration of all total injective morphisms; all partial morphisms; all isomorphisms; all morphisms that extend a given base morphism). Furthermore, given a span, it is possible to enumerate all jointly epi squares. Enumerator implements Iterable, and hence can be used in loops (e.g. for (Morphism i : Morphism.getIsomorphisms(g1, g2)) { ... } to executesome code for all isomorphisms between two graphs g_1, g_2), or as Streams (Morphism.getIsomorphisms(g1, g2).stream().map(i -> ...)). All of these enumerators compute their results lazily and so also work when the total number of possible morphisms is very large.

The following example code creates objects for a pair of graph morphisms $g_L \xleftarrow{m_{TL}} g_T \xrightarrow{m_{TR}} g_R$, where $g_T = \overset{1}{\bullet}\ \overset{2}{\bullet}\ \overset{3}{\bullet}$ (three isolated nodes), $g_L = \overset{1}{\bullet}\!\!\overset{A}{\longrightarrow}\!\!\overset{2\,3}{\bullet}$, $g_R = \overset{1\,2}{\bullet}\!\!\overset{B}{\longrightarrow}\!\!\overset{3}{\bullet}$, and morphisms m_{TL}, m_{TR} merge nodes $2,3$ and $1,2$ respectively. Then their pushout is computed and the result is printed to standard output:

```
String sgfContent = """
  gT = graph { node n1; node n2; node n3; };
  gL = graph { n1 --ea:A-> n23; };
  gR = graph { n12 --eb:B-> n3; };
  mTL = morphism from gT to gL { n1 => n1; n2 => n23; n3 => n23; };
  mTR = morphism from gT to gR { n1 => n12; n2 => n12; n3 => n3; };
  """;
Map<String, Object> sgfMap = SgfParser.parseSgfString(sgfContent);
Morphism mTL = (Morphism) sgfMap.get("mTL");
Morphism mTR = (Morphism) sgfMap.get("mTR");
Square po = Pushout.compute(mTL, mTR);
System.out.println(po);
```

As an application of the fundamental constructions, graph transformation systems using the single-pushout and double-pushout approaches can be directly described and processed by the library (...ti.graph.transformation). So far we restrict to injective match and rule morphisms. In SGF, if a rule morphism is not explicitly specified, then elements on the left and right hand sides are automatically related if they have the same name. For instance, in the rule `r = rule { { n1 --A-> n2 --B-> n3 } => { n1 --C-> n3 } }`, nodes n_1, n_3 are mapped to their counterparts on the right hand side, the c-edge is created, and node n_2 and the two edges are deleted at rule application. To enumerate all possible results of rewriting a `Graph g` using `Rule r`:

```
for (Morphism match : r.getMatches(g)) {
    Transition t = r.applyToMatch(match);
    Graph rewrittenGraph = t.getTarget();
    // process rewrittenGraph somehow
}
```

3   Existing Tools Using the Library

In this section, we describe some of the existing tools that are currently using the library. Notably, we present: DPOdactic (a didactic tool for graph transformation), DrAGoM (multiply annotated type graphs for abstract graph rewriting), and Grez (termination analysis). Additionally, we give a quick overview of tools that are currently under development. These tools demonstrate that the library can be used in a variety of different application areas.

3.1   DPOdactic

DPOdactic [12] is a tool that walks the user through the process of applying double-pushout (DPO) graph transformation rules. In this setting, a rule states that the occurence of some subgraph L is to be replaced by another graph R. The relationship between L and R is established via an interface graph I and two injective morphisms that map I to L, R respectively. A rule is applied by locating a match of L – where DPOdactic also allows non-injective matches – removing parts of L, but keeping I, and then adding the missing parts of R.

In the tool (Fig. 1), the user is presented with a rule and a graph G that the rule should be applied to. First, they select one of (possibly) multiple occurences of L in G. Then, they input the context graph, followed by the morphisms that relate it to the other graphs. Finally, they input the result of the transformation step and the related morphisms. The tool checks all intermediate results for inputs and provides direct feedback to the user, including hints on where to look for mistakes. Optionally, the tool can also simply compute the result of each step.

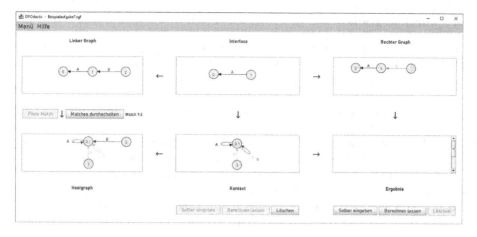

Fig. 1. Main window of DPOdactic after the user has provided the correct context graph, with the result graph yet to be computed (by the user or by the tool).

3.2 DrAGoM

DrAGoM [14] is a prototype tool to handle and manipulate so-called multiply annotated type graphs. The main application of DrAGoM is to automatically compute strongest postconditions in order to check invariants of graph transformation systems, in the framework of abstract graph rewriting.

DrAGoM uses a materialization construction to extract concrete instances of a left-hand side graph out of an abstract graph. Then, it can be used to automatically compute the strongest postcondition of the materialization, i.e. an annotated type graph, specifying exactly the language of all graphs which are reachable in one rewriting step.

3.3 Grez

Grez [5] is a tool to automatically produce proofs of uniform (non-)termination of graph transformation systems, i.e. whether it is possible to obtain an infinite sequence of rule applications from some start graph or not. Grez uses various approaches for analysis: some are simple (e.g. if all rules reduce the number of nodes, then rewriting must terminate at some point), while others are more complicated (e.g. termination arguments based on weighted type graphs [4]).

Typically, algorithms classify rules as decreasing, non-increasing, or possibly-increasing with respect to some order. Grez can then combine the results of multiple algorithms using a relative termination argument: if one algorithm can only prove a subset of the rules as decreasing (thus terminating) and the remaining rules as non-increasing, then termination of the remaining rules (for which a different algorithm can be used) implies termination of the original system.

3.4   Further Tools

Numerous other, smaller tools that are currently in alpha stage are being developed using the library, with areas of application being the analysis of (conditional) graph transformation systems, satisfiability checking of graph conditions, and tools that automate various basic tasks.

As an example for the automation of basic tasks, we have implemented a tool (Podmineny) that, given a pair of cospans (typically corresponding to the left-hand side of a rule and a graph with interfaces), computes all borrowed context diagrams [7]; a task that is tedious and error-prone when done by hand.

As a case study, we have partially re-implemented the tableau resolution algorithm for graph properties as described in [11]. While this tool only implements part of the functionality, it encouraged us to start work on another prototype tool, RSsat, for both model finding and unsatisfiability proofs in the more generic setting of reactive systems.

Table 1. Overview of tools that are currently using the library. The columns indicate which components (Java-Graph, VisiGraph, VisiGraphJS, VxToolbox, CatLib) are currently used (●), will (○) or could ( ) be used in future versions.

Tool	Description	Jg	Vx	Js	Tb	Cl
Grez	Termination analysis for graph transformation systems [5]	●	●	○	○	
DrAGoM	Manipulation of multiply annotated typegraphs [14]	●	●			
DPOdactic	A didactic tool for double-pushout graph transformation systems [12]	●	●			
Podmineny	Enumeration of all borrowed context diagrams, given two graph cospans	●	●	○	○	○
RSsat	Prototype tool for model finding and unsatisfiability proofs for conditions in reactive systems	●	●	●	●	●
TGC	A partial implementation of tableau resolution [11] for graph properties	●	●		○	
Your tool here	:-)	?	?	?	?	?

Table 1 gives a quick overview of current and future tools.

4   Future Work

In addition to the existing documentation for classes and methods, we plan to provide an introductory user guide for getting started with the library. As supporting material, we will implement several smaller tools that can serve as examples or templates for the development of other tools.

Naturally, we also plan to lift the restrictions on the injectivity of match and rule morphisms and to extend the functionality in general.

Our library currently supports SGF as a custom text-based data interchange format. We feel that the simple syntax of SGF goes well with the design goal of facilitating the development of prototype tools. Future versions of the library will additionally support the Graph eXchange Language (GXL) [10], an XML-based interchange format that is used by other tools. Note that GXL is not primarily designed to be hand-written by users (as SGF is), but to be generated by tools.

While the VisiGraph library has no strong dependency on Swing and support for other toolkits can be easily added if needed, we plan to provide interfaces to additional common GUI toolkits directly in our library. For the generation of mechanical proofs (e.g. (non-)termination proofs for graph transformation systems in Grez) we will also add direct generation of LATEX code.

Furthermore, we plan to use the library to develop further tools to demonstrate applicability of our own future research, such as the analysis of reactive systems conditions.

References

1. Akehurst, D.H., Bordbar, B., Evans, M.J., Howells, W.G.J., McDonald-Maier, K.D.: SiTra: Simple transformations in Java. In: Nierstrasz, O., Whittle, J., Harel, D., Reggio, G. (eds.) Model Driven Engineering Languages and Systems, pp. 351–364. Springer, Heidelberg (2006). https://doi.org/10.1007/11880240_25
2. Azab, K., Habel, A., Pennemann, K.H., Zuckschwerdt, C.: ENFORCe: A System for ensuring formal correctness of high-level programs. In: Proceedings 3rd International Workshop on Graph Based Tools (GraBaTs 2006). vol. 1, pp. 82–93. Electronic Communications of the EASST (2007)
3. Bruggink, H.J.S., Cauderlier, R., Hülsbusch, M., König, B.: Conditional reactive systems. In: Proceedings of FSTTCS 2011. LIPIcs, vol. 13. Schloss Dagstuhl - Leibniz Center for Informatics (2011)
4. Bruggink, H.J.S., König, B., Nolte, D., Zantema, H.: Proving termination of graph transformation systems using weighted type graphs over semirings. In: Parisi-Presicce, F., Westfechtel, B. (eds.) ICGT 2015. LNCS, vol. 9151, pp. 52–68. Springer, Cham (2015). https://doi.org/10.1007/978-3-319-21145-9_4
5. Bruggink, H.J.S., König, B., Zantema, H.: Termination analysis for graph transformation systems. In: Diaz, J., Lanese, I., Sangiorgi, D. (eds.) TCS 2014. LNCS, vol. 8705, pp. 179–194. Springer, Heidelberg (2014). https://doi.org/10.1007/978-3-662-44602-7_15
6. Busatto, G.: GraJ: A System for executing graph programs in Java. Berichte aus dem Fachbereich Informatik 3/04, Universität Oldenburg (2004)
7. Ehrig, H., König, B.: Deriving bisimulation congruences in the DPO approach to graph rewriting with borrowed contexts. Math. Struct. Comput. Sci. **16**(6), 1133–1163 (2006)
8. Ghamarian, A.H., de Mol, M., Rensink, A., Zambon, E., Zimakova, M.: Modelling and analysis using GROOVE. Int. J. Softw. Tools Technol. Transfer **14**(1), 15–40 (2012)
9. Habel, A., Pennemann, K.H.: Correctness of high-level transformation systems relative to nested conditions. Math. Struct. Comput. Sci. **19**(2), 245–296 (2009)
10. Holt, R.C., Schürr, A., Sim, S.E., Winter, A.: GXL: A Graph-based standard exchange format for reengineering. Sci. Comput. Program. **60**(2), 149–170 (2006)
11. Lambers, L., Orejas, F.: Tableau-based reasoning for graph properties. In: Giese, H., König, B. (eds.) Graph Transformation, pp. 17–32. Springer, Cham (2014)
12. Matjeka, M.: Ein didaktisches Tool zur Anwendung von Graphtransformation. Bachelor's thesis, Universität Duisburg-Essen, June 2019

13. Minas, M., Schneider, H.J.: Graph transformation by computational category theory. In: Engels, G., Lewerentz, C., Schäfer, W., Schürr, A., Westfechtel, B. (eds.) Graph Transformations and Model-Driven Engineering: Essays Dedicated to Manfred Nagl on the Occasion of his 65th Birthday, pp. 33–58. Springer, Heidelberg (2010)
14. Nolte, D.: Analysis and abstraction of graph transformation systems via type graphs. Ph.D. thesis, Universität Duisburg-Essen, August 2019
15. Runge, O., Ermel, C., Taentzer, G.: AGG 2.0 - new features for specifying and analyzing algebraic graph transformations. In: Schurr, A., Varro, D., Varro, G. (eds) Proceedings of AGTIVE 2011, vol. 7233, pp. 81–88. Springer, Heidelberg (2012). https://doi.org/10.1007/978-3-642-34176-2_8
16. Schürr, A., Winter, A.J., Zündorf, A.: The PROGRES approach: Language and environment. In: Handbook of Graph Grammars and Computing by Graph Transformation, vol. 2: Applications, Languages and Tools, pp. 487–550. World Scientific (1999)

Multiscale Graph Grammars Can Generate Cayley Graphs of Groups and Monoids

Winfried Kurth[(✉)]

Universität Göttingen, Abteilung Ökoinformatik,
Biometrie und Waldwachstum, Büsgenweg 4,
37085 Göttingen, Germany
`wk@informatik.uni-goettingen.de`

Abstract. A graph grammar with parallel replacement of subgraphs, based on the single-pushout approach in graph rewriting, was designed which constructs Cayley graphs of monoids of transformations of a finite set, with permutation groups as a special case. As input, graph-based representations of a finite number of generating transformations have to be specified; they will then correspond to the edge types of the Cayley graph which is the final result of the rewriting process. The grammar has $7 + d$ rules, where d is the number of generators, and operates at two scale levels. The fine-scale level is the level of elements on which the transformations act and where their composition is calculated by parallel subgraph replacement. The coarse-scale level corresponds to the transformations themselves which are organized in the Cayley graph in a sequential rule application process. Both scale levels are represented in a single graph. The graph grammar was implemented in the programming language XL on the software platform GroIMP, a graph rewriting tool which was originally designed for simulating the growth of plants.

Keywords: Graph grammar · Cayley graph · Permutation groups · Transformation semigroups · GroIMP

1 Introduction

Cayley graphs have been used for a long time to visualize the structure of discrete groups (see, e.g., [2]). They can be generalized to other algebraic structures [1,8]. Particularly, for a monoid M (i.e., semigroup with neutral element), the Cayley graph of M w.r.t. a set of generators G has the node set M, and for each node $n \in M$ and each generator $g \in G$ there is a directed edge $(n, n \circ g)$. In this small study we will demonstrate that the construction of the Cayley graph of a monoid which is defined by a generating set G of transformations of a finite set can be conceived as an application of a graph rewriting system. See [6] for an introduction to finite transformation semigroups. Permutation groups, which have been intensively studied for long time [3], and their Cayley graphs emerge as a special case when all members of G are bijective.

© Springer Nature Switzerland AG 2020
F. Gadducci and T. Kehrer (Eds.): ICGT 2020, LNCS 12150, pp. 307–315, 2020.
https://doi.org/10.1007/978-3-030-51372-6_18

As our graph grammar formalism, we used the "relational growth grammars" (RGGs) which enable a parallel replacement of subgraphs and are based on the single-pushout approach [4]. "Parallel replacement" means here that in every discrete time step all subgraphs which are instances of a left-hand side of a grammar rule are replaced by the graph given on the respective right-hand side. (For a discussion of the advantages of the single-pushout approach see [9], p. 104.) RGGs are supported by the programming language XL and are available on the platform GroIMP [9,10]. They can be seen as generalizations of Lindenmayer systems from strings to graphs and were primarily used to simulate the 3-dimensional architecture and development of plants (see, e.g., [14]). The software GroIMP contains also a powerful graph-drawing algorithm combining an energy-based layout with a tree layout [5]. Furthermore, XL was recently extended to support rewriting at several scale-levels simultaneously [12,13]. These were our main reasons for choosing the RGG approach and GroIMP for our purpose. We think other graph-grammar formalisms and software tools with a suitable expressive power could solve our task as well.

It was not our intention to create yet another calculation tool for transformation semigroups or permutation groups. Several such tools, devoted to this special field of application, exist already (see, e.g., [11]). We rather wanted to show that an existing graph rewriting approach, originally introduced for a quite different purpose, can solve our construction task. Our method is not optimized in terms of computation time or memory efficiency.

In the following, we will specify our used graph model and all the rules of our grammar. The description of the rules will also contain the proof that they indeed build the Cayley graph defined by the generating transformations given as input. Finally we will discuss some weaknesses of our method and also the graph layout provided for the final results by GroIMP.

2   Method and Result

2.1   The Graph Model

The graphs supported by the language XL are finite, connected, directed, rooted graphs with attributed, typed nodes with inheritance and with typed edges [9]. That means, each node belongs to a node type, with a class hierarchy among the types like in object-oriented programming, and each node can optionally have parameters (attributes), their number depending on the node type. In our case, all parameters will be integers. There is always a single node of the distinguished type "Root". Edges belong to one of a finite number of edge types (without hierarchy). There are no loops and no multiple edges of the same type and direction. XL provides three standard edge types called *successor (s)*, *ramification (r)* and *decomposition (d)* which were taken from a graph model designed for multiscale descriptions of plant architecture in [7]. Beyond that, additional edge types can be defined by the user.

The concept of multiple scales within a graph has been introduced to represent several levels of spatial resolution in vegetation models [12]. Although there is no geometry and hence no "spatial" scaling in our application here, we retain the concept of scale-level by denoting the transformations (i.e., the elements of our monoid; node type T) as "coarse-scale" and the n elements on which they operate (node type E), which form a finite set S, as "fine-scale". Without loss of generality, we can assume $S = \{1, 2, \ldots, n\}$. At the coarse-scale level we will additionally need *intermediate nodes* (type I) of indegree 1 and outdegree 1 between any connected pair of T nodes—this is particularly necessary in the case that the Cayley graph has loops, because our graph model does not permit loops, and by inserting an intermediate node each loop will be expanded to a cycle of length 2 –, and *connector nodes* (type C) which will provide the linkage to the fine-scale level.

The T nodes will have an index parameter which will help to control the step-by-step buildup of the Cayley graph. The I nodes will have a parameter i, initially 0, which indicates if the target T node of the corresponding edge was already used for constructing a subgraph at fine-scale level representing the transformation for which this T node stands ($i = 1$) and if this transformation was already compared, in a subsequent step, with all previously generated transformations ($i = 2$). The E nodes will have a parameter k between 1 and $2n$ indicating which element of S they represent, either as a preimage ($1 \leq k \leq n$) or as an image element ($n + 1 \leq k \leq 2n$) with respect to the transformation under consideration.

As edge types, we utilize the standard types s, r and d as well as one edge type g_i for each generator ($1 \leq i \leq d$), and additionally two edge types x and y which will be used at the fine-scale level to calculate the composition of two transformations. In our graph, a transformation $f : S \to S$ will be represented by a subgraph consisting of $2n$ nodes of type E, with edges of type x between nodes E_k and $E_{f(k)+n}$ where the subscript denotes the parameter of the node. Each element of S is thus represented twice. We could have used a single set of n nodes instead, with edges between E_k and $E_{f(k)}$, but because our graph rewriting mechanism does not allow non-injective embeddings of the left-hand side of a rule into the host graph, this would have made the calculation of the composition of two transformations more complicated.

2.2   The Grammar

Rule 1 (Initialization): From the start node (in XL called *Axiom*) a graph is created which corresponds to the identical transformation *id*, shown in Fig. 1. The nodes I, T and C represent the coarse-scale level. The C node is connected by decomposition edges with the E nodes at fine-scale level. The interconnections of the E nodes by edges of type x form the pattern of *id* on an n-element set. In XL, the initialization rule is coded as follows:

```
Axiom ==>> I(2) T(0) +> c:C /> E(1) -x-> E(n+1) </ c,
           for (int k = 2; k <= n; k++)
              ( c /> E(k) -x-> E(n+k) </ c );
```

where "+>" stands for a type r edge, a blank for a type s edge, "/>" for a type d edge, and "-x->" for a type x edge. The parenthesized part in the third line is iterated $n - 1$ times. The syntax "c:" in the first line introduces an identifier for the particular instance of a node of type C generated at this place and allows to refer to this instance by using the label "c" in the rest of the rule.

Rule 2 (Copy): To build up the Cayley graph, the transformations which exist already in the graph have to be multiplied with the d generators. In the first step, we simply make d copies of a transformation which is already represented in our graph. (This rule will then be applied iteratively to all nodes of type T.) The mapping pattern at fine-scale level of a copied transformation is still identical with that of the original, i.e., the generators have not yet been applied (Fig. 2). This copy rule application is controlled by a global variable m, initially 0, which is incremented by 1 in each application. For the sake of simplicity, we display here only the code for the case $d = 2$, with generators g and h. The generalization to the general case, using the remainder of m mod d, is straightforward. On the right-hand side of the rule, the XL method cloneSubgraph(c) is used which creates a copy of the whole subgraph accessible from the node marked with label c:

```
t:T(i) +> c:C, (i == (int)(m/2)) ==>>
    if (m % 2 == 0)
        ( t [ c ] -g-> I(0) T(m+1) [ cloneSubgraph(c) ] )
    else
        ( t [ c ] -h-> I(0) T(m+1) [ cloneSubgraph(c) ] )
    {m++;};
```

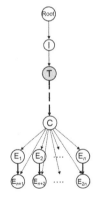

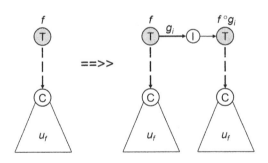

Fig. 1. The graph which is generated after the application of the initialization rule (rule 1). Edge types: s = thin unbroken, r = thick broken, d = thin dotted, x = thick unbroken. The fine-scale representation of the transformation id forms the bottom part of the graph, consisting of type E nodes.

Fig. 2. The copy rule for generator g_i. The type T node represents the transformation f. u_f is the subgraph at fine-scale level corresponding to f. On the right-hand side of the rule, the Cayley graph is extended by adding a g_i edge and the node representing $f \circ g_i$. The subgraphs at fine-scale level are not yet updated. Edge signatures are the same as in Fig. 1, with an additional g_i edge (thick unbroken arrow).

Rule Family 3 (Insertion): For each generator g_i, there is a rule which adds a third layer of type E nodes to the existing two layers. The new nodes are parameterized with $n + 1, n + 2, ..., 2n$ like the nodes of the second layer (cf. bottom part of Fig. 1). Connections are established between the second and the third layer using edges of type y which mimic the application of the transformation g_i on S. These connection patterns are the input of our algorithm. The code of each of these d rules thus depends on the action of the corresponding generator. As an example, we show the rule code for the permutation $g = (3\ 4)$ on a two-element set. It makes use of a context condition, enclosed in starred parentheses, which ensures that the rule is only applied to T nodes with an incoming path consisting of a type g edge followed by a type s edge. The additional condition "i.used == 0" ensures that this insertion rule is applied only once to each transformation node in the Cayley graph:

```
(* T -g-> i:I T +> c:C /> *) r:E(j), (* c /> *) s:E(k),
(i.used == 0 && j == 3 && k == 4)
     ==>> r -y-> E(4) </ c, s -y-> E(3) </ c { i.used = 1; };
```

Note that in the third line the order of the parameters of the type E nodes, 4 and 3, has switched according to the represented transposition g which exchanges 3 and 4.

Rules 4 and 5 (Composition): Like the insertion rules, the two composition rules work exclusively at fine-scale level. They perform the composition of the two transformations represented between the first and the second layer and between the second and third layer of type E nodes, respectively. At the end, the subgraph where they have been applied has again only two layers of nodes, and the type y edges have disappeared (Fig. 3). Rule 4 is the proper composition rule.

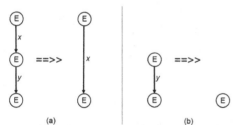

(a)   (b)

Fig. 3. (a) Rule 4, (b) rule 5 for composing two transformations. Both rules are applied in parallel mode, but all applications of (a) have to be finished before the (b) applications start.

Rule 5 serves to remove unused E nodes of the second layer after rule 4 has already been applied. Rule 5 is superfluous in the case that all transformations are permutations, because then all layer-2 nodes will be consumed by the applications of rule 4.

In the XL code we have to ensure that rule 5 is only applied when all possible applications of rule 4 have already been done.

To this purpose, XL provides the method call "`derive()`" enforcing to carry out all preceding parallel rule applications listed before:

```
[ a:E -x-> E -y-> b:E ==>> a -x-> b; ]
derive();
[ E -y-> b:E ==> b; ]
```

Rule 6 (Redirect): After copying the subgraph representing an already existing transformation f, inserting the connection pattern for generator g_i at fine scale level and calculating the composition $f \circ g_i$ at this level, we have to check if this newly determined transformation occurs already in our graph. In this case, the corresponding type g_i edge at coarse-scale level which led to the T node under consideration has to be redirected to the existing T node (which has the identical pattern at fine-scale level below it), and the new T node, together with the attached subgraph below it, has to be deleted (Fig. 4). The search for identical subgraphs has to be done systematically, so in the XL code the rule application is embedded into a control structure with graph queries (syntactically marked by starred parentheses). The flag "**used**" guarantees that the redirect rule is applied only once to a transformation node.

```
for ((* i:I y:T +> c:C *))
    if (i[used] == 1) {
        for ((* j:I z:T +> d:C *))
            if ((j[used] == 2) && equalTransf(c, d))
                { [ i y ==>> i z; ] }
        i[used] = 2;    }
```

The method "equalTransf", called in line 5 of this code, checks if the fine-scale subgraphs below the nodes c and d represent the same transformation (code not shown).

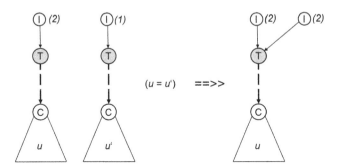

Fig. 4. The redirect rule (rule 6). The italicized numbers in parentheses show the values of the parameter "used" of the type I nodes, which ensure that the rule is applied only once to each newly created type T node (here: the T node below the I node with value 1). The rule is applied only if the condition $(u = u')$ is valid for the indicated subgraphs. Edge signatures as in Fig. 1.

Global control. After the growing Cayley graph was (possibly) modified by the redirect rule, the rule application process starts again with the copy rule. The indexing of the type T nodes and the application condition in the copy rule ensure that each transformation is processed only as often as there are generators (i.e., d times). In XL, the correct order of rule applications is specified by a control structure where the rule blocks are called by their respective names ("compose" stands for rules 4 and 5):

```
for (int i = 1; i <= d; i++) {
    copy(); derive();
    insert_gi(); derive();
    compose(); derive();
    redirect(); derive();
    }
```

The execution of this loop of rule applications has itself to be iterated until there are no more applicable rules (i.e., all transformations have been processed). The result is then a graph consisting of the Cayley graph of the monoid generated by $g_1, ..., g_d$ at coarse-scale level (however, including insertion and connector nodes) and with all generated transformation patterns at fine-scale level. As an example, Fig. 5 shows the generated graph in the case of the Klein four-group $Z_2 \times Z_2$.

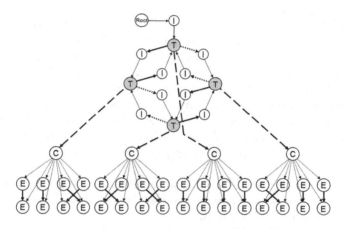

Fig. 5. The full graph representing the Klein four-group, generated by the permutations (1 2) and (3 4). The Cayley graph is spanned by the four nodes of type T (shaded). Edge signatures as in Figs. 1 and 2, additionally thick dotted arrows for the second generator.

Rules 7 and 8 (Simplification): To obtain the proper Cayley graph, we can first get rid of the fine-scale level by applying the rule C ==>>; which deletes all connector nodes. Since in the single-pushout approach, all "dangling edges" are deleted after removing a part of the host graph by rule application, this disconnects all type E nodes from the root and thus, because of the requirement of connectedness, deletes them, too. Only the coarse-level part of the generated graph (without the C nodes) will be retained after this (parallel) rule application. Finally, we can simplify the two-edge paths between two transformation nodes by merging them into one edge: t:T -gi-> I u:T ==>> t -gi-> u; for all generators, $i = 1, ..., d$. This last step should only be done if the Cayley graph has no loops (which is always the case for groups).

3   Discussion

The presented graph-rewriting method follows closely the way how one would construct the Cayley graph of a given transformation monoid manually. To this end, a combination of parallel subgraph replacement rule applications and procedural control structures (for ensuring the systematic, sequential visiting of all type T nodes during the construction process, and for comparing their transformations with all previously generated ones) turned out to be useful. The language XL provides an appropriate environment for combining these programming paradigms. However, some steps still required workarounds. Particularly, for testing the identity of two (graphically-encoded) transformations we had to write our own method "equalTransf" for using it in rule 6. The copying method "cloneSubgraph", used in rule 2, was already provided by XL, but, as for equality checking, a more intuitive and short-hand notation would have made the rules even more transparent.

In terms of calculation time, our method is far from being optimized. Particularly the search pattern in the insertion rule family tends to produce a lot of mismatches which are then sorted out during testing the additional parameter conditions, especially for larger values of n. A more intelligent pattern matching strategy for the transformation-encoding subgraphs would probably be possible. On the other hand, our calculation of the composition of transformations by parallel graph-grammar rule application (rule 4) is potentially very efficient. To make use of this parallelism, however, requires a splitting of the execution into several threads and their processing by properly parallel devices (e.g., on the GPU).

Cayley graphs are used to display an algebraic structure graphically and thereby to visualize its symmetries (see, e.g., [2] for groups). This requires an appropriate graph layout algorithm which is able to map abstract symmetries into geometrical ones. GroIMP provides already a "standard" graph layout which is a combination of an energy-based layout (using simulated annealing for optimization) and a classical tree layout [5]. When applied to the full graph (before simplification), it has the advantage to automatically separate the fine-scale from the coarse-scale level. From the examples of final Cayley graphs (after simplification) which we tested with our algorithm, we could see that this layout tends to produce already quite appropriate results in the case of groups, whereas in the more general case of monoids the existing patterns of similarity in the Cayley graphs are often not fully reflected in the layouts. It will be a special challenge in graph drawing to provide an improved graph layout algorithm for Cayley graphs of finite monoids' which could then be used in combination with our graph-rewriting approach.

Acknowledgments. I thank Ole Kniemeyer and Octave Etard for their ingenious programming work. Some prerequisites of this study were developed in the project

"Multiscale functional-structural plant modelling at the example of apple trees", funded by DFG under grant number KU 847/11-1. All support is gratefully acknowledged.

References

1. Caucal, D.: On Cayley graphs of basic algebraic structures. arXiv preprint arXiv: 1903.06521
2. Coxeter, H.S.M., Moser, W.O.J.: Generators and Relations for Discrete Groups, 4th edn. Springer, Berlin (1980). https://doi.org/10.1007/978-3-662-21946-1
3. Dixon, J.D., Mortimer, B.: Permutation Groups. Springer, Heidelberg (1996). https://doi.org/10.1007/978-1-4612-0731-3
4. Ehrig, H., et al.: Algebraic approaches to graph transformation II: single pushout approach and comparison with double pushout approach. In: Rozenberg, G. (ed.) Handbook of Graph Grammars and Computing by Graph Transformations, Vol. 1: Foundations, pp. 247–312. World Scientific, Singapore (1997)
5. Etard, O.: General-purpose graph drawing algorithm for GroIMP. Internship work presentation, University of Göttingen (2011, Unpublished)
6. Ganyushkin, O., Mazorchuk, V.: Classical Finite Transformation Semigroups: An Introduction. Springer, London (2009). https://doi.org/10.1007/978-1-84800-281-4
7. Godin, C., Caraglio, Y.: A multiscale model of plant topological structures. J. Theor. Biol. **191**, 1–46 (1998)
8. Kelarev, A.V., Praeger, C.E.: On transitive Cayley graphs of groups and semigroups. Eur. J. Comb. **24**, 59–72 (2003). https://doi.org/10.1016/S0195-6698(02)00120-8
9. Kniemeyer, O.: Design and implementation of a graph grammar based language for functional-structural plant modelling. Ph.D. thesis, University of Technology at Cottbus (2008). http://nbn-resolving.de/urn/resolver.pl?urn=urn:nbn:de:kobv:co1-opus-5937
10. Kurth, W., Kniemeyer, O., Buck-Sorlin, G.: Relational growth grammars – a graph rewriting approach to dynamical systems with a dynamical structure. In: Banâtre, J.-P., Fradet, P., Giavitto, J.-L., Michel, O. (eds.) UPP 2004. LNCS, vol. 3566, pp. 56–72. Springer, Heidelberg (2005). https://doi.org/10.1007/11527800_5
11. Linton, S.A., Pfeiffer, G., Robertson, E.F., Ruškuc, N.: Computing transformation semigroups. J. Symb. Comput. **33**, 145–162 (2002). https://doi.org/10.1006/jsco.2000.0406
12. Ong, Y.: Multi-scale rule-based graph transformation using the programming language XL. In: Ehrig, H., Engels, G., Kreowski, H.-J., Rozenberg, G. (eds.) ICGT 2012. LNCS, vol. 7562, pp. 417–419. Springer, Heidelberg (2012). https://doi.org/10.1007/978-3-642-33654-6_29
13. Ong, Y., Streit, K., Henke, M., Kurth, W.: An approach to multiscale modelling with graph grammars. Ann. Bot. **114**, 813–827 (2014). https://doi.org/10.1093/aob/mcu155
14. Smoleňová, K., Kurth, W., Cournède, P.-H.: Parallel graph grammars with instantiation rules allow efficient structural factorization of virtual vegetation. Electron. Commun. EASST **61**, 1–17 (2013). https://doi.org/10.14279/tuj.eceasst.61.830.825

The Glasgow Subgraph Solver: Using Constraint Programming to Tackle Hard Subgraph Isomorphism Problem Variants

Ciaran McCreesh$^{(\boxtimes)}$ ⓘ, Patrick Prosser ⓘ, and James Trimble ⓘ

University of Glasgow, Glasgow, Scotland
`ciaran.mccreesh@glasgow.ac.uk`

Abstract. The Glasgow Subgraph Solver provides an implementation of state of the art algorithms for subgraph isomorphism problems. It combines constraint programming concepts with a variety of strong but fast domain-specific search and inference techniques, and is suitable for use on a wide range of graphs, including many that are found to be computationally hard by other solvers. It can also be equipped with side constraints, and can easily be adapted to solve other subgraph matching problem variants. We outline its key features from the view of both users and algorithm developers, and discuss future directions.

1 Introduction

The subgraph isomorphism family of problems involves finding a small "pattern" graph inside a larger "target" graph, or establishing that the pattern does not occur. When the pattern graph is part of the input, these problems are NP-complete; despite this, subgraph isomorphism algorithms are widely used in practice, including for model checking [23], for law enforcement [9], in biological applications [1,6,20], for compiler implementation [5], in designing mechanical locks [27], and inside graph databases [19]. This has encouraged the development of practical subgraph isomorphism algorithms, which fall into two categories: those based upon backtracking and connectivity [6–8], and those based upon constraint programming [3,4,15,18,25]. Presently, the constraint programming approaches give spectacularly better performance on hard instances [19,26], although simple backtrackers will often (but inconsistently) run faster on some very easy instances due to lower overheads and faster startup costs.

This paper gives an overview of the Glasgow Subgraph Solver, which is the current state of the art in subgraph solving for hard instances [26]. First, we will discuss the range of subgraph isomorphism problems that people sometimes wish to solve, and then describe the main techniques the solver uses to solve these problems. We finish with a list of potential future directions.

© Springer Nature Switzerland AG 2020
F. Gadducci and T. Kehrer (Eds.): ICGT 2020, LNCS 12150, pp. 316–324, 2020.
https://doi.org/10.1007/978-3-030-51372-6_19

Fig. 1. The arrows show a non-induced subgraph isomorphism from the pattern graph on the left to the target graph on the right. This subgraph isomorphism is not induced, due to the extra edge between vertices 4 and 6 when c and d are not adjacent.

2   Subgraph Isomorphism Problems and Variants

Figure 1 illustrates a basic subgraph isomorphism problem: we have a small *pattern* graph and a large *target* graph (both of which are inputs to the problem), and we wish to decide whether the pattern graph occurs inside the target graph. Usually this is expressed in terms of finding a mapping from the vertices of the pattern graph to the vertices of the target graph, as shown using the dotted arrows. Beyond this, different applications have different views of what exactly the problem to be solved is—we therefore give a brief overview of the common problem variants.

Adjacency, Loops, and Directed Edges. It is generally agreed that for a mapping to be a valid subgraph isomorphism, adjacent vertices must be mapped to adjacent vertices. However, authors (particularly in application-oriented papers) disagree over whether non-adjacent vertices must be mapped to non-adjacent vertices. We use the term *induced* if non-adjacency must be preserved, and *non-induced* otherwise; when we do not qualify our terms, we are talking about both variants. A further question is on how to handle loops (that is, vertices which are adjacent to themselves). We take the view that loops may only be mapped to loops, and for induced problems, additionally that non-loops may only be mapped to non-loops; some other solver authors disagree or have not considered this question, and may handle this differently. Finally, in the case of graphs with directed edges (which could potentially go in both directions), we treat non-induced as meaning "the edges mapped to must be equal to or be a superset of the pattern edges", and induced as meaning "exactly equal to".

Vertices and Injectivity. In the classical subgraph isomorphism problem, the mapping is required to be injective—that is, each pattern vertex must be mapped to a different target vertex. In some applications this restriction can be relaxed: for example, we may prefer local injectivity (no two vertices that share a neighbour are mapped to the same vertex) [12], or even to find a homomorphism, where there are no injectivity requirements at all.

Labels. In some applications, either vertices, edges or both have labels, and may only be mapped to vertices or edges with matching labels—for example, in chemistry problems, labels may represent different atoms in a molecule, and we may not map a carbon atom to a hydrogen atom. Richer labelling rules may be necessary in some applications, such as in temporal graphs when we care about "before/after" labels rather than exact matches [21].

Deciding, Enumerating, and Counting. Instead of simply asking whether a subgraph isomorphism exists, some applications want to find all such mappings. They may require that these be explicitly enumerated, but sometimes a count is sufficient—and counting can be exponentially faster than enumerating in some situations. A further complication is that the number of mappings and the number of *images* of mappings are not the same, and different applications assume different definitions—sometimes, the number of mappings are called "labelled" countings, whilst the number of images of mappings are called "unlabelled".

Performance. Finally, we briefly discuss the common misconception that subgraph isomorphism being NP-complete somehow means that it is not viable to solve the problem in practice, or that every instance will exhibit exponential complexity. In fact, with good algorithms, instances that are actually hard to solve in practice are rare. We caution that benchmarking algorithms for NP-complete problems is challenging, that the size of the inputs is not an indicator of difficulty, and that only comparing performance on a few easy instances can lead to design flaws in applications built on top of these algorithms [19, 26].

3   The Glasgow Subgraph Solver

The Glasgow Subgraph Solver provides a high quality implementation of algorithms for many subgraph isomorphism problem variants. It is open source software, released under the MIT licence (which allows for commercial and closed source reuse). It may be downloaded from https://github.com/ciaranm/glasgow-subgraph-solver. It is implemented in C++, using the Boost libraries. It supports a variety of input file formats, but given the subtle and often undocumented differences in meanings of inputs in supposedly common file formats (e.g. whether edges are explicitly listed in both directions for undirected graphs), the solver has been designed to make it easy to add new parsers. The solver is primarily intended to be run from the command line or as a separate process, and its output is easy to parse for use with other tools.

3.1   Algorithmic Details

The Glasgow Subgraph Solver is based upon ideas from constraint programming. In a general constraint programming problem, we have a set of *variables*, each of which has a *domain* of possible *values*. We also have a set of *constraints*, which restrict valid combinations of values for subsets of the variables. The goal is to give each variable a value from its domain, respecting all constraints; usually this is done using a combination of inference and intelligent backtracking search.

To model subgraph isomorphism using constraint programming, we have a variable for each pattern vertex, and the domains are all of the target vertices. The constraints depend upon the exact variant being modelled, but we will usually have one constraint to deal with injectivity, and then a set of constraints to deal with edge and adjacency rules. A key strength of constraint programming is in the ability to add additional *implied* constraints, which we will now discuss— these can vastly speed up the solving process.

Degree Filtering. In an injective mapping, it is easy to see that a pattern vertex of degree d can never be mapped to a target vertex of degree less than d. This often allows many values to be eliminated from domains before any search starts. The solver uses even stronger filtering, based upon a result by Zampelli et al. [28], which looks at the neighbourhood degree sequence of vertices.

Distances and Paths. Another source of additional constraints comes from reasoning about distances or paths, rather than just adjacency. Audemard et al. [4] observed that the fact that subgraph isomorphisms preserve or reduce distances can be used to provide additional filtering during search. An early precursor to the Glasgow Subgraph Solver [18] strengthened this result, using instead the fact that subgraph isomorphisms preserve paths: if there are exactly k paths of length exactly ℓ between two vertices in a pattern graph, then there must be at least k paths of length exactly ℓ between wherever these two vertices are mapped in the target graph. This is exploited through the use of supplemental graphs, as follows.

We define a *supplemental graph* to be a graph with two distinguished vertices, that is subgraph isomorphic to itself under the interchange of these vertices. Letting G be a graph, and S a supplemental graph, we define a new graph G^S as follows: the vertex set of G^S is the same as the vertex set of G. Meanwhile, there is an edge between vertices v and w in G^S if there exists a non-induced subgraph isomorphism i from S to G which maps the two distinguished vertices of S to v and w respectively. It is reasonably straightforward to prove that any subgraph isomorphism $i : P \rightarrow T$ also defines a subgraph isomorphism $i^S : P^S \rightarrow T^S$, where $i^S(v) = i(v)$. The Glasgow Subgraph Solver uses this result to generate additional degree and adjacency-like constraints. This is sometimes extremely powerful, as illustrated by the example in Fig. 2. Currently the choice of supplemental graphs is hard-coded, based upon performance on a range of standard benchmark instances, but we believe it may be possible to automatically make different choices for different families of problem instance.

All-Different Filtering. Suppose a pattern graph and a target graph both have exactly five vertices of degree five or higher, then those five vertices in the target graph cannot be mapped to by any other pattern vertex in an injective mapping. This is an example of all-different reasoning: more generally, if any n undecided pattern vertices have less than n available target vertices between them, we have found a contradiction, and if they have exactly n available target vertices between them then those target vertices must all be used only for those

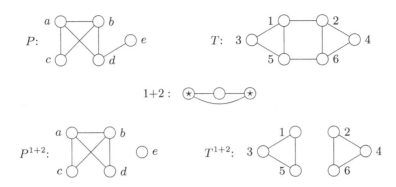

Fig. 2. On top, a pair of graphs P and T. In the middle, the supplemental graph $1+2$. On the bottom, the modified graphs P^{1+2} and T^{1+2}. These modified graphs make it immediately clear that no subgraph isomorphism exists between P and T.

pattern vertices. Deciding exactly how to filter all-different constraints is one of the big differences between constraint programming approaches for subgraph isomorphism [4,22,25]. Currently, the Glasgow Subgraph Solver uses a special bit-parallel propagator, which gives a good tradeoff between performance and filtering power [18].

The other major contributing aspect to a constraint programming solver's performance is how it carries out backtracking search.

Search Order. When performing a backtracking search, the choices of which variable to branch on, and which value to try first, can make a staggering difference to performance in practice. The Glasgow Subgraph Solver uses carefully chosen strategies to decide how to direct its search [19], including always branching on whichever vertex has fewest possibilities available to it (tie-breaking on highest degree). This has interesting implications, which are not yet fully understood. For example, in the absence of other filtering, this will cause the solver to always grow connected components, which is the optimal behaviour for certain kinds of pattern graph—but it is not clear whether exploiting additional filtering could theoretically lose us performance guarantees in some cases.

Restarts and Nogood Recording. Rather than using simple backtracking, the solver employs restarts and nogood recording [16,17]: the solver runs for a small amount of time, and then restarts from the beginning, remembering not to revisit any part of the search space which has already been explored. Combined with a small amount of heavily biased randomness in how branching is carried out, this avoids a strong commitment to early branching choices, which are most difficult for a heuristic to get right [3].

Parallelism. Modern hardware provides a range of opportunities for parallelism. The Glasgow Subgraph Solver exploits this in two ways: by using bit-parallel data structures and algorithms to carry out inference as quickly as possible [18], and by using threads to explore multiple parts of the search space in parallel [3]. These parallel search capabilities scale at least as far as thirty-six cores.

3.2 Future Directions

We finish with a discussion of possible future directions for the solver, and with ideas for research and engineering challenges which may be of broader interest.

Problem Variants. There are other problems involving finding mappings between subgraphs, such as a surjective variant [13]. Some problems also involve wildcards, not just on labels, but on pattern vertices; work on k-less subgraph isomorphism [14] may prove useful for continuing to allow powerful inference when wildcards are present. More generally, we have experimental support for connecting the solver to an external constraint programming solver, to handle arbitrary side constraints (a bit like Satisfiability Modulo Theories). This could be useful, for example, for temporal graphs [21]; we would be interested in exploring this direction further to tackle suitable real-world applications. Another potential application area is inside graph rewriting systems [10]. Here, the pattern graphs are considered "fixed", rather than being part of the input, which has implications for the theoretical complexity of the problem. However, when patterns are numerous or large and complex, or when side constraints are involved [2], it may be more practical to use a general purpose solver than a dedicated algorithm for each special case.

Symmetries. Some applications involve heavily symmetric pattern and target graphs [27]. Handling such symmetries in constraint programming is, in principle, a well-understood problem. However, a practical difficulty is that because the symmetries vary on an instance by instance basis, symmetry-breaking constraints must be computed for each individual input rather than for a model as a whole. An implementation of the Schreier-Sims algorithm [24] which has no costly external dependencies would make this approach more practical.

Faster Counting. Currently, the solver handles the counting problem by explicit enumeration, except that for non-induced isomorphisms, any isolated vertices in the pattern graph are treated specially. Although counting and enumeration are equally difficult in general, we believe there are further opportunities for speeding up counting, for example by decomposing the pattern graph into nearly-unconnected components, or by handling pattern vertices of degree one and two specially. We would also be interested in implementing approximate counting as an option, as well as seeing whether uniform sampling of solutions can be carried out more efficiently in practice than by explicit enumeration.

Special Classes of Pattern. Certain special classes of pattern may be counted efficiently—for example, if the pattern is a star graph. Some applications involve counting occurrences of many different kinds of small graph [1,9,20], and so it would be useful if solvers could detect when they were in an "easy" case and switch algorithms, rather than relying upon end users to do this. There are also classes of pattern graph where decision and counting are still NP-hard, but where more efficient solving techniques are available—the solver currently switches to a different dedicated algorithm if the input graph is a clique, for example.

Proof Logging. Given the increasing complexity of both the theory and implementations of subgraph isomorphism algorithms, we should be concerned as to whether the outputs produced are correct. In the Boolean satisfiability community, proof logging is the standard solution to this problem: solvers that claim unsatisfiability are expected to be able to output a machine-verifiable proof of this fact. Recently, Elffers et al. [11] introduced a more flexible form of proof logging, that we believe is better suited for algorithms that perform strong inference. The Glasgow Subgraph Solver includes experimental support for producing proofs in this format, and we hope to see further research in this direction.

Automatic Configuration. The solver supports a wide range of filtering options. Its default configuration is designed to reduce the chances of poor performance on hard instances, rather than to do well on very easy instances—for example, it will create supplemental graphs before attempting any search, which is a relatively expensive one-time cost that is not necessary for solving many instances. We have previously shown that it can be beneficial to employ a simple connectivity-based algorithm as a presolver [15]. However, it may be possible to take automatic algorithm configuration further, for example by selecting the set of supplemental graphs to use on an instance by instance basis.

Benchmarking. Finally, given the importance of having good instances for benchmarking and for informing algorithm design, we would be very interested in collecting sets of instances from other applications. The instances by Solnon[1] originally used for algorithm portfolios [15] give a good starting point, but having more instances from a diverse range of applications would be very beneficial—even if those instances are all either very easy for all solvers, or are too hard for any current solver to solve at all. We would very much welcome contributions from the community.

Acknowledgements. We would like to thank Blair Archibald, Fraser Dunlop, Jan Elffers, Stephan Gocht, Ruth Hoffmann, Jakob Nordström, and Christine Solnon for their contributions to the design and implementation of the solver. This work was supported by the Engineering and Physical Sciences Research Council (grant numbers EP/P026842/1, EP/M508056/1, and EP/N007565).

References

1. Alon, N., Dao, P., Hajirasouliha, I., Hormozdiari, F., Sahinalp, S.C.: Biomolecular network motif counting and discovery by color coding. Bioinformatics (Oxford, England) **24**(13), i241–i249 (2008)

---

[1] https://perso.liris.cnrs.fr/christine.solnon/SIP.html.

2. Archibald, B., Calder, M., Sevegnani, M.: Conditional bigraphs. In: 13th International Conference on Graph Transformation (ICGT 2020), Bergen, Norway, 25–26 June 2020 (2020)

3. Archibald, B., Dunlop, F., Hoffmann, R., McCreesh, C., Prosser, P., Trimble, J.: Sequential and parallel solution-biased search for subgraph algorithms. In: Rousseau, L.-M., Stergiou, K. (eds.) CPAIOR 2019. LNCS, vol. 11494, pp. 20–38. Springer, Cham (2019). https://doi.org/10.1007/978-3-030-19212-9_2

4. Audemard, G., Lecoutre, C., Samy-Modeliar, M., Goncalves, G., Porumbel, D.: Scoring-based neighborhood dominance for the subgraph isomorphism problem. In: O'Sullivan, B. (ed.) Principles and Practice of Constraint Programming, CP 2014. Lecture Notes in Computer Science, vol. 8656, pp. 125–141. Springer, Cham (2014). https://doi.org/10.1007/978-3-319-10428-7_12

5. Blindell, G.H., Lozano, R.C., Carlsson, M., Schulte, C.: Modeling universal instruction selection. In: Proceedings of Principles and Practice of Constraint Programming - 21st International Conference, CP 2015, Cork, Ireland, 31 August–4 September 2015, pp. 609–626 (2015)

6. Bonnici, V., Giugno, R., Pulvirenti, A., Shasha, D.E., Ferro, A.: A subgraph isomorphism algorithm and its application to biochemical data. BMC Bioinf. **14**(S–7), S13 (2013)

7. Carletti, V., Foggia, P., Saggese, A., Vento, M.: Introducing VF3: a new algorithm for subgraph isomorphism. In: Foggia, P., Liu, C.-L., Vento, M. (eds.) GbRPR 2017. LNCS, vol. 10310, pp. 128–139. Springer, Cham (2017). https://doi.org/10.1007/978-3-319-58961-9_12

8. Cordella, L.P., Foggia, P., Sansone, C., Vento, M.: A (sub)graph isomorphism algorithm for matching large graphs. IEEE Trans. Pattern Anal. Mach. Intell. **26**(10), 1367–1372 (2004)

9. Davies, T., Marchione, E.: Event networks and the identification of crime pattern motifs. PLOS ONE **10**(11), 1–19 (2015)

10. Dörr, H. (ed.): Efficient Graph Rewriting and Its Implementation. LNCS, vol. 922. Springer, Heidelberg (1995). https://doi.org/10.1007/BFb0031909

11. Elffers, J., Gocht, S., McCreesh, C., Nordström, J.: Justifying all differences using Pseudo-Boolean reasoning. In: Proceedings of AAAI (2020). in press

12. Fiala, J., Kratochvíl, J.: Locally constrained graph homomorphisms-structure, complexity, and applications. Comput. Sci. Rev. **2**(2), 97–111 (2008)

13. Gay, S., Fages, F., Martinez, T., Soliman, S., Solnon, C.: On the subgraph epimorphism problem. Discret. Appl. Math. **162**, 214–228 (2014)

14. Hoffmann, R., McCreesh, C., Reilly, C.: Between subgraph isomorphism and maximum common subgraph. In: Proceedings of the Thirty-First AAAI Conference on Artificial Intelligence, San Francisco, California, USA, 4–9 February 2017, pp. 3907–3914 (2017)

15. Kotthoff, L., McCreesh, C., Solnon, C.: Portfolios of subgraph isomorphism algorithms. In: Festa, P., Sellmann, M., Vanschoren, J. (eds.) LION 2016. LNCS, vol. 10079, pp. 107–122. Springer, Cham (2016). https://doi.org/10.1007/978-3-319-50349-3_8

16. Lecoutre, C., Sais, L., Tabary, S., Vidal, V.: Nogood recording from restarts. In: IJCAI 2007, Proceedings of the 20th International Joint Conference on Artificial Intelligence, Hyderabad, India, 6–12 January 2007, pp. 131–136 (2007)

17. Lee, J.H.M., Schulte, C., Zhu, Z.: Increasing nogoods in restart-based search. In: Proceedings of the Thirtieth AAAI Conference on Artificial Intelligence, Phoenix, Arizona, USA, 12–17 February 2016, pp. 3426–3433 (2016)

18. McCreesh, C., Prosser, P.: A parallel, backjumping subgraph isomorphism algorithm using supplemental graphs. In: Proceedings of Principles and Practice of Constraint Programming - 21st International Conference, CP 2015, Cork, Ireland, 31 August–4 September 2015, pp. 295–312 (2015)

19. McCreesh, C., Prosser, P., Solnon, C., Trimble, J.: When subgraph isomorphism is really hard, and why this matters for graph databases. J. Artif. Intell. Res. **61**, 723–759 (2018)

20. Mukherjee, K., Hasan, M.M., Boucher, C., Kahveci, T.: Counting motifs in dynamic networks. BMC Syst. Biol. **12**(1), 6 (2018)

21. Redmond, U., Cunningham, P.: Temporal subgraph isomorphism. In: Advances in Social Networks Analysis and Mining 2013, ASONAM 2013, Niagara, ON, Canada, 25–29 August 2013, pp. 1451–1452 (2013)

22. Régin, J.: A filtering algorithm for constraints of difference in CSPs. In: Proceedings of the 12th National Conference on Artificial Intelligence, Seattle, WA, USA, 31 July–4 August 1994, vol. 1, pp. 362–367 (1994)

23. Sevegnani, M., Calder, M.: Bigraphs with sharing. Theor. Comput. Sci. **577**, 43–73 (2015)

24. Sims, C.C.: Computational methods in the study of permutation groups. In: Leech, J. (ed.) Computational Problems in Abstract Algebra, pp. 169–183. Pergamon (1970)

25. Solnon, C.: Alldifferent-based filtering for subgraph isomorphism. Artif. Intell. **174**(12–13), 850–864 (2010)

26. Solnon, C.: Experimental evaluation of subgraph isomorphism solvers. In: Conte, D., Ramel, J.-Y., Foggia, P. (eds.) GbRPR 2019. LNCS, vol. 11510, pp. 1–13. Springer, Cham (2019). https://doi.org/10.1007/978-3-030-20081-7_1

27. Vömel, C., de Lorenzi, F., Beer, S., Fuchs, E.: The secret life of keys: on the calculation of mechanical lock systems. SIAM Rev. **59**(2), 393–422 (2017)

28. Zampelli, S., Deville, Y., Solnon, C.: Solving subgraph isomorphism problems with constraint programming. Constraints **15**(3), 327–353 (2010)

A Simulator for Probabilistic Timed Graph Transformation Systems with Complex Large-Scale Topologies

Christian Zöllner$^{(\boxtimes)}$, Matthias Barkowsky , Maria Maximova ,
Melanie Schneider, and Holger Giese

Hasso Plattner Institute at the University of Potsdam, Potsdam, Germany
{christian.zoellner,matthias.barkowsky,maria.maximova,melanie.schneider,
holger.giese}@hpi.de

Abstract. Future cyber-physical systems, like networks of autonomous vehicles, will result in a huge number of collaborating systems acting together on large-scale topologies. Modeling them requires capturing timed and probabilistic behavior as well as structure dynamics. In [9], we introduced Probabilistic Timed Graph Transformation Systems (PTGTSs) as a means of modeling a high-level view of these systems of systems and provided model checking support. However, given the scale of emerging systems of systems and their often complex topologies, analyzing only small or medium size models using model checking is insufficient. To close this gap, we developed a simulator for PTGTSs that can import real-world topologies, automatically detect violations of state properties, and handle the graph pattern matching as well as time and probabilities efficiently so that complex large-scale topologies can be considered.

1 Introduction

In future large-scale cyber-physical systems, such as networks of autonomous vehicles, the interconnection of the autonomous systems via complex software and networking will result in massive systems of systems where a huge number of systems collaborate and act together on complex large-scale topologies.

Since these systems of systems are often real-time critical and exhibit probabilistic phenomena like failures, modeling them requires capturing timed and probabilistic behavior. In addition, structure dynamics needs to be taken into account since the interconnections between autonomous subsystems may change at runtime. Finally, given the scale of emerging systems of systems and their complex topologies, the modeling must also allow for capturing the complex large-scale topologies in which these systems will operate.

In [9], we introduced Probabilistic Timed Graph Transformation Systems (PTGTSs) as a means for modeling a high-level view of these systems of systems

Funded by the Deutsche Forschungsgemeinschaft (DFG, German Research Foundation) - 241885098.

F. Gadducci and T. Kehrer (Eds.): ICGT 2020, LNCS 12150, pp. 325–334, 2020.
https://doi.org/10.1007/978-3-030-51372-6_20

and provided model checking support. However, with model checking, only small or medium size models could be analyzed, which is insufficient since the small models will (1) likely not exhibit all characteristics of complex topologies that can lead to failures and (2) likely will not allow to study emergent phenomena and failures that result from the interaction of many autonomous systems.

To close this gap and to enable the analysis of large-scale systems of systems, we developed a simulator for PTGTSs that can import complex real-world topologies, can automatically detect violations of state properties, and handles the graph pattern matching as well as the concepts of time and probabilities efficiently. The simulator maps the application of rules of a PTGTS to the probabilistic application of graph transformation (GT) rules and a dedicated time management. Scalability is achieved by exploiting the local nature of changes and by managing time in a way that avoids global updates.

Employing graph transformation systems (GTSs) and incremental graph pattern matching techniques for the simulation of complex systems has been proposed in [13]. A link between GTSs and discrete event simulation has been considered in [14]. Also, an extension of GTSs with stochastic behavior and related simulators like GraSS [15] and SimSG [4] have been developed. However, to the best of our knowledge, no simulator for GTS variants that support timed and probabilistic behavior (like PTGTSs [9]) has been presented so far.

This tool paper is structured as follows. The preliminaries, such as a running example and the PTGTS formalism, are introduced in Sect. 2. The simulator's concept is outlined in Sect. 3. An evaluation in Sect. 4 shows that the tool can import complex real-world topologies, can automatically detect violations of state properties, and can handle graph pattern matching as well as time and probabilities so efficiently that complex large-scale topologies can be considered. The paper is closed with a conclusion and an outlook on future work in Sect. 5.

2   Preliminaries

In this section, we introduce our running example, briefly recall the framework of GTSs, and recap the formalism of PTGTSs. As a running example, we model a scenario inspired by the RailCab project [12] where autonomous shuttles on a track topology form a system of systems.

In PTGTSs, we use the formalism of typed graphs [5] to describe the states of the systems and their structure. A *graph* $G = (G_V, G_E, s_G, t_G)$ is given by a set G_V of nodes, a set G_E of edges, and source and target functions $s_G, t_G : G_E \to G_V$. Let $G = (G_V, G_E, s_G, t_G)$ and $H = (H_V, H_E, s_H, t_H)$ be two graphs, then a *graph morphism* $f : G \to H$ is defined as a pair of mappings $f_V : G_V \to H_V$, $f_E : G_E \to H_E$ that are compatible with the source and target functions, i.e., $f_V \circ s_G = s_H \circ f_E$ and $f_V \circ t_G = t_H \circ f_E$.

Let TG be a distinguished graph, called a *type graph*. Then a *typed graph* $(G, type)$ consists of a graph G and a graph morphism $type : G \to TG$. For two given typed graphs $G'_1 = (G_1, type_1)$ and $G'_2 = (G_2, type_2)$, a *typed graph morphism* $f : G'_1 \to G'_2$ is a graph morphism $f : G_1 \to G_2$ that is compatible with the typing functions, i.e., $type_2 \circ f = type_1$.

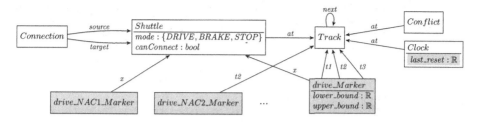

Fig. 1. Shuttle scenario type graph and generated extensions (grey, see Subsect. 3.2).

The type graph of the running example is given in Fig. 1 (without the grey extensions). In the context of this scenario, track nodes are connected to the adjacent tracks by *next* edges. *Shuttle* nodes are located on tracks, which is represented by *at* edges. Shuttles can move forward on tracks being in *DRIVE* mode or can stop resp. brake by changing into *STOP* resp. *BRAKE* mode. To avoid collisions and unnecessary braking maneuvers, shuttles can communicate and establish connections. For this, adjacent tracks are marked by *Conflict* nodes.

PTGTSs are typed over some type graph TG containing at least a type node *Clock*. Furthermore, for every graph G we use the function $CN(G) = \{n \mid n \in G_V \wedge type_V(n) = Clock\}$ to identify in every graph the nodes used for time measurement only. In the following, we call such identified nodes simply *clocks*.

The type graph in Fig. 1 thus equips tracks with *clocks* needed for time measurement to be able to control the time for rule applications.

The adaptation of graphs is realized using GT rules, which are to be understood as local rule-based modifications defining additions and removals of substructures. A *rule* $\rho = L \xleftarrow{l} K \xrightarrow{r} R$ is given by a span of injective typed graph morphisms with the graphs L and R called the left-hand side and the right-hand side of the rule, respectively. A *match* for a rule is a graph morphism from L to the current graph G describing one option where the rule could be applied in G. The transformation procedure defining a GT step is formally introduced by the *DPO approach* [5].

According to [9], PTGTSs are a combination of Probabilistic Graph Transformation Systems (PGTSs) and Timed Graph Transformation Systems (TGTSs). Similarly to PGTSs, transformation rules in PTGTSs can have multiple right-hand sides where each of them is annotated with a probability. While the choice for a rule match remains nondeterministic, the effect of a rule becomes probabilistic. Similarly to TGTSs, each probabilistic timed graph transformation (PTGT) rule has a guard formulated over clocks contained in the left-hand side of the rule, which is used to control the rule application. Moreover, each rule contains the information about clocks that have to be reset during the rule application.

A *probabilistic timed graph transformation (PTGT) rule* R is a tuple (L, P, μ, ϕ, r_C) where L is a common left-hand side graph, P is a finite set of graph transformation rules with the left-hand side L, $\mu \in Dist(P)$ is a probability distribution, $\phi \in \Phi(CN(L))$ is a guard over nodes of the type *Clock* contained in L, and $r_C \subseteq CN(L)$ is a set of nodes of the type *Clock* in L to be reset (see [9]).

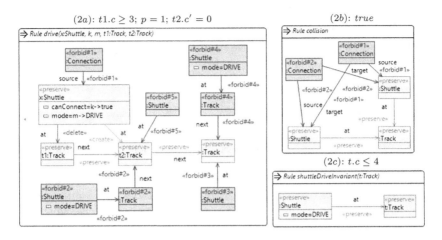

Fig. 2. PTGT rule *drive (a)*, atomic proposition *collision (b)*, and invariant *shuttleDriveInvariant (c)* of the shuttle scenario PTGTS in HENSHIN syntax.

In PTGTSs, we also employ negative application conditions (NACs) [7] and attributes. They allow to increase the descriptive expressiveness of the rules and can be added straightforwardly to the presented formalization.

The behavior of the shuttle scenario is modeled using 14 PTGT rules in HENSHIN [3]. In the following, we only discuss one of them in more detail and give an intuition for the other rules due to space restrictions (see more details in [11]). Shuttles can drive alone or can build convoys to reduce the energy consumption. The rule *drive* (see Fig. 2a) allows a shuttle leading a convoy or a shuttle driving without a convoy to move forward if there are no shuttles located too close in front of it. The restrictions for the location of other shuttles are given by NACs of the rule. To reflect real-time behavior, we require that moving on a single track can take between 3 and 4 time units, which we express using the corresponding guards and invariants, respectively, formulated over the track clocks for the driving rules. For the rule *drive* in Fig. 2a, the corresponding guard is given by the annotation $t1.c \geq 3$. For brevity, we refer to a clock c linked to an element e as $e.c$ and omit the extra node c. After rule application, we refer to c as $e.c'$. To measure the time spent on a track, we reset the clock of the track to which a shuttle is moving when applying the rule *drive* (annotation $t2.c' = 0$). Other rules of the scenario handle the connection attempts between shuttles as well as situations when shuttles have to brake or stop. Some rules, such as the rules for connection attempts, have higher priorities to ensure their timely application. Furthermore, probabilistic effects are used to model connection failures.

State properties in the form of invariants and atomic propositions are both given for PTGTSs as conditions (non-changing rules) over clocks, the satisfaction of which can be checked for a given state. In the context of our shuttle scenario, we consider an atomic proposition *collision* that is depicted in Fig. 2b and that identifies a collision whenever two shuttles are at the same track without being

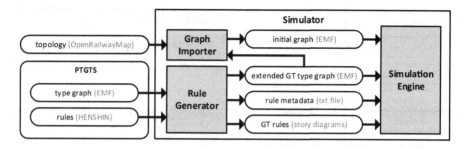

Fig. 3. Architecture of the PTGTS simulator.

connected. The invariant *shuttleDriveInvariant* in Fig. 2c ensures that a shuttle in mode DRIVE should not remain longer than 4 time units on a track ($t.c <= 4$).

A *probabilistic timed graph transformation system (PTGTS)* S is then a tuple $(TG, G_0, v_0, \Pi, I, AP, prio)$ where TG is a finite type graph including the type node *Clock*, G_0 is a finite initial graph over TG, $v_0 : CN(G_0) \to \mathbb{R}$ is the initial clock valuation assigning the clock value 0 to every clock, Π is a finite set of PTGT rules, I is a finite set of probabilistic timed invariants, AP is a finite set of probabilistic timed atomic propositions, and $prio : \Pi \to \mathbb{N}$ is a priority function assigning a priority to each rule (see [9]).

3   Simulator

In this section, we present the concepts behind our PTGTS simulator [11]. Each PTGT rule is translated into multiple typed GT rules. During the simulation, only specific GT rules must be applied to specific subgraphs. Structural matches are marked to avoid searching large parts of the graph after a local change.

Our simulator consists of three active components highlighted in Fig. 3. The rule generator creates GT rules from a PTGTS and the simulation engine selects and applies these GT rules. The graph importer constructs input graphs based on real-world public transport network topologies from OpenRailwayMap [10].

3.1   Simulation Engine

The simulation engine's algorithm for applying GT rules is sketched in Fig. 4. To select applicable rules and affected subgraphs, the engine keeps track of so-called *markers*. The engine is implemented in Java. It uses the Eclipse Modeling Framework (EMF) and an interpreter for story diagrams [6]. The interpreter allows for graph pattern matching starting with a fixed partial match, which, together with the engine's marker bookkeeping, makes the algorithm incremental.

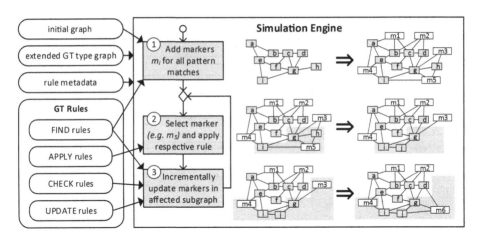

Fig. 4. PTGTS simulation algorithm based on marking pattern matches.

Step 1: Add initial markers Patterns occurring in the input graph are marked by generated FIND rules for PTGT rules, invariants, atomic propositions, and NACs. Since the rules for NACs are applied first, NAC markers can be used in other rules [2]. In the top example in Fig. 4, markers m_1–m_5 are created.

Step 2: Apply a rule Out of the created markers, the engine selects one that represents an enabled rule application with highest available priority and satisfied time bounds. Afterwards, it computes a new global time t'_g s.t. no invariants are violated. Then, the engine uses a generated APPLY rule to apply the actual PTGT rule at the marked pattern, and, finally, resets clocks. In the middle example in Fig. 4, node h is deleted while node j and two edges are created.

Step 3: Update affected subgraph After a rule application, the subgraph affected by the application (incl. all markers) is determined so that the necessary updates to the markers can be conducted incrementally. As can be seen in the bottom example in Fig. 4, CHECK rules remove markers that became invalid (e.g. m_3), FIND rules mark new patterns with new markers (e.g. m_6), and UPDATE rules update the time constraints of remaining markers (e.g. m_4).

Termination. The simulation engine stops when no rules are applicable (due to a lack of markers or due to violated invariants or time constraints) or when an atomic proposition (e.g. *collision* from Fig. 2b) is matched.

Handling of Timed Behavior. Simulating the timed behavior of a PTGTS requires according to PTGTS semantics the advancement of all clock values whenever time elapses. To avoid changing a potentially huge number of clock values each time, our simulation engine only maintains a global simulation time t_g. Instead of a time value $t(c)$, each clock c in the model has a last reset time value $t_r(c)$. Whenever a rule mandates a clock reset for c, the last reset time value $t_r(c)$ is set to the global simulation time t_g. Whenever the time value $t(c)$ is needed to evaluate a guard or invariant, it can be computed as $t(c) = t_g - t_r(c)$.

To handle guards and invariants even more efficiently, they are translated into lower and upper bounds, respectively. For example, the guard $t1.c \geq 3$ of the rule *drive* (see Fig. 2a) is translated into $lower\_bound = t_r(t1.c) + 3$, which can then be compared to the current global simulation time t_g.

3.2   Generation of GT Type Graph and Rules

In this subsection, we describe the generation of the type graph and GT rules based on the running example of the PTGT rule *drive* (see Fig. 2a). The GT rules are generated once and stored in the form of story diagrams [6].

Extended Type Graph. Markers for all possible pattern matches are added to the type graph. Moreover, a *last_reset* attribute is added to the *Clock* node in order to store the values of t_r as well as lower and upper bound attributes to marker types. Fig. 1 shows the type graph extensions for the PTGT rule *drive* (see Fig. 2a). Similar extensions are made for all other rules but omitted here.

FIND: Identifying Pattern Matches. The FIND rules create markers for pattern matches. Their left-hand side is equal to that of the respective PTGT rule, with the exception that instead of NAC patterns, NAC markers are employed. Similar FIND rules are generated for the NAC patterns themselves. To ensure that NACs are found first, the ordering of FIND rules is stored in the rule metadata. FIND rules also assign lower resp. upper time bounds to a marker, which are computed from guards resp. invariants as described above.

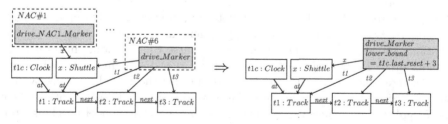

APPLY: Applying a Rule. The APPLY rules are similar to the PTGT rules, with the exception that they require a marker on the left-hand side and perform clock resets. If a PTGT rule has more than one right-hand side, multiple APPLY rules are created. Their probabilities are stored in the rule metadata.

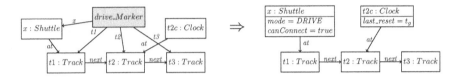

CHECK: Checking Completeness of Pattern Matches. The CHECK rules remove markers for matches that have become invalid after a rule application. A NAC of the whole original pattern ensures that unless the complete pattern is found, the marker is deleted.

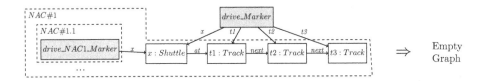

UPDATE: Updating Time Bounds. The UPDATE rules recompute lower and upper bounds of markers affected by the update of *last_reset* attributes.

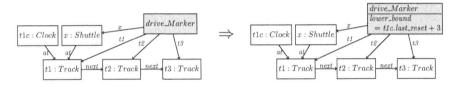

4   Evaluation

For evaluation, we constructed input graphs from the tram networks of four different German cities, including Europe's largest connected tram network in Berlin, which we modeled with 9184 track nodes. We assumed a density of one shuttle per 10 tracks and, in case of Potsdam, created an additional topology with doubled density. For each topology, we generated three sets of initial shuttle positions and ran each of these experiments three times, leading in total to 45 runs for up to 25.000 steps each (most ended earlier due to invariant violations).

We were able to use the simulator to improve the PTGTS by discovering and analyzing situations where invariants were violated. These situations were too complex to be efficiently discovered by our previous model checking approach in [9] using PRISM [8] e.g. when a violation is caused by three shuttles approaching two subsequent crossroads with a specific timing.

Also, we tested whether the average runtime for a simulation step does not change according to a trend (i.e., it is stationary) after an initial interval. For that, we ran three different stationarity tests (ADF, KPSS and PP, see [1]). All tests showed statistically significant results (i.e., $p\text{-}value < 0.05$), except for a single simulator run in Frankfurt where one of the three tests had a $p\text{-}value$ of 0.09.

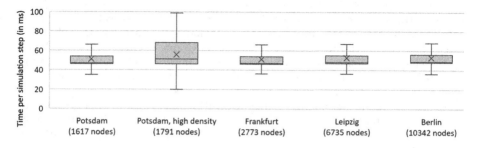

Fig. 5. Distribution of runtime per simulation step after non-stationary first interval.

As can be seen in Fig. 5, when excluding the non-stationary first interval, the size of our example models has no significant impact on the average runtime per simulation step. However, the higher shuttle density appears to have an influence on the runtime, which can be explained by a higher rate of rule applications for the connection of shuttles that affect a larger subgraph.

5   Conclusion and Future Work

We presented a simulator for PTGTSs [9] and demonstrated that it can import complex real-world topologies, automatically detect violations of state properties, and handle the graph pattern matching as well as the concepts of time and probabilities so efficiently that complex large-scale topologies can be considered. As future work, we plan to formally analyze and further improve the efficiency of our tool, provide more mature tool support covering, in particular, the transition to model checking, and support checking for more than state properties.

Acknowledgments. We thank our colleague Christian Medeiros Adriano who supported us in the statistical evaluation of the experiment results.

References

1. Banerjee, A., Dolado, J.J., Galbraith, J.W., et al.: Co-integration, Error Correction, and the Econometric Analysis of Non-stationary Data. OUP Catalogue (1993)
2. Beyhl, T., Blouin, D., Giese, H., Lambers, L.: On the operationalization of graph queries with generalized discrimination networks. In: Echahed, R., Minas, M. (eds.) ICGT 2016. LNCS, vol. 9761, pp. 170–186. Springer, Cham (2016). https://doi. org/10.1007/978-3-319-40530-8_11
3. The Eclipse Foundation: EMF Henshin (2013). https://www.eclipse.org/modeling/ emft/henshin
4. Ehmes, S., Fritsche, L., Schürr, A.: SimSG: rule-based simulation using stochastic graph transformation. J. Object Technol. **18**, 1:1–17 (2019). The 12th International Conference on Model Transformations
5. Ehrig, H., Ehrig, K., Prange, U., Taentzer, G.: Fundamentals of Algebraic Graph Transformation. MTCSAES. Springer, Heidelberg (2006). https://doi.org/ 10.1007/3-540-31188-2_15

6. Giese, H., Hildebrandt, S., Seibel, A.: Improved flexibility and scalability by interpreting story diagrams. In: Magaria, T., Padberg, J., Taentzer, G. (eds.) Proceedings of the 8th International Workshop on Graph Transformation and Visual Modeling Techniques, vol. 18. Electronic Communications of the EASST (2009)
7. Habel, A., Heckel, R., Taentzer, G.: Graph grammars with negative application conditions. Fundamenta Informaticae **26**(3, 4) (1996)
8. Kwiatkowska, M., Norman, G., Parker, D.: PRISM 4.0: verification of probabilistic real-time systems. In: Gopalakrishnan, G., Qadeer, S. (eds.) CAV 2011. LNCS, vol. 6806, pp. 585–591. Springer, Heidelberg (2011). https://doi.org/10.1007/978-3-642-22110-1_47
9. Maximova, M., Giese, H., Krause, C.: Probabilistic timed graph transformation systems. J. Log. Algebraic Methods Program. **101**, 110–131 (2018)
10. OpenRailwayMap. https://www.openrailwaymap.org
11. PTGTS Simulator Project Website. https://mdelab.de/ptgts-simulator
12. RailCab project. https://www.hni.uni-paderborn.de/cim/projekte/railcab
13. Ráth, I., Vago, D., Varró, D.: Design-time simulation of domain-specific models by incremental pattern matching. In: 2008 IEEE Symposium on Visual Languages and Human-Centric Computing, pp. 219–222 (2008)
14. Syriani, E., Vangheluwe, H.: A modular timed graph transformation language for simulation-based design. Softw. Syst. Modeling **12**(2) (2013)
15. Torrini, P., Heckel, R., Ráth, I.: Stochastic simulation of graph transformation systems. In: Rosenblum, D.S., Taentzer, G. (eds.) FASE 2010. LNCS, vol. 6013, pp. 154–157. Springer, Heidelberg (2010). https://doi.org/10.1007/978-3-642-12029-9_11

Author Index

Printed in the United States
by Baker & Taylor Publisher Services